The Square Root of Two to One Million Digits

edited by

David E. McAdams

Author's website is http://www.demcadams.com.

This book is for educational and entertainment purposes only. The publisher and author are not offering it for mathematical advice.

Other Books by David E. McAdams

Parrot Colors – An introduction to the concept of colors. For preschoolers.

Flower Colors – An introduction to the concept of colors. For preschoolers.

Space Colors – An introduction to the concept of colors. For preschoolers.

Shapes – An introduction to shapes. For preschoolers.

Numbers – An introduction to the concept of numbers. For grades K-2.

What is Bigger Than Anything? (Infinity) – An introduction to the concept of infinity. For grades 3-6.

Swing sets (Sets) – An introduction to set theory. For grades 2-4.

One Penny, Two – If Sig's penny doubles each day, how long until he can buy a dark green sports car? For grades 3-6.

Learning With Money Activity Kit – Teach large numbers and counting with over $1,000,000 in play money.

My Favorite Fractals (volumes 1, 2) – Picture books of wondrous fractals presented as high resolution images. For all ages.

All Math Words Dictionary – A math dictionary for students of pre-algebra, algebra, geometry, and pre-calculus.

The First Million Digits of Pi – The first million digits of pi. For all ages.

e to One Million Digits – The first million digits of the Euler's constant e. For all ages.

The Square Root of 2 to One Million Digits – The first million digits of the square root of 2. For all ages.

The First Hundred Thousand Prime Numbers – The first hundred thousand prime numbers. For all ages.

Orders of Ten – A book that illustrates orders of ten with dots (1, 10, 100, … dots). For ages 10-15.

Geometric Nets Project Book – 80 geometric nets to copy, cut out, and tape together into 3 dimensional polyhedra. For ages 9 and up.

Geometric Nets Mega Project Book – 253 geometric nets to copy, cut out, and tape together into 3 dimensional polyhedra. For ages 9 and up

For an up to date list, see www.demcadams.com.

$$\sqrt{2} \approx$$

```
1.4142135623730950488016887242096980785696718753769480731766797379
9073247846210703885038753432764157273501384623091229702492483605 58
5073721264412149709993583141322266592750559275579995050115278206 05
7147010955997160597027453459686201472851741864088919860955232923 04
8430871432145083976260362799525140798968725339654633180882964062 06
1525835239505474575028775996172983557522033753185701135437460340 84
9884716038689997069900481503054402779031645424782306849293691862 15
8057846311159666871301301561856898723723528850926486124949771542 18
3342042856860601468247207714358548741556570696776537202264854470 15
8588016207584749226572260020855844665214583988939443709265918003 11
3882464681570826301005948587040031864803421948972782906410450726 36
8813137398552561173220402450912277002269411275736272804957381089 67
5040183698683684507257993647290607629969413804756548237289971803 26
8024744206292691248590521810044598421505911202494413417285314781 05
8036033710773091828693147101711116839165817268894197587165821521 28
2295184884720896946338628915628827659526351405422676532396946175 11
2916024087155101351504553812875600526314680171274026539694702403 00
5174953188629256313851881634780015693691768818523786840522878376 29
3892143006558695686859645951555016447245098368960368873231143894 15
5766510408839142923381132060524336294853170499157717562285497414 38
9991880217624309652065642118273167262575395947172559346372386322 61
4827426222086711558395999265211762526989175409881593486400834570 85
1814722318142040704265090565323333984364578657967965192672923998 75
3666171215982578860263363617827495994219403777753681426217738799 194
5513973127406689832998989538672882285637869774966251996658353525 776
1989393228453447356647354694295216889148549253890478538883452609 6524
0965428893945386646257449275563819641031697983306185201937938494 0
0571563337205480685405758679996701213722394758214263065851322174 08
8323829472876173936474678374319600015921888073478576172522118674 90
4249773669292073110963697216089337086611567345853348332952546758 51
6447107578486024636000834449114818587655554286455123314219926311 332
5179706084365597043528564100879185007603610091594656706768836055 71
7400767569050961367194013249356052401859991050621081635977264313 80
6054670102935699710424251057817495310572559349844511269227803449 13
5066375687477602831628296055324224269575345290288387684464291732 82
7708883180870253398523381227499908123718925407264753678503048215 91
8018861671089728692292011975998807038185433325364602110822992792 93
0728717807998880991674177410898306080032631181642798823117154363 8
6966170299993416161487868601804550555398691311518601038637532500 45
5818604480407502411951843056745336836136745973744239885532851793 08
9603738989151731958741344288178421250219169518755934443873961893 14
5499999061075870490902608835176362247497578588583680374579311573 39
8020999866221869499225959132764236194105921003280261498745665996 88
8740679561673918595728886424734635858866449682238600698335264279 9
0562831656139139425576490620651860216472630333629750756978706066 06
8564981600927187092921531323682813569889370974165044745909605374 72
7965244770940992412387106144705439867436473384774548191100872886 222
1495895295911878921491798339810837882781530655623158103606486758 73
0360145022732088293513413872276841766784369052942869849083845574 45
7940959862607424995491680285307739893829603621335398753205091998 93
6075139064444957684569934712763645071632791547015977335486389394 23
2572775400382602747856741725809514163071595978498180094435603793 90
9855901682721540345815815210049366629534488271072923966023216382 38
2666126268305025727811694510353793715688233659322978231929860646 79
7898640920856095558142614363631004615594332550744493975933999125 419
5323009321753044765339647066276116617535187546462096763455873861 64
8801988484974792640450654448969100407942118169257968575637848814 98
9864168549949163576144840470210339892153423770372333531156459443 89
```

703653166721949049351882905806307401346862641672470110653463493916
4071462855567980177933814424045269137066609777638784866238003392324
370474115331872531906019165996455381157888413808433232105337674618
121780142960928324113627525408873729051294073394794330619439569367
020794295158782283493219316664111301549594698378977674344435393377
099571349884078908508158923660700886581054709497904657229888808924
612828160131337010290802909997456478495815456146487155163905024198
579061310934587833062002622073724716766854554999049940857108099257
599288932366154382719550057816251330381531465779079268685008069844
284791524242754410268057563215653220618857512251130639370253629271
619682512591920252160587011895967322442392674237344907646467273753
479645988191480793171800242385545388603836831080077918246664627531
174442500187277795181643834514634612990207633430179685543856316677
235183893366670422221109391449302879638128398893117313084300421255
501854985065294556377660314612559091046113847682823595924772286290
426427361632645854433928772638603431498048963973633297548859256811
492968361267258985738332164366634870234773026101061305072986115341
299488087744731112295426527516536659117301423606265258690771982170
370981046443604772267392829874152593069562063847108274082184906737
233058743029709242899481739244078693752844010443990485208788519141
935415129006817351703069386970590047425157655248078447362144105016
200845444122255956202984725940352801906798068098300396453985685930
458625260637797453559927747299064888745451242496076378010863900191
058092874764720751109238605950195432281602088796215162338521612875
228518025292876183257037172857406763944909825464422184654308806610
580201584728406712630254593798906508168571371656685941300533197036
596403376667141610495637651030836613489310947802681293557331890551 9
705201845150399690986631525124116111925940552808564989319589834562
331983683494880806171562439112866312797848371978953369015277600549
805516635019785557110140555297633841275044686046476631832661165182
067501204766991098721910444744032689436415959427921994423553718704
299559240314091712848158543866005385713583639816309452407557009325
168243441682408361979273372825215462246961533217026829950979089034
594858878349439616204358422497397187113958927305092197054917176961
600445580899427878880369169432894595147226722926124850696173163809
410821860045286102696547576304310256027152313969482135519821409716
549097319992834925674097490392297126348693414574933198041718076111
963902278664075922434167762466236238913110270343304576368141128321
326308582239456219598086612939996201234156176318174312420089014983
848560480879864608393596492366514296812577314322914568716827621996
118278269531574983802624651759054103976181287604216386134502213262
727756612441133610775195557749508656360673786650623185640699122801
875741785494661253275997697960597760590756489106661015838417202818
530432119044657752554277543798726054881736198267581686283295260789
932226683602838513512281059318591028641508157056319717315183136250
243590414632122392176633982689368253150530059891547029095371932662
073411234947433678846902013904978428521634144292145895582878476693
946464267812219049785636355263368278051860098699248937786002398769
169807656621943898544370805946433362333810587458162354756001365924
352426571430834655457680023708146757325254702550747637471635067851
599173693793251032682760628645914618204721486370370771926926823623
334720379245964691810526139153086280291440965482563873092730426544
662929045896063751918711469345361973324789527070315309309901921199
199993615765003503984054067425387927527922724733566770607837911384
488936261367657060263600315132952095395202854897384486256134924414
708607086602676349978793420875836121947116994223848482595914304528
107062601508969135303017720062717054402090669514915274597719705947
695474095210287872557856880022193717743558110793930883384558648277
291008629554566141306721230848740227121058686323388237413884428938
155444647105755651468435702946635062893873569868688376480326519528

```
4146535173953027361201374203000986739838514321900436028982698293529
399414129230580384565022707216815161941011449826301364900877048398
48838609065336859905458389520318564801493272142390865164999431659
207965953569430723112911629286797517156688905439322035691293324570
208067194440497304943981408227829602799424541083166675921424835182
723817205041039274288801556223380796147512433514731021284545944899
44499600075243751957011668341744749079588209951783768023236517674
97230148745774272599476096219843271483529861119027873584905217975
908374197486026706053746231530039375212367867752848692195857137554
269684827836317861109933680143915905974842858054516130230143979057
016108898627779610750673332676048654929251399781390535882276893732
204941483940135560356560442140176120605131806891989962606184831853
40183623782172663758045524719626617492542285204571442048578342113
2280085287042054889923412785548123676153770710425446968685219911228
354266349997127483660762462418207364666171283947484732804744304033
441072004287271275670279567582429262719454580530026664899650795697
781786219421720052371653694677041951119127046248360511302890464377
511486948878496151188414719100012558838366606772084112351535588112
67789571558590412576261601067513135358021242733187100006358249545040
995794072547989003168265123731190556682915194305370848930786919742
829049038603723116099283424317122250994547150192866648787107951995
180054633883844315481724635480244518030845273431000621371034625733
060012349737443558180965678464641533905146569193245623531405779193
698988423647183525375805257713311200797104068315492665402026046806
81839143782721476906324246951712863673844313983337176176159418699934
662623453734523567940124168092291163609563721674528391709909146648
507392051516056047378710615470216996074656930979442612146925615934
256494019122989514732544715181263258368897282262833295240359700727
863364604594707124174729468775705958157349962848099567839255474240
448991887071069675242507745201229360810574142653234724064162141033
353340551104521261750359028403745459186450472762434207177092979354
010214096464502836834180407586081001407216192477179809859681115404
46443728568959286683197779777869346415984697451339177415379048778808
300220583350467465553230285873258351570859964906867287596729503872
547570879169554736691708701241333922148466851743706661548819529332
272737436041082542596603039869326542235052369108595126300831846755
5034597583955050840356701558879777364438048182138707003440236180412
002114837279422740787378933162708101362649828962927256244580539713
414221451109999544582142923783881026483948233951418767468967831862
86817882755582573193951815531695164501494357263106045694929670986
252043393852078220762219100344692696633425908530581604497802577632
5448937080006267787317954852985668394869467335696300140293131419025
780775816945815272529343422590519791831662164448751781696775276770
913043157342564054922938187395110844166830924911159785773327363884
141850737936300263921806800194982396664712313171902523703199058771
977410007132407519204181221413242532729491860004200841548511547411
573059872196212988541663720877522483769485974767293301868390522500
148690382610848248198167593107772702648826209072384775290587650403
2667275884825218516231074544988758827465678094971230876614426414824
157903570393312256518933356281836185405746706380618398489466284245
736564564213907216305295529359284877555242754559513382771500178401
655305485442285011988365575680159346450558994424849627412711989883
15804769181416769185321657169645222594594712469319957116419861884
797789121142681164383772384836318673186075647785369993038705466322
969807567584682123028077261006969174078202479498821095473343011265
454421701958523758807853480037372471187611100087719035538815731922
513338424947745031188119474559536533660920641929344003507856422343
292324929727084724823557671740589500126876360081245211244875643428
094659313361856432414855780791931151265097295891605299303077105635
245451483457209224551984890588904219806543973353757599824858037546
```

39273653764196748062696838271292001434956674852247241454863603621 1
58472323173699806171993642113631458071198839681295705611588124620 5
88579665056221507482089747764177083787052924202880290044002480686 8
12542207579059424347046448957544023873693604740130860360759917438 7
61563529677605801833493087946627011608050737610718002215525191 99
37962007091613832272801773133201900597804820796075803249946223853 8
58035734780187138028403981200468123707909245727285765451048971703 1
02370548678793364378157807400767774215280311849815576981656151626
11572020454026441299316117077331253846128936763791838537050094206 3
06091032540258476822203676824927947340006177512952630726563785309 7
36864200077666658899328454612246507300220956287272622278080395483
40381096280576492897465184363194984026129976189004678190927370964 7
82787243577522066846540002468330746087835876558905305694257499098 9
03922046300471457205905371209131427588653769314804000087179138456 9
09936299878478854217781540735051706253205095144782206672526086204 1
07996222703480818013800661007192268140291976835488424399162809803 6
18597719358892265485872816327690542861746632308136287764990073775 9
93244175214767760469369622332151759264505564525638405467004045215 8
00754543796810384355851479430922963521978522832957454572715647931 8
50418896070128059492295921835949370745803903214104366016376509554 8
94454190263391196074110066949778024695409365628127538496323601062 5
84653667050765177029695130396858702367912875413588064402634238235 6
80607640745176119088333712091415762805652237901273564193534565267 6
29624402660228245426196034228352400205032950531908532014968045135 6
43341034313292235896972831087395694381318094316669133905264891483 3
28798827628525630451206376149000452186427171115089762827528671466 3
61173898285874253172165962476433238400349004962987894870010518844 9
41186604397391074937573495289347707396386659332554385899935379941 4
38406624221022683285116625113683447328966132105267508937948344634 9
30352785321301278211526859429843756517451093039924958664609423868 4
70021535501803780001870111315193787540109149588908076473345500264
09805683214381160075146182788449046812481468930974300001090198432 0
86663092251381121115994812796367839081222437819101877799403407652 7
40603823415053271741627867488808575410121428667466310361088001818 8
43540182368653221687750411978076525811538417365621835675013034456 5
95936590974690077656309515636628348639979754937563840529672328356 3
40303159165495886112229959996868270142840723914623001617354408314 3
64380458922055411017953513558852713479849378761337910756559954145 2
89177701575813487576801862492222977666211542249711334173960319676
39093505123209476166427534743883333886999799164638367503241863248 6
28418784699609638082751299633817393742209534758616322163052027035 1
70374902985685255958141929549951765525821234731081976633013417508 1
51236775231516073208818295640726347645058857576136189361870128904 02
67922647049496787237402581300834763975644632633549675285749537015 1
27100694464420624617536442894986049205218232133843262753351988294 9
20864907096059216545737690959513893789976626877086832350596409805
01885698199740566005441532138407349781540809435407596815613389643 2
08404153151024324324765063558240978534681151056323898040138038148 4
97352874444290693934437338180109010178859205640690769726390933 91116
13684666699318068382346740389223672292551660244685974607635829037 2
85294971583690637290950336859511823872403865668038440985913599965 8
83006227997529136849456170519932915994923434041372725380763684029
51873698179735366570959942204751059644075907078002039362324718238 0
41377992837396735880963895473938471289506230448447324704438233902 5
13149138594475212792714610672252683555520931014098250324104135681 1
88893441706348018188790385243728434504139952675083907492931559489 9
27997402060166861010605738362034369613992370502059136912247881448 2
19700455646296061801529572677466540252240321520106268059246929418 4
14651692694297031644748922553356819470105586075395031257487792718 2
20198068050655134718926265099870403872393615262809117150163983918 3

820803710766444723112559429793084157485754971284956770768913053139
151928316074937226046483741211241052740458076907749903216761531996
566097430089028478792209894555140349566776329368418912928228889379
139256579030617042195174642667176286007376548254894908236704904178
982794694813371005437573522926259359568093716797717773842816611959
931987815037440223252940166513596488398918771266667644592828072477
419805118340357726301941505622270926688151008740810216304551111936
897033987589916343676655245969002296639061823459922443716156818871
678501955219269047700887628817012435907238421885909432302524000872
836341134600247463505407631743028561032883144639525995577714162497
515992886034410104753346774530437278688576119622685894813897884351
225166906761879132344620772426389891117519375536755089771173608077
985499293374857587940796948901185382605111362359173403913986090018
722454028726512923507251346336039947797211253440796964196584329248
583896153707862544624052734183729616587128099621567514167788885218
201782685794750886056191761433450530724225794421504401189938032821
176942753515500503593840261927124840735344805764151349209066433260
871886931878391137249135429063143277314147565234427698926410729192
964778315226765309633771907887021210173628892801332066553846887522
709821416994745346783973061884338063688567887509348371281299459471
416740210647944623047509596911213284185737450768802174200091903786
114928999321369828255050439412523429387891529294488067290453371558
685893911940586799267968019751929463531321204605827301365246354919
747717843125514719561089448171687369595009755149090580423770550765
831660455263178819159288580151410990335159992764926020916753796585
654071721490272772072079533046409492679296980145647407586168417518
270355419152328590131991897564442720919580664737853965474943503366
098455694220541232209149476985226606686931349412846052436006261919
200954559599299203576635844725208888438770109848509614553662505648
222331082774877124964592394034410384880456557209153720836923704220
390308166921534433655552965914773759520794595970591492130243833379
570937471630364094522401198254550375439726080376366587365259895269
116799601027835888111571584115744794740352868900094824133918451378
059992251898473594116542190094366985029180072615270895483247691079
054750239576659419788818441052018288716411670528264469474464709886
880658944170090145701739592379880631201342950834144109670046000697
066301139883806544103595903638542889053395979476135555393092353500
102274640257399654926031871210054395169315102736251469580843669483
749113385315323780426247949617769295622506132578266616587528170615
484688178446049181585002235785076884446778740147504422625751101736
486183263373253354192130963105322132562419461409302789335825311735
548312186188185808274496642586804545788804160409610003849873310755
633890347244188014704960144521546446352768745300463429370465787461
230411428584139315470696445598862702341355336961468894130579524344
828578813842913133400443882694562774572044648927503718896002502 14
810243226623247446675922220957579680398758330118809412348594105539
793115016614982410947734045594477362140729819067342271221329828201
893551814807116129360475393545191644885191976819894324688266584699
908103487433057154252114898292494064135204869933170852436546029080
065024215298598119251209628332680767252801429242638106620922006716
242390535394287757500791358700376806504069329046236701998064642381
662780745584542797281363649612815600633612819117180777498572128490
965404563806025279792900141167180888216727632369573144529953790935
999635720195691053823183323228796036615493435619556357778872038051
407147028627486665772728496380606820908743374755778931220716618329
099796807740695830545505816946391053802547129809501916405917729214
706120537973758318816904034573730892287048764298832763431061565604
142353102111260771177930873317131713383405539907714874603978634100
181548943060212764145175368556781241048762789184459864887507256919
070705612618906511541537434331704771073512089749664585871232326061

```
2277676488017419041705326219011292994134240685947797064845451230437003620444460307001911450415420812728658532659862596583412919318743374681425409919891908009325715225950256975763625837854206876078753364238902453524223147811695030742299688266236189437975463431486940473705466022052365853522433491893414953731854617756390850630149805930744131272144306096868121549167999336020911246342068160464235634526699224710694338071397510765670422960906083000214645676075669544782233742030122313638536548811896867622119656961068380955681107416168288872914897089291727202031664475298155649165099329642475935553094342723473525567423358885410408162561808385896759248129675817885803743143764480510225754841455195600309544841537831759812222780836261003829621913668746990601355910987901789394509507080084701971030094963485673126176537990696366534238478394751006091980236622402583791667733659818620969012370485544178833081743224152709618583973205487774714196202102091912200354193993755848965796596777487966649833738446464628849232209159662520817710945518557849193422490391126440162540353415291859606278442584979907537650351143662301268745038748418190800021735115212731055425489544564305352174412647603739881432392237995520803725631538278957675603269534153713162286842231199071736627159252166809502575991355368041140314363749285705000161825898070568736394803223614046040114806234062268309640850645651849478620873085379933150826863834559299180029565026007129052213461451425916832384997363792953241667672636608451393291270052813463426544459202882461427621775375422965611776774352683633779277270530866888712035544787678456074403451818739133581638413567183639815500757116010247023062209136180213041332617816241751224462126905137587603315297832296688192798031720614092427303623900983851487008360779814038114398492015100688229636700554973943285334706986769952794008280800408871510193767467730882920960774142894924319076599523269541480873863532377880542082237256367753706673147303496863970704898031245446088106656204930116583699242859143171957326639117940460873354170410387477676568434175416519619874924892507420893760141194424029639425417630165604729567661334499190693680735551658503150396913387483435296368054125218883032166460594553739445123752178397465350332767285694918193988259952860975698800611916256873292306230585674189853838045570809358377914628586246872311025994015876757690831677174283925303011600838846048071478361609641715594546399221829104927969229251112793795664013381991163506963539326360985956462057610408705683800302694519711248638111853378910981402240524583922160456345211221741701040567973307539584647058474754960443223342630066921057776700574771792447903388417399895395625218314754341828854229277388521031765033181685116423984842459090668564084864712227534729176002772087091119916253536523207304127403457281143076790748835435004321724871067033312039256058908763836947478428601451134482256881522505295704764411457698138038591192713356349551780814736321077680156756794279876371559640348376284440532510433462372579811524905785545455452128459612304154962148839747200361422992055663045249896473658280189451430478607908479278319769496910731263109830485633506795521367459452740637830001391251206426019247343307523851977600345083075542671038439897795080702918698409431888452283895341047458008310255934322333268376672915092233641315967653120611587458495988481505442424328431684072867512430343206581773398533408669746044667571456675222004330063396653272711639216282843359363915970535605351461391484568462400484626935209063545366851485497344595022174912956216820534347169458903066650112239953739250769149599980731164875683223107356012908889074480041139114929364660679542826773578999601714474747822597197369102206176629747963359413820508653835366913170891009758629322982345585218809771609131415612283794230140742857517219609708392468807241064334663820378880194299968923340577439945587187721671949412173978424845551124239495108231334190009549809749441168544692712681046436762979
```

```
951628949214864249196131714911299164723580230000506478446110811914
449115812299764633424470996174148529445175829914157131372881744978
663466979834683063377226946211202454420349057367761082493287387912
706838889328012987679170081110139014998227247474545924421841219503
017784199339710292780971003172800480512407455154973275048680372399
013584337669127305599571328206218815992141130876955806638951579665
775394244413241315371071369661381734700574482745460028185122194881
094099093203083696286639456596762750124811505090895014507857953164
711383535236783150519442462603033024233866607777623963532367792520
543071072814179089963458476941987851049288285046330076222359001391
963015771950482636735933640454911225862198229883550564611200430090
520331006969844714235105314071504276373299915834406869863675333351
079658921179232619884213541214952878833669486524486147483513338067
666140739939721171174258077014776173413160355802526005061346050994
514494808115176759222658322435066901650708586969301664407829476134
100953558555025555973395101109712982941416813316290912997166393986
770253895589191479403266826999918493493521145641673825912822446260
474686998341002689357087169821589951314472624674435142671224348360
464877244577905302925494885818976837409053632700690035935815381883
280740828133022036509060649110278848197191631340902967476872679485
708198499821462735902524243518033251794029613213416361860160670646
726579833801692963268988520060136260127308122721574835345068943834
941229354390242537575276935242785398862908394926459788524969063109
016368095320411338520466963037354159339352275611976274730671323084
914276213000530951932312884148955410680616740648799311628819724340
996737600093836031980956229275123373419530042271675265748542474 8294
603373865038667525655809964547552269265897588852563853292193801975
692542773496925611476888086411726702213493152855142211570520058969
666638052038104514692230017952607407386746595628401458292793638099
004324083574265721606100589276934909467829254643104382358817121534
163957942829172179362408197768819899019978733140361218579680658635
836814653558955738187837879210025324208016567419832526439504211082
153922017378503914351674199577152156214636359058908552855660974755
528542340303264688623171663272306314113253065883613291776807543214
046263688827594471614338928326902599705018230149037097799114932182
970034345553862403942080577861275758494151568256422080453230079309
019116615353434782211794369044850125888462028698089681465461734 9737
831554789210426289448390506854412032720652374931405325170693575800
090799417928559317281056994376439412694664934570902480722236062517
133594884428904861879125690145136706459235092312563008362115495698
642742822983298049515800422796155220772188169501772040152352682132
562384768196467946508310757531703922608556146933191076828363656585
285066867708773314799646708576262754660296271676307593189843573386
709657017379322828234819216178187700704734967289663187703448331207
245745902134879582560494827184614765808417211499946854803681654737
055739119000597890021034377469279468790761859868970990466907028633
273646321478565924562227186274617896242617308191565710953780887906
210137634562553455427872498576451110514792642181602336671418980139
700712972986873282822161863841631129139087937144100443787731724496
829823520214297586440824917781147307608372773515153045369225956908
916303290224998830272931041640064920140282253288875902793130857765
623765546576499418464267103698691707059191385989330859375437465778
744024340421767226124563578400296687454724507237893772604841729537
531149249611862975908219854645299621591789872203301708655899813711
247509633047309722806506025812936326292023721769898118542181621497
036652310506492195857613634072272973464467527473852366778039742256
265274288738163396279152344894750071035704323055086330738705595 04
709704223112376212193195753906049935667463010742768459394995465955
497215492094314313600228526551330882682568446783672070869302727333
326543140936959799774757251648656093889015974966204772135307231257
```

```
76395662827622901153667810443564268355205959032732041101941591 0623
13224980121075141619321442104168538720916349784817562178617762 2288
90105867147994106015776350936484541015472003561105423001876393 7194
62379052951539126455419411099300801282463142808366035865498594 3824
15741276615019556596477471231145521988678134484244364073854528 3514
09647126347426960370617561065388275867938635884085367748880421 7287
68315437910382402724164103510769580691860549061336294668741293 1799
16557774775664852417361469670995689731625551315071526377319131 7175
16989497832401686871488313669781158182326385822972697878024287 8221
21295698816233680033879843352183322680767487176431408751324746 2340
52349879184847668268295616438928794038967706478767283588109660 6481
62531582424385915943578234529335797618609442793133303422963677 3288
21148012516066324362292892965570524145729258190733355159023087 1119
57026902387019554455972339436581286903114329458125014251884569 3967
76835149554430476026363981924230113099136607386910994818916701 6589
18535224610165524399123034154691509650613337574536443198759207 3348
23839043618145230004162856240178989661872993243359237891279256 6089
49415876734705551490720707611671730965073513646926538883863019 9300
30541130154220297873483229683717362289165329531827179977341842 4230
73500560498706316137883700547174126357960794405301742138827466 6841
18222343231694126094014895249875267394863904890764212721169711 7220
01376013590455819175168779717618660293497250710388274361157014 1066
44578105100135067350691284817103249133295514721029682665729358 3776
00776585271406735718234186091576190979074852627061960974327839 7440
62195281371868901178670869015790015505061922360571445739706041 8587
43996318822577354587091346279727560768912856748610370590980611 3218
76045048742172796985312160568894934224324078228305238271344062 9583
69171976697184275805197218719577589349082463036176217101228587 1913
61637847857196761183924798254488891446680529057852098644781989 3689
29078732480501714397606548833506889116194809729553239596084126 26285
60342743775801058132345724648829660857216592741563562814964100 8049
54256817567964768010548889433267801048640465724291697524024598 5799
43231020276941986194545670508140448615912446109549533985036181 8922
67470044724119922280866415538338996517840392327919638032934090 6520
06414740476650917711590990708088204104346134836618874309123514 8789
56003809594186791967316693219121047834208947318434885713285636 1440
80344037386737479499390765225963002337807256201081672843774094 7482
04430163722131953928089072365430666440098227698510338321679833 09404
06099575235990494712319398719599051631265725722617896465674770 425
38713251064353097458323097084107331189812014476464577914235732 4887
64319711258271029798752633428433557953569922565057949092308669 3770
02790666364522037314515749656148162618588528566810640204720142 4332
22213898082335243400026257218104923740342933520854088533181454 1802
87778125463001936557949148442799953029441147175167081886444216 7070
23391688783428021896673288216515803065115988169156137091318045 0124
81261833398057451020283579255536354601188115754159547114922870 8473
35885068942708517824011738597558630393366332930143668899742102 3173
45796658515711559591969533870407520797373330628808526843124917 4317
27930361698319604976090563017785959731984177147632175931492540 4713
98181234408341869060733107306870228133005451668137079990879890 8963
19546202343993187247995643712325767418916631641520305278194069 5740
48089437747208541996957462535350882405618255957108021743597126 7
05069402546421153273914034817493582344460288491903529717034361 862
71496582233771781042198935335106366316511744501656381651924049 6156
25150790861123581699291613239493471905931734777040136846628050 9806
83375370187721517051788910354822136744744507147985432589265397 1791
54182200674665152085339698698557357884979377970446007658652551 169634
24515704282934552014699955648618896440313524095881591474720300 2141
38468364419662076817723310317917887052884598193251871105991350 3525
30179848772285579328648747510119395405150337615279010491080759 8042
```

```
2869140039331902826334954729532342060192780949379984756527098415918
0738523388418078142377053699013487160743211314885094546739362748792
7656294224381737690756304188919928601138564573088018861331632924097
7497100070846351948207918454076634335102757004011619401162271610395
0657266740557225600955942894373864738247050130431656666130451464655
0593711233555171795797372241171206870206535219152026527849027407531
1026265343043315754436881070612837812844732501052153518010772445257
7458385730783648224018800322616203232907208329809769383668685800069
9735187473972029591946690234506436755781023849929697274553414183066
0934498993121577722419060268006235706742692497722261207437890584076
1110602576661587455868756225825500357340899237490420880992681826277
1485046403102990438670497613047782813172983754347584685622855153168
9577746946387006403775492624370371012909543270543316033372828978255
4754444442270075597793743712978253655635992289769581596616805701963
1466828041969376265749532036758395016725071557075650006269070761679
0422965605952710360566741912039358481154279888794065527558958744552
7240924981267747274714523089388434180073890087243377596470098193649
5522545073441552275209843293401914171541498330928881006369427454728
1280789229402486723336101369756372840819922663761854682236519361763
4976478451374737566169676288557891706670822511710109432328821940163
5175287938873634954626113660562653921987701504418064664471678417048
4103012500577844094006298771448112187534232558819875778652128925470
3753355233263969256048851581997712443529685525012576800575465133654
7160530299192825913021561030900505994487359078573038349557135093066
3153180517931158961867762547087754711590881373036345572061857715124
1051502735643895563727276609900121463749378365699843697242034171875
4180098664821318361296867251676149629666456466934877854923677598627
2015762841692251062456553394374917413118580922685713815582165894450
0202706375824312115684562967095555271529098565312333672562403364419
7964506373652252082364270848450282230747408499307537629218713608335
1595193920219186192671193701358125212484822791952216151439719087370
5058300125943547914250503856428918542436153167130534758489799395103
4668713876226236237253854867755690017467543782794662973938677387494
5640043196206581695997083207077665032120855635374157575882558312929
7837718215123197859910440863396292730764608859399303287343066103077
8989234534533196297244687315502858747540107704119551610317085711629
2144663557254973801299238863770568941557612546709782437991333337796
4873736012161213902941507629441970311220004144089656360853166155685
2295230732916130184152208112712112277341449799704226751145566895102
8081280058263499734188941797236366054390777750164790754951428012581
7646962364117033609865560475961247925538624995714282292955679692746
2608847324610363220750236694712555665528693719752296941077821627762
1457198546752878491178355064181775413505305964542918836892738872619
0237757955877284911783550641817754135053059979169097939697924565400
8791081455139845839676614557724568939555315073946669653502650996685
0271443051262931097998079520353299643002444634280240274736843076796
9533194918042344568101563119928948343984604686898727422421778149493
8860501475488437997093973498185291352533830145315182879608645995232
6988283347100236401838291725278341273738640666245802148727756428972
2995440605975135770019869916284725549700160711802045789679681585646
5731426222487754366947651946369383650656678934306239232666827511127
5772430334941335651369199105559241378214475688185781780077186548780
1842221493223906182413560177994219991753876511615785766151551010278
0246280574131623372664281261390015621266261922300371215233379407660
7281811997397841248780786280472984172909617911950490054527478842891
4746834379353740978007239550551429239052453494913696022031391753109
2117398384927473343632030950911822555561599146928761124663047656336
4050521360428470145364228385272067558789690790405386115281275577337
5471983220323309897611927640075987280463182211498731429200813670349
10867265693969157425343450
```

```
70213579000070031696886368027300595763575239160114955002117871042 5
87197128809222305905743457885861012482476647303858561110317798451 4
35659416239682506612220734704433659690571495163939290724189904597 1
56504340511480824810215705437251430642280446175082136556694682
18845397273062639036121831890362753976560136913970052841569734854 9
19035023490451602585364379385478817639655917963803486228938713533 0
96155087418574107461096881806667535070560182644715189747984686603 9
11002762680380730817785478385096636245047271387897025405364299964 2
11314233619832199795409136576000393375079861518000175643580478329 0
33283471843561266708710954866363043225866343793454780051780011756 1
66014492176528084010101440369775459722443768278944292549208167967 7
64647197217257545147797075211863098424510087860132651431648987799 9
47854943873553254379024786728025848361510402694442455156323601139 3
68422462509851532099476470448560512625993913709189705852386235637 5
62692855231081896326880675289493668599793184604134237684440336901 8
49227468856083911956169695876401290430511628127821705206972452779 5
49504095761619333598589572277102102887149373490806249597738391376
66339228330184067875379162021012647111774030977087930222108946883 9
37295481893909854437708501418358182991808305774515064776260450541 9
94774782158464281318349971792158319205591128414048616145269365955 8
61855464844493564688907781478169958123011218965259351683133112142 3
40453646388211350328896987524840571604491077086777802178182613136 3
83399390163577456756513329802954798276111986432116358744014065462 9
93561840710021577568164299571406401295906901768433327643063232085 2
17085355735886569296877887589645482173884335540903166206051304434 9
39375711227965527548574120764798615024076571465819844366219118382 5
62648520564336574690135042580985766778124052019290114713099710492 1
95311483750874457215436533840579277426291932045801313717202040712 4
48663189994721389623376545899837362761587357815853603314227084459 7
92805042400693459843042630083662968532617864045721873192446699638 6
96454586680008134507336594803012423262312133786375162043111134116 2
38211119783069551958914027187876592649601616857888476222074761532 0
77193952960225140206240821732843378393514233838017678394252283483 8
85434209224077105610958918457376838528575612566732358885920947474 9
36431785650705531298667991193175998658992092891921781617575202113 6
52786114767825751839802921408203900012760728976858600811752319802 2
11535122940876574297746368670930301533354042505876419710508686326 5
25802333436233049391551201677671326659274888072479646189506345714 2
67569003107657792403740183793650784135113625589549081262229441687 2
86614538937620864080410509153653122798204347981903510116881290149
52981278293406133559720783871832834088196015929992991210960330733
73712169024641636446304874881968464564757094641689688460871409599 1
19519973254965726406483843922572258412511918769939622931453201641 0
95088963852064915799311485402280581550241179261559887325584878513 1
72605144302425726069475882436481259595591540465302362984078996353 8
07707650948558516135354146792570105749963139986912371978904265479 1
56242614010007994823832489417782772271882740652836291392018233281 5
78858145033518522318538874421508888669101204815843705404920469111 0
27400170278408702706485799844978158759769305805439458747805448319 2
44201977444492101700884184024293611319537332618885234142417406213 6
82357180373988155185695944249736273441280874592245748768241843922 28
16669227541594855008049451382268726790362850241656065982805107741 4
60640500054848784878194182423214846678122311690563230281722262857 5
75946307593532477265942060542960523988692857374789307916898837007 8
66446739050576653805329709977873116913720190027659226188534167486 9
15203363948587646776288492694099495288715908780627806987802906139 4
33747600048361453496032824189890282094520716524778600924948587110 8
86065783034906823611518317458702866204703590700071818222989617616 1
04520552122083050429482802786645547333041418294918179647407872227 4
09447128655909107702357938704193439657117639818889595944792196292 47
```

3840238417170401560069957035654575045013293386787871351643031624 36
4401978928150319374801217192232827989490260443521220210131876146 67
3680470333262966195861544927633524872579684301440136622280872421 88
7991036563096802886695557226651604825961913861407434105417679505 70
4709553156527674047832445921514135868197441735611598345832084084 34
4907750814255803855761152992014799823106094806549578398107791069 96
8526534395370798191311323826931111787097508631804617471544662353 4
9759561234403806863010798862767381382910777335923984781054809303 69
8335134666601107313832505289766868774592093583817647165641264151 77
2731965298912217369956970562103224503858375434527718783154676505 50
4468577493195564724037411806028519235620471092580936470766384550 28
3280675503302045862495677594044142652536941494997208289999458426 83
7707179121519311960934012422901648588197496347411904774008885351 12
4170915759259122719122000235950386050904234324610509462259337272 62
1341393843675364692759959541052984000108841326444717476452012328 27
3621758989214119399994694844136951436961976384273325628523252921 95
1898917284524312385225429771089250876033978037048533448779558315 59
3667321034718897329984538801194738813365572524297519344287007200 14
4885967129553680412748230593461189971788069173854467145770011536 96
1231618059509781783415339900139848145535409937740499242699519686 18
3269926994632553715586922876186050606050706184073400056367527533 48
8603674202347515637368970388674310820986994049690502669479231473 6
4078711585810550603846251556323526324028254556483459933185587661 22
9766981456741954451069068664115283619163749142296258456740251678 2
9332444571393554384836295200022162297372847238078395312547695329 69
7841224676476689775050377870795458509946573476610595982648479162 1
2726514733371497478606872487465646845799651682381804748620863939 97
0211438400362403127695315013453253186574408194072813625425735688 7
7226023176024442054537836650472173564005771250830095413949038635 15
7495643380517325786996478932729971949602626450335988784177035733 03
0002343206549165036608321381933528878014044331877146112961299949 59
7438947395635344895104947848976831296173449650652596840608924934 76
4770899420728202707014614871969807413291500568197160097872130332 3093
6449111778173744090215271944017370233250077397024706959185138767 6
5144754537590796466789330168775972148919030441622356457819719488 22
1526540748384121785491795152429488278402012107708165585019139828 72
5276567598069894515500242293174903206502627021836811658044844764 50
8973174141367145920832617354040123670082851414247398928550675383 34
3107753066552631684379041994424741780930260210003173294539408512 62
7908876322943319380034226576464484102189646773557017813703606193 75
4833525859904988387033038419637679544848367914001803816374728481 68
0076766475494039497239686539022479402421540453890027027616967288 87
1242785794795238403751777314660833733946302387672935517544514977 21
2438559274663767369439396485706619298086379956710744012585026965 63
0458930787880532438866673038140754973411719656042220287963377918 21
0228805320967368150975189952089110659986698885263530837619302157 46
8901153766860899679470409418062656550135286874081599388612032203 01
1159406118244639930048265609456117035910776279625330417374812787 90
5180836185460529855589629232000900730542585838683256857533951325 91
0780492421176804152483135188747160476658866969164308104819396613 93
3602389061947597559949774851087586679615407585727814193721620311 18
8724910605586404616099981771172067288199480751576927162892964300 78
6293460144671441856000893148084450095967231038530416150498680448 06
1658397971688694173936467841076744044868779690120070321004282737 44
7479621183533262324320473580788806327282315854970470827249187721 47
4521550400859426040451091208837343252915292653247959604257313856 35
8868322807014003026408061288006000889864191633905998244868218719 09
6493984899233830149471017508049692688882437917867061575934464944 26
1647257186623828761399086968715607308285652128059571847987956018 31
1977383027055992317882344038451460235407226300704578724852698490 89

```
4152609475958303377522393476787527410442043706561260937141372654988047067382493931237911865259512468358280447735099458072922393113097279398739236181263194309648189853685994482001115326184222906627472654059177362570543376108538124744220049124963273318695876009656893115453235922138887682882454580651675358447270195928193480550637295523133903311340767347423415390357795782907452389109401299267987179841169862341954818305702062658224846785242252436920135797576596623705195199831059811253304419437017070854095177972071003190321025453984009957070942503495150746856825980488147179013195437493294399761098025368238696911921186015299676513178634812953731989724093306756560955924741143521960670690556920536104484244769384241107340114278316700302003015720229251048693235061688787988969301675060274646004080375056756535803338066107458990020233373740299070189902028605763751232647218642412537402434910892814966098167808853565253937850054761967825513451930242937933658119101886946363849445012294396359384398370881217563652996761776075431925954975718196352433073047754533528496174798515452568060805976165223267351645319250326185637617955431931660513667169056027137983054868737838614958230647405707900055095614094148846646892516754902088056624848504548278924376071069610068178033701718533779405456967726082477046081330935274810370127029477327489274967622163267185884772368074416620126647917396977166943779859700682687364538617004490235759286375303570179540375491673164960133258686874457567031896575675812847816273181844956202742753471149364542573327681432438939484704170121614743916434295703517455006492339273189753185442216417129969425228032293004442589577706722997943828242301565275899905088901270972491725795179603430027908759536936374942315788139099243210459633657286189410039768953962504579309721764800839180643341684234559596063605140764089546204553894014532820227843158604855652349218114981875538424592728130968057352907443432386422067369905249986915463269359258921088697643636653953043316936964017728699374713199142194666990188196697929507352626178660980578845864764830822714119822044574875520082959672083953696519347304814528654444941028203353655824206922079114652126512379567325304406442379543334821772596401045040552152959052371112839746180348884903390326196229563405843217266259091872438043745553937279542011816164296406991236138911467050840045969769664037411496707338013580469408536577792630930879792136465498503838637841145415843769910612781475640292011584695306813986984532289610168793263478338876889693953526579427398779547470782735451209487021061585757351160236800128034370557255625628136319012179681222929269118918021141720927752996926560410082331193447219766611841373699520782459591169068990193077092069948814403436140507458296828125875278272639945513864136292080871688130706863338330854547869605583603432740471299417082864517904256443350944847645743112317048178036452077024087774733749263104317507860133169367130127357010859769247216665734301044339057977426569205584780409818108407733336140020516333981124319584817819861339523803974993276983590221969704659930364551109698236136845714968140930715263184937825998116719460774917159333173695518691040162381274478781444679099598450990105397322033012235439745285280640806709888386840835171818490211197897594836725096842939280047649454653794697598385911139823390122193866891492010815357765522498565703094058519395681997171691353579840844267086540750698988333848166559548976692500671475854403353961845164435748740811532588069232409471555444223103379511376956704623279392250007706025119580019429886442116421620329314077052939545395250691810147502526231287593839906983594569908481673840939550592494060584844701837224673206591092904536639407414105334304051450275049946264432995162753193979427723699185718024325079924404555533697823341354432641342560206353430010389981604290292175532054322504650029194341686140498878586688119207039528953193443334871768241700998962859352542402410701184621765941752268136830818831765200
```

87385758029661017971089549127570947046489759129639007254667188 3421
47605130167826403409147120125602598837513243335844131697171767 4156
29859478458077885114464255059160272700568192111366269662812553 5583
40331687855546902774658573359081661542195251366174312006329379 4206
01306543017582598239468956494648116545055390624462725406525479 4795
20219701418780042911357850041499690353504628814180305514470449 9175
03150195660058811212783334496747320124448032344802415393768109 6744
82458241312048590792681414597406673446850306102774783160416556 876
22785246875058825171361852564847224279604285954702104738132988 579
64535405357603697315889368419165369030202305736836086700426790 93624
23796784531756321205149225137647815617416945610416541243837594 8739
90931592744759832350597374956768131105471847279354166515395579 4820
36289678773485797629754502544180155935890854389844222603048640 1243
18039638378885276773370192062902845451597025581202504361006001 1125
31944426879016772637322785841600719933913185599409074310795418 5932
62199463893176001227990537893419265036013681605468564329550360 1957
33799901686338997528893727234215754501570469769720205152135293 7362
95242968882748141252048116650708513692920803795632514256690878 0256
34323429587433403882136167496786404155767494467602605282041289 2879
62848285758752341071805593570366530432349131209079594106357008 4350
34219422627649478365640786969808554542081740120758828848210566 140
21730273552277066888500781041121548046161417849974165352504936 1357
94776622784715986124764760047228960130941509901662467221849188 974
34467121597437223325499296049770726302139173065441729802795980 3490
84083968689158617693995858034740709860552853425403750735096292 8456
03705860686840495194231816971707615567104584517241361922202643 0537
54028443119193828048336531921493320922072547282506580404252101 6967
35962650496651073419586745688796717488418213932617116579362931 3370
38775652158908792452762358734202425423140152449334603162278308 2427
01173785145399042133334722508677016724357299122652803938725712 4032
30922229537968296464216212785215379601136627303094078844958905 7703
56652779293806364682136117769294951902200134673919809779027300 3482
86119813484546377912772389286614197852030271658363892153890385 3664
94876123770645703230255802861929586723503282627688897726101265 7477
49494480826571174723974756628052551858982567229825928322231387 3588
45915710606585966550588358874432169046454886803341334813106611 8488
25000070160640513375523188953746504609479920730008226393564093 1044
27957490663780119645396079426600626973342925246176438600440721 7669
50864969268391001088420896521097467350634685962265583283912523 5456
45272987766200762220371487732537070400155528825970102275280159 6943
40899625067953723547753999695910112369758569177202965673610612 9758
08050773887480337838161904464500674830561380385278213603474559 6011
81012788527955584874209957099753233799842699695489209017419661 6941
82498693048400010174476727654651333965743064219836533659962808 6516
29686035001589764756451087762072682963738867020702851539641909 2401
61428236551624029650653572587665192464711535810871295539339854 0699
41033002330634579855310655236896611325592313459648329118567672 3039
87658897687749701847989976051218812346326165330263812386365977 796
74033218196650540961237063293716393797098273098580710232572012 7774
66350089759975104880602383527650773792309294417805333322727796 5774
31198584012256682418802192089628319963930036038411146713369697 6928
83934484912602643489065656225866057754305087290995914211323424 7415
74521721099726489807368890170132471888351212088296396370612147 7926
61230055051710298455668792986382063016098973104006337670839080 5125
45036901493804379108248508281594436013584745840929625587151841 2176
34410197320044553719129215488827160303317100705888096860077741 7400
45533448000790761133472081969235273640713238527458709819672647 5975
12236925439442477598890292096224000368959283427301680829481082 2831
01985970282376390800208691203359588971802640208471924301693547 3303
65414618962400317081137392230856016735056262924146462893386205 0330

```
552200979538959383536318097955997708688363570126870314531980181206
254516410065118121690833521506794941645173040395957204680026729505
890255151418654661138406568213727380092844267258927761475821432159
083838108325688200027771555348943747671302251419506010389213896395
669029313148892399142685841599917428266827305403977451428318952646
229366714484495738640291467612297750018104617279590650657565133276
381016619047817586644566576952148650187685610264254197311264100545
329300683976747073133363591844291314229894108034435360656456635829
404850487331709823240172711447489727703595681906320291182104232616
505955565782133323546654482884359061766605790588614248813346357 7650
590414668928857764403291770424617153919437978562515807951484779669
474821402330746784641397846628354250386235486309926434804102944456
735296987604609063817809653069826903867344408835380167199393689207
794046857518381512365447286840167040270758845473725619532434134485
260990355710206505867908519137971586214495398498402350041123635368
078381302750691815461735738247624757608321009623636855869762977870
095345565871397398632446309132942042736962425350234878975343551932
814785656836126483249809219652251694822665983111598350229169 1319168
500767272203713642274079617997881037120482859600314332684314168446
961258365213703499617029470673383529492193843255632077859650145932
175930499231573581117162029901810116080291259736532699748993578117
890926453937695106960835636964666908632133261369418802954442725957
036758427803349634401574136767064815209147423378181013304209477479
603031281914245492488931706073435123341575415823991861915009487681
270693664076932568487035256589626194849271556671689039365798226017
000651861844156593812177985717355804409156075986829157259752227205
393440982101817789122843068592282949069589653369796636933509675093
566416810053010218058991844283084544177516147468041428961544543952
329090105800093094365259910801197140023518394513396121349641381946
888980019761200714423284974551915971352136717259079743100799256777
862693105827705007927328057219130264258627625438226447256398317579
672160608737722707000211383634470426836275780864913835914402751480
434592898122034547973385301745180364729483484180893408909754492306
224183962125687078152848871559519929491655770428252749226828350056
233295675340984337759820430551727471439465633616787527407428762750
431745454457020402401252976164436649771893311823574438586583115314
899505808029502067644353594485116812791408139966578732298849318213
999238289018453372945652410348441434556407827131462559475948024849
358397924242341648181701621085629510856860862854488298064661183419
365886457632109904168712029069324966634820415991689866437980848440
031413423221721816588472601324356712558524510581693072846569177314
291305312477189440396949896369974477897648609233072266451704562537
181861273035840595451636648956104373300737712722792001565679953780
312472059742971314129812594167703085271810057454304123955875464065
960326911946097951513448828080847313734544451122987393090658853204
458961489329057930452158175223990889528793877553188106996587993204
339705348999869698208767374943265836499721544855267749965769677 9353
282746956513157735420665082981806885819351151272338270936915963092
626219846068117989951037772265383401890711111988867376398054469146
749241853019378628912764358199547234052453122115015587142011411374
972171926059633275477219118436855933995321993692122068219052716822
095132194104553921005906743148571423144596303171710005924264004820
075398286668975411773473108593139331501837998492264158546951466 6941
537005433593796461143682055312126651162727692102387308636872620555
056729294175313712034779625567845049020940257675630584297124679994
304462305910173994192049501943426578201385854063888319578434163691
881222433892383511437432883462883595907460577093781626611997342538
515516976845985028460355924847783918179869129025446416990685428988
369319611820644335057368707024281107021668172609320522435330267075
330569203332144651770638887253973735002715832329020543641761557879
```

```
0858061118085839758480348581087965949020284403326988507888504106045
6722309171254776339771368572306124349389856991502567226050841297235
6999700483925060656204091896952637869678798039201415048287379058876
6256578055000524290627346396237562305467911862000163560419899497955
6262430927209123299235754466535776989381703079170253241509891565875
0272382249408621560007695606110697411268949685547104256516734171866
0595238442317248773770846238777219355127920059889236997228415004826
6104324807747865599206997539757463650379444969836317508737545051325
4726448748275362306060796615228674697259356751538157514952256575996
4213538783307152133883812878385960373729952001091605335838752764499
7476198635816212531228150181290598395301482452541455745452621273816
8857332313392846397935640814202697292349012839202580905149126183905
9067506188453197533702064061229430444427580506477915974933657340155
3392639762767396333624833229357444011880831116944148793169006977485
8853935373119717330895839664455152304824508124284162793149454322365
2058931917886830586995950312203552521456144611884421692457051548326
9391612612199982750525033051713666089422962932556199458794439031896
7663094484551657534562561057959219575553599984412424328843143977255
5559589075525842268145194486945122853586639349646490981969865208725
1134450091482814575588969036632680491316675661118532103201778056645
3420253234631147694278791231386635417387358695733618678606453048275
5794051814778720733980792620300872687015118536151881684187956767315
6174341165061888152624745762334738306523446007337262281716288921856
3899512837042310997268324824046897607126107729276738667485360521215
0835442644240834674890726121983208131366251285792093793494479842145
5529809923102009456786285159433526413736429886205184199044485242215
4306547560418929083688281759775337108794523791704896873899637547935
0105858351721624112267767128161847447975446757824262325662088509935
7467006731076414668035938809978296861139821433754520542243944555155
2245524110567428850699388449116180513413380829894631195694471159195
1789778684086958853613364652720684334671280159543248400595674643165
6021332618049022546576027528512588847320213286122465807284837099116
8177198117734030314264853565700758587719192824024330180244489425785
4827799446959295597221246860677731041188849203345938704231535180235
3122289583582557009421832811155898240120992711169828439870694491865
6075852532682272024464928896027290691817243468326399823713580426215
2460864610597183339900487004750418704965110204935355114692070232826
0485890422268996329515587867337167445672507594538552336889139645835
9881143569650792751072572818297642587857494666118795789747773346335
5938851878355343315795342629547986724358087817013695430411996215985
5303921882563768746332996737178849532406289469138896654004863122105
3814123660078667045282945615307397123840905847630129373126442596735
7451684545232820087731908616066162711043757003637536732873874570406
8637117902815999987344604440020963749298030795213343965109259228565
9753787538041990901269225083749125520786126288299644937751486613085
7880304945593671218119520279145982374429902140669676787582617848575
1163705258764110179601900694234641052336006717256153989870650677785
7546967150366013341871869750086743567771420588422870728904770135875
7513445537269747102568935937738871242435412468286561822015895005735
6314874708459461459186151620377292877504435242048054491230106682985
0469261300948122957667519727736061909026177966586990069097664213895
8757624388014291394847634479680888156013466821247743405414877825635
0926484735038005759232331400448742948707199797051932422556560455115
0918234549357000770355653184597844681479651382181153077263054887525
8850842519973083805499312622728773567734156283998814698112125939465
2594667985690495514133950634842921140820295885516672168210805642985
2453242077806998565288311132833131054230475978269120112138511206725
5008727890533472041916428468239490569260613372535105901909711685375
0117423495562118271127425557000227021133751568494873686540476160415
1382706297854801526911163481686852825623146909822636593841346334445
```

```
2454441717049755976078397733888525524571612170078079563106667373630
5853440719144197598222959028368675098518812583376293636836655510998
4615657604577896635137338226745822968001717959376274018061677784902
2861160170640648470704388864366022155818017629878412123453927 0109
3200577554067051757414825546875651384978088464299015144139551 36600
6281134856542542106876639363925879501362353044397168953837831 40839
3910775099279093048615336401839844768346402538576878152019397 52103
2536275948118557889965272734367457645558484584777350058425997 63092
5475611432098247534360914816117740060592238255275192566333488 28676
7471207846228376710101623690294364660467922507585368106003284 46711
1883707600140059758493736308636543337419773677552469925865305 02773
5026660671419155666686468788657982220901895278833316701242616 97490
9430077372995238044414949490902783531086644489237721374900681 92050
3141762786587215559483649838673360450399153225589719031928239 45475
8479713187048376283510413259110594272757807676190847400611388 03007
2851608958919240960677911223623065396205129793475212295899617 64351
6935349453004715318486366841324389400952539863265200447711789 24283
2010377986534721990457284541585181598408728616930132214940592 19286
8799709794190021375310032583166640684670226160286590038643562 55053
5479607193713661727322879645701402509930502791464251786168348 60501
2985370482891856016773577760549779185864321265765328542216745 86247
8208580295025127454898876217015784179860648091252586898800241 12785
2221921620683471693167562641674142894052871294580848481573564 89635
5076578924270246229329958298823911845888962033242792694776648 25498
2040525937751962165299012687287344390905393667875900811148464 72765
8778967610031921026991013216206230101271004537108315063806563 5340
5686812621741158493393984779679949485218749682274227239065884 55097
7838775546995706180124856657299318060793931618043590093478733 36456
2828112868431270642211114888123424277047497437527979413888105 075124
7757005439727943388288013038658151401723613786875015635593536 15878
1049333846158063872571216933219833031314127335465475598239025 93587
8775229050268730777012090784852665946138165134233512529326817 89229
7237621570339450742172115956587377598245705259779162149185578 76911
5081473273125754227471616555924841848420056255402139719929699 41053
9021364818350148529131218912429367183037724208657137936253976 25519
5976361587983228443739973573703562192975938676083752885392672 80921
8639717509636043781103907716631388799532918319589493859224432 03860
6852379456720328880037686695118933578471516570897037410309450 86774
9712349351834767413781301217604278993653385363303717477990232 55937
3103747974578848867053519148554969774032743686238295304901992 77539
6295890931820239940565363488763720304196544385998752678713841 4954
6842652894716203170834000366088263240091689446555317195378765 34741
1196696641450393376632730226704975470351009824092301934238829 38232
1479722970582613050815575181906263850672243414220022362443166 87167
1440395034321438238835570880573367248057095643299705456777993 27598
7115586533232889215653248147202617993700769764863592918534350 02067
6363231394569472251393307693774983007534169398159666446205009 13645
4464888334441413145520212660807471798034147251492319897373062 03672
3754170225365934640418880462093374655818942874710761076151790 14654
2705811403737635835983088787739196381890297514645415475075180 95416
2321430838110672634905696335981044045851089043600875269584182 08072
5925276214534445697292591875802260308499451187678465513583908 13469
7858279395508613642030286836185343574289713064948164565172166 09361
6608278090287613832655842271333030999107504595267900380516588 25192
9459617499297570649930255021222265503893886669361110270543291 19601
0816992828332905123359453875606234125976163171924164621710779 6923
4783041918945870562324205466170021540805798683023194603798445 52246
9175247501450211268293172444531654022626177592154267033290940 3493
1489666540076557302384459529408511148272790038534119560578548 77165
3549753877311492830955307063249702325868732070741391429499147 26976
```

```
1096604687887233414323099571691004779411920057438778576530291 03793
6891403076197120044781817869734924078791037322439245492881107 8543
9754983718641023650491176144640773748923034503488374940235093 84012
1730149096714189978560550637003295160546014722073839440134972 32842
9506841929609113964918061687379271136644810406104052575862053 27060
0362839228819166740852562667303949384846876701571546115648249 0717
0138900414941939625562464622270782007340433246722360462437144 61142
3567166899105694217076371890942929224560218755286575146883635 83494
3055032876344462544820995432504009880460250686697003326946331 45583
1715279619762462080018954263344305644144817215367901389705164 34914
4312873376281963969490168934628169901673586293718444628718644 44801
6998546709803854204782163857439475481910234694842548601663301 98087
1183687298005610592455420728867723239573168610859164839323585 48407
6158339492109223702422096305113023687435328476421531272187108 13030
6653854744245113020572818123976076542195488372241651067846705 12829
4472875003380370679829152933739582848802387882493067546392737 20996
5780111945749194589520687322346524498799179731923660526332764 56286
4539912860979202961469062796025830668024963157038240039299911 03013
5106200702892276831672386801441671418313209186307181668453627 95085
2401493436021553404459442658967046488689286222945774327145054 80588
4572831113942680925308010584527596399103557273909644283928699 63855
8794506272092025661359169871083928390805234869392922829940297 09677
4838475911016134129206570666019715298210900112472239105930658 46220
8881497047753214131243750590709941583119333114884492590366987 63968
3839543069003161395741345642731139284069807349998417828736855 50085
6451042755884309114864777778129541864771393329595770659848290 45952
2843872510439603306030296023259230250862828573278000382698461 81052
9301074673541174081421058973321988086991066613887250523102103 46916
3219505070240368784825494052941609789653924378662138673748823 84561
4226272131446355612057036658065154059016400805262689316841685 93759
6161407821120289226571600088097642650962273108864843561881618 95857
6844927466677663304556409432324511493376754985622372834752492 85210
1495881544345914951767302148424953871743818459007985645999681 22468
8396682599950586936338670057195669791996440910989329245588532 38401
0149875289930290896558116153336317490716708743175007762182142 85669
8592982867299382607157132213023031393435181683423498251963152 34693
3988604901899056876106362743151290094379876788877763662313396 51111
2830743544471189814360445927826692138208779671225441131857649 22771
3158414094066199111635564834674101915306453716085722756609230 81302
0558304891871281104738912671065431602195275005165571403239325 89135
2064152754113561710730609855539213317853847677274962449367076 55750
1335763297193449184342528714658072757771118420584749302827694 605919
6030663197910339312556627179492361157020036046584710825972408 64139
7989373787212613924737285278636782989069197908722632764638968 26110
5244783129623381372761373801947768916431110971016234837224594 92546
5524110651840086464146523224679362573818078014364910538821011 18129
3567041831993607120627681634425815760773938960038669792188273 59632
0363007419087287741649056528795112064259952880191506432436503 0609
2424448519238788192034698215962628574135099360203892874021268 16050
7615129862490756632817463565382749327318703128680177693913872 54113
3876246595500143814501709989326714889622056923802385566573248 07108
8648027490989871355211454734365647445240806515479558821678333 1642
2089340825507159932130002446344428174778589304273835005553142 05237
5938792565515910156597043460507859617798974694400874113772179 36933
9924215266488796031603179666265062061426381856913940396060643 88375
9358583784639904410627042097337016027313084493827711025307449 74243
4523816265879979193324779713614682553450951218915819350899365 35864
8651573798856163653042014008713153763117598187858424946771001 37846
6077061512797010900481726366460241232390426502788339065790080 49863
4772664074050672475475733443681170960518153899109072623370885 47565
```

```
354893812835431464156212438324395941339186775437445354268995609915
418939065874871334520440107206269319781879322293198089075019008013
815769819166508641104389385255484700672151581525092231928543671062
640730495505824326271256703634364762106669935854616067417372529671
275660797509529615926096120354970769357084816301215035832988013792
224455776597734421742697621885420209265661782766529872428262916534
934378544272322475531311966239439724391957484654406086988204730 54
867274019570385267689352748142002115676830238550999426678590782385
606010491526874137525778293785031424510534511431032374560145056059
086797228200648222403981440260476164691363400441805373308919517231
509982810369515597926732505529462119278728183910936203809933999283
857980309888346558800156727116410917154610430758134772510668038058
934798964182026627218885295272881736879948536581397633344537431277
774357829553517547822562312639990869205929775270376103945382624858
333023293065917836116119024875524186335328454045130848968387693636
827389672560221196408321221693801474898703597371652888587086495348
748766880006809609874782238544009509883409161540636762250494156126
329519263134573203663622717828917597040030424418772834965416287159
372132631975620808318647207226521754947436756635803894716635981326
906931646801836233003484094210158268295153001966407338849213705893
859609200842389996924917669201150067918046533762559389178422015659
755312368527871271070962821136560367634648332331128413724245316100
181824544158850024390673112370786899675212970675041162786200416865
311322548588398440967654056300915033722755379566574974163794734922
486144211356137261620405459650594808449381284113204848633711192201
324064783708139261194594869273106468379217729576075311289150238706
591658277045634035628787325571215939879725009603645512870234793741
645714504698048089991734577889599026771746706321184944036870684107
564734453015108067508419942032239103756751209588737435923721913108
839791570550994360314281715562897171420014423351400761441651875811
121951057769749660332928207121257415649531570867089698105436525844
918647327304122603186693473481407128343277073204639503051038441250
130676583505943728936447590117171551870603655034272166144863756629
511546235401878714626634189846375820626477491175621127356844678074
424043275567119168273446469411404778258055138031130536065567527554
103333041962127110621035417589220644787939252221159714574974915567
414089741732988899058336436805825620316272148526835067188378604000
863892589033850283743078871371081664191050940051524132071826988755
924805495135101967896696464633972560475129350802139353979201297771889
473544291189041480086363440225096369442363934896189800464089835766
212291548935877092733510851138612661429992464243590597893144210322
098288569705833635601234099214534795783230133759681041465266015210
093927435593227952442553013703620415610042419598608406139270335410
735434006252045633299892805430614838823536534252358071003893231605
854691630761985109515287052387531861268385072392998741553809306851
534551552533284184966102761914675040714631351705305288035419425560
303067011913004471358675464496475415497508503815544384030856160117
081505346878218400480551805444179808696573397351735646390878976087
299733868703870385069176289928961332415738546163320553537003503052
899958618151442677115164286428357869657428895568599889992883611411
929421000790621865240492339437145108376424233854265824400866981470
086570759997412336403870346002237757482120509305874636318768011968
144314663880616024056632473777168572740211560895636420841168522765
674964292893977414662394104754511467555736790695411864644115114949
025888914750145249954729418383940109618809637159569650704554780224
014378686636939548110121442467493620785935691997710075325239021402
663221366377517772617172568214133836899846027950384239563984348431
109223848746741356389009972324997968078363652965442721364058282946
918187063308329281662632216019235403217627253506032940597704393617
742542690045382334436298972505667736513346008337071149974094079613
```

5419630688183906709167600340979416249498031187082185864014437526 63
8084531783095118043287294683818825542083754547701682994552982434 15
0314271285864029279458417825822031608829580253633723869235699894 72
6757812074662941304655440261159875056456866914010183205266254946 208
8362869465147038783741970519515471253394510250342413144630314282 62
9413500534696081470977153900935707315912916261538779126390611696 89
5425817227965922998967667636882135772028734562019109965099034466 44
9256617886613685428293533945477367173716989531241966392266683632 26
4904230705791612153634748356765316125211642010951437073305207425 93
9960220250676762188972688953830490896383981062432141192575816608 23
1725559572078784969375053604523387984891130046934122858965358174 62
4338568629529992602591950048610048349575065235433051615199073281 41
0724644583726178124786757987781721943186315647930287219816518253 59
4931144543828837736006364494009875135421618438785966874805471702 46
9371223203655556349566706605704696979081464939434889029447252233 28
2759375267568169888353671259246119810050595030892530107114126635 11
5816010081525448452792275827144178200059000707289216634207469399 72
7656849335914911760825841378854021412426390559444263133225283820 9
3855202503302044677815005862869768523383995527965448237904142095 57
6304521776906329641700597926069349508009804229444901770541555550 28
9500800051134989379370285541258364427973066691294478372126673696 52
3534142074156735808889865126577390501014452438055498709485814452 99
2539881852974831664243309135363028044512113611809871684431496547 66
6817093860090186358645588906188145736604195435772757127397031103 78
2432532495037455896860303546905980458009052480521947515206522192 63
5809678198765665289518630944001320885718427694251530926673103542 94
6610724226420124173281249827302447150889383109799379910419182336 2
7604212329571209664898004419581443666261356678834451163564121050 2
6349912123730975104650367194847958018157568018645696117528607487 48
7068530458469843660644626513756594067941706303537951903956089293 18
7993344515397921067518598097495673393328142422209081042745477115 87
7607658783086090220298196793674221397562353439354808252579199846 024
4514502101395068191007536918185023466186644110197304412419085810 28
5927064571008762536383186388273148178784850461378426273539968447 33
1043279775550773610116548589618738567718190731349935386296159630 29
9055312954461343622639217165979561773906044078750781375494521197 3
2084783027632247985954565905559723792980567982884624777450319821 25
3260027237627388582022598711951504734474089081002146122086816527 85
3250321099739070891445429687910406347890637323559932821921379099 35
7163031074253776445247748810067064588133120535527388057150808580 39
6146916851167042409555883734939095883611823873481115894135138352 93
0780540672165575326400369391350842867360532916480909778937076789 99
1312888234964331427854358118019584928287537725450431831377539410 16
4617982683825319381153619166131727928051813950726797268712428901 87
2250769884640525961991630592474734635781644625844607399168302217 92
9263190314526648237904409430893916385793283640822141162618719629 23
9548432674993224847875432749619035546848866957455872392270254924 15
6031455385238352635080875238973012907096512881957859451655781605 33
3727822503840754016393486298652703140159073858897235351170428531 38
4240785034973523146548631006785260052159528145350834093300839041 32
7565777588672134490636402881732120192091357497333251704692219415 001
1447132690635523697965060956195141566589213908966249704323528289 91
2980502451265506467621311643626388790597060997887041316518848370 60
4888730050747860188897866723091188393113970658333023496213397282 04
9621868861622009402970901676858386205046493478134522494847836130 55
1407543953060433246588418394885561718615418933503769784714818787 07
8618520093960486626311599630198767654593257820535652864010829724 51
3695290984109087154890963085110179855498819183000839464500105304 448
5254821041097069083191001258119649049314306573580744922989790064 60
7132499978409730648381576045872570673542932134059778889862370585 52

```
172372052056143500473157313501598234350726893415340602708378476532
675793464801910488308685467770611171458649509076620849504701685425
522710889373356784730209798472986028550216365546289580385076683886
766770022293816221494634697935735488963342068079540103477113287395
702090657532461445512500710373049125346262226777757242596072807298
481380996573893077709645518059552440606976380707282773800739929665
723916997901790484538398772684135340772759911540102065939534115201
429207990657805550524391062456472669508160166940521470513528271635
119446311241414226604370034965059326393604089572386378302467738004
113798401473263610012634311263901409243754773178135531349962015973
068489437745662697805643005025511842706828816995640996773071371794
076298295654410357615196833086585390871098908572158514893979334011
393716377392947876781367077659768715495470475543397450844593525770
360187091449084343002556657590746227491817861484622443653505112916
087818738017266940366518391277161614277856012970306041944802636057
144062848820903049100874609890764550275173388819318879857480311682
930965594050197791408366431363839136360657430412976718754288687736
370018617848937995736892537595010323245546710330777418617765122150
757854850156006004604038088217679161339588881534853905368358356573
390241123654233976580885456737483050504020425584747888979290739954
188267225664515599934978992800888484958149401181152115527508941684
054976831715849911517342609772250120153393846598719058258228680581
337546183045434405978344515414835468122506451978577597618036886858
913232088331224798880732602522465987976871020397812344751824793667
684366929965814654090887823084106118409288968160071712549517908365
048854719991879976444479088555317371564463965633215036771533530063
880269511300620641248805912355313294397187124976789814680123049044
248294824536796577443839202390756453322067763543853879465711591288
967022701643680709817241698280145657072147598669858065170022191594
666549407141017339115490121352471352786312388148766142720132064222
524548421187180105785237504792361380061458729999350993143642905906
165159387048755156102905017525790236236480539154631958686723847892
037566923129155651355379091575744539779840678827873704504373117828
571058708693260059714821857029844574434463623395618523427676912525
342920269065497877383322871342775826987537430368613247671333377253
498843062125308614887453701587917321644809103887259668227305918358
451756614873025344364318290838208660163042706369280711195499400761
984041535415932932349526793760399350014811985977776804603514098812
881682743211203127555326359119596819317604370002233957772278227896
767767542937471756655703237166465325218413676353978637981152544169
193604166664217440141349349640942839724862617069048210774657463946
808337682164782055589093550852946100291851793044907705587256255377
572120524897879291479336006146936784623649990755512473514853196373
285871566229159881873314976325785177210948254087129953767935276586
937432121237834116756743306492303680458717490378909996726513749270
382954481530318183224241723089144736716902447169911245337722323139
072122684696234389961354256846237270205169150415523650394954013907
067919629402418703430548422048936039388770140133666482029358100390
207613117976263199869407596628098517998192357168172972795817297158
416519480847797554233461881233753825476519793865967656046944806883
586750927561765524097327144335577689997899094547512072958624637837
180873660206561399449338936801773622772537274642739151213198741058
968148752877422614083336765301746040732265183776600426090333 6597167
693360292202099770197681756480137965379521830735628737916911538739
615486325857127181750951633648710399132704274165470535154858496545
944443528680125665217495269796913899454531037433079246751429804456
629705049788153075164147463982117405779962995514553087414799551817
186674764309200795479585327263659021871299408679299762465457086842
347330015326249530254764736907456509761210322718351676039871020189
782773679187241159253927019128950824348351148722907761179822674334
```

357669794073606038289124694231304835992765126630421654535217633212
786632332597120583836662754678387659214066996938081519909567181314
592623081221640039853298815465827207563118613016786301873368547939
077258856536843319856528061877165640640846005626819437737589959596
198806979288224445710779753720512351399670103414542665571714263979
042746749977503517188490284200084733689560447914987590301979425422
659962421944501798290665007299736698795587021362020825679895029487
092063242012672117319121447994926500569557598554609723990180436327
544324219025587245267365341472201985428975429545465159208862602699
271402296804771226844822860989339142079704235511465029560416491740
348805791442102730784348601012685703348872738704969788724282471511
588138618594251966864502153635314878908810201023670060412897255999
74693082592483870380641416377166677309365770204873195255476282537
5376054103821030367996185905295370821790710104198511105716823024
218038539937148310945959667179790119786966554264228194456924589004
569022233348927096848870327670007717610705245503710960153006546029
84905100701031103481079918586367258432140180822740683850841310420
547983774613879683657934736083098632914454751117702517380091562661
598838245664873072395105611755990591617177616345798573623034615259
796351843475942934676052870169914045274329264375840416846467190792
000311849547862922401257375085090444764300449532528621073103603959
17374505424098004528775558081860827798038587381800725315572075864
398951257429541746250269781494996363296101295145658559421709530698
229725189004831694843635825672315744829540008470195824491638174567
728783217349446297575661062810403234901944580436630163685456640239
853958802135751999320956731987508587690841600724047423761801158328
205265046238163901132972070377176818967652772464891967667461209253
903680520340691042353023935597024286057225046817522528545877189934
148088961137386409058871123607794045319242359410280987599918406236
069466151666836780465882227491841931580149190318191698802888487348
030440623089653295476058508321508001240353292000493516446182456589
754132224556732180264056143618541761803962951442171788377206795
604118159420803325179436982251100851490205957157061510626230132971
96702890508625012629234722523269517928634792796949413410000423364
695349692487126641571924374609542804752626734365891817077029172145
713768812554103317351482946881989007918702265736018707384681649143
699987106984600191689096593720241885047972435728376389138752753136
002982458931858887692561219882516729402714166274110109057051670404
402408071406498828829856959197346104377331011447691475262910698853
825137038535780974653904779067255016467803360181679329887431471630
986480579478362549657833764176886790879835274602335676250018178656
294031767036627373023092372650150318692641076974688586843741406926
217164346625532916146295533259182009186877965261372564356330656930
328000571565435358092330765657957720130631938875793257086924428051
465871768974300069205330722032513886201409797823700948963305597855
85878493114238138600390173786175522090623568100329374408534126400
387815396520551675568639044844723503541766917548061538046414744450
765548385189949264172483381716509416349088134945512114655369978995
145141230301364830886559843341576778303064076509992913759659377328
528489009510586178016536326383603460339580911178329293138602013565
601390250579862845514783907063407148491454161331524588764228310105
560062277714107826625389670360564548197317753107336837589035160500
915367236132513812801871611930792012258359764383730370592117054257
953423358198015318515126042948982620100785924031693828033960472680
258501293249560560424638640064333216748278397221360514840268680599
942827359841128760884110893507804050858006132343738800839809840590
054293030177183800775190281710525966503917790960190831974926023203
560291684927003465470315057004893775563711126409434156167157552241
495001627356967672537219929973631027739754928302666627426325254605
840336874603732532129366191747964004550638388639389829482369837875

```
0924533977824317875197117984743193636816414301472705131405862057 21
33494425048536528260864446164422843850298557364319120633565493935 6
052822275137658330074985953051872917953824709460422985071101416790
1785555877764898382610804393073106004521591657873080959078276329880
5606422923212023207785561531050450270731776024483138549340474379 7
03990426472194245081810695586486573927624178613863402401106041352 0
26402662727135022116124838758540952400872014214462003929370583436 1
97576481019267736340022154316600814160611606409952708962292182885 9
30991787560903171083557497351817308836315172918737386698062090547 0
24289635589233454592571183695513439603353613464486035196493890255 2
79396853650534027981286803044996293254654711754305876107969995472 3
43611613493541234088929457834056858566597253673176557256497694857 1
33054142273797582984878705336243101326395879860008655569101460056 3
43697818915923790484504172272211386176827208260163149345105317896 2
30034308562997973575201294229120391135746597393937081862784548327 0
39849487109509873607449216379654270600143612989064171600578857760 7
33042307674963549648599293139697135128483938233910949167588127609 3
18503436730019811259069796487490901250549746976828472000553837348 2
41411931946144035624003740911007021800262208954655465412745127067 0
62169714158728110866892283179740356698924610404876326200621906771 3
33627520401636770973844571534830495891916508809275590114017100518 5
75248121483504767085020112920564624109509701245676160173843285733 2
72082773188172306247453368386893615738339752968578025025033317197 2
37773991740399857825983521179543913002740409667091053284265071339 8
05230340014103416831698768638549554703951339378965439153850025036 5
67010030672218112617115715652431966179249286065748786228769724774 2
00779128836498967094980388641728130423589811192782824748257469110
17632630117113528812801038314324802813493684523986741331047795742 3
68355088755495752734885269651357615831493756710070034901352776049 5
66457901416465319959574974469904335987971996950532050259547425 1
19525724480804009247760329669175706463163447530864685474801740459 1
02398530145544357791297359610903939709449797803870632953189898664 2
47359223668789143745559594694968069281187299469127784323704933772 9
82719772601713197095363315172659040397132878904056511964061802291 4
27136220442007053539697287187384030654689749771293566238982663862 7
80082681900471412422003154158250170198907990430390291849927767740 2
31149383515996132359915896028897527030394581548554686058925716687 7
84365599773020685427333836251795736435488710764330866849761643393 5
95832018354245360808482085732918191056364111745463879494518834943 0
82316284773978565207368969334325050250436900909851852888749279872 9
47282241015695303879640216684095715413326539361378451395527352942 0
70203087005459922745754523678873686946530890764964283904057658095 2
29327706620352520332187582701937202329146321104395636163450956902 5
57655681605356623516437808930696373287785381023284864626846687266 3
25698198360245534746345397303483817702775174023146485590687136362 6
66347792755931475900447884862052529926428599587241086470392659387 0
16740293072075223624763013717247798201345514618159578089178901272 7
00189166373095118427306152744669942657918888891004820762432551044 8
13147161553342512488947117141460427499128365115774922007707947259 86
09864961859113433296317553943351109304512890757363062664246214278
08791690166672347976788105603255886001036023066865884136930226697 3
99430491456042174685451712842941904983586549824881549289230213551 3
29325156730699100312584637670123844499518775598994218928084451700 7
82150373470241370410190901850072082058227262399872123807410197388 0
23939320585419615743863954886476480466406836821662162955090624672 9
92775498684792551449284507062165229807118061373068357835524421515 8
70422601289795491552231355128311452666922735734568074852693886830 0
72796091432020066141139042393865249370363161112216012927657841053
77312417069744162566703894475617282362956728431447731529433667004 2
78779017769427383111105590067980832107160918994198885070821162027 8
```

```
852381370134306585094427192751855373256728119190147762363835935582
817811773041664180545304563097845136336138091634703879096755541301
073708098665019736648403052508169431912843956462437153928852666965
934993598341808997484084637774249351131282116157423068531785489493
607056281934694375954996717298258501610015024676578412642212682217
658109101129003956282556114331899516450270248330700747553654072247
602649902361050453233097146711328159799639433471852799121594805 15
906073044606236374634188967670329214956153440578256662816028005644
923227622338673112775727098708365441661018481422933532988826492111
649091008743694921683602082034669625292658620591848141252715838918
693039129081270240872915926628039756356334399454314219461623542738
131287526814095614872224550458935817508911406285606102817357811107
439867231122332105994120119721236180675912354679401140026354569843
481267098280260432627593664621233123362108938871247808114613074828
381816583444719047220533122464448637432189896213644148703137443511
202510404442174578407407509281152308743238466528478876674176145 95
971252975297151814451322497526738471337253596115516352313862031971
590017228429250069329123052315873290295184553726154798914016999584
676698510909400813266910612508193558097366628199654193584465774076
889163096882304996791530800535342427729281401576187467757182319073
659093587436895644484615176244700466869612359223621114738492436649
713583349472057884273861654627852054769551662598241910661104413503
879822793805573735903301590819929297939948013333587477675592267662
597393221138211519965835177471444457261687315890325250870687321937
336775096122737572868054927973397290545545645436089950911173138558
705917551422948852285105001083734608565790029548118662961829997876
067297908752009459370548949359906569759535231618991122931174581660
344711105537997531595907177223728983534140421459789165565963112 3815
470116701172230674820579061378091908916394209968249006750969 49803
572007589835261776009107167637097191388912818357972124874354205685
169235565886245951047824661072835461854056340735525188315758635762
151648873932239030655171426138997179683242928858270700952690547330
614737849111650248513094545091260330530549739644349170480990616788
446256767862007730077123260967538290593047293608966125133534456361
327907603159765861590101588440546086259330647952646337791879422529
002171845655502488072168429354673749029287220190690865545039609552
925976379534378894424994204806013393924287140347282014898387621020
883010654991802648170984845379785905252868611431601625257534584365
338274924205194392102290623004607365865497462922377007004745059948
466626752219072970743287585984669070928174667993446865230538642992
139713708083927437673121830831829451238641919475057263400735063892
469182711840584776390345505087969998833300440987918607218465357801
468285608419628135423348911437473969436353478564236917262978995421
270155332632843257236875018729457549643820264424317411526789396007
076848388601546355279893667582152343945023868109317982749075824869
622360611767111890871804342058292895506492775928982028552540268818
878592213177126118964115876216388284590901389318309486408786619308
170867966760211576900642006513975026046461163026036059743338757 54
933295380822388618318246874265456305537601525846354442172303782549
105853467271136477905797268903722403321025838075919122795243804093
365072597877593090926338595008321182333415073444095853656184391234
223032060489868776007711121714602939663688393967191538659473154528
726896334164855636942811638115334847228313243448131392994345515193
015589835957055859290007466799426361167921449154595219617372012 95
830225133417017195138870849901020760396924117949339475546946711764
529142140255642970376440434725992865973666261734627865400149386039
234793277288177113728517027414891921705077971065654302954736853151
878335749534928042388057889935402774490884786971241913848313069057
333595085718185409609513062323571982658487488317968856359452585390
453938831344333902946146826987360918034386975281657157343121270703
```

```
32897910206841434706402579350111982430409072961860774158707964861670705848350284881674753873599568268565335647313573522294854346387296997985989891775909214993299692627985712522018609317243034600614761188215107145425184736508942790341374239278988744752855203158502318978480791464302328077640417736253936737111961595382027059631440114478760201273848485020768344277141043336221994596262561641079066011515993186681769183719298879162076480058621428954272242080188259367261454134045309355722678425711367719434831032412801609265182166888310526939418429236590238107933968663925968184677838187428006181551188275820630027240883446535424860946681615209301308156428707434491680750108000809902486603784183169687589604044612227317821328979825427220273711353527767676105439871713579365915021743735398902382785092385390121007711097173838265972336232330728886404183175276642610829497099276144377388765109505342607099031130707428445279399819142676244439665879805613047427878132037459035380338235159654812783578259172856014323547135910777618206818790552794502499331738649477429387375779139550135675815958161334658238512307614133274495668846663382696172890440637167571533268853116347093043023363213788723108006945584534526972255513773943264605035213531677701309191680502394414918743604219677208101568191592158883270756691728157325450039099391497124260339887120211295623518589001094344650730686775691141135176241368261569821195568108222599218650517329480263653302471480931755842380501778016419976072734614619098985000891902631717428714519927816978645285109886078755146714188045630667875853936140028155977962748670576235206310676527678472039469210130912818396710543644964452556440547307639476175553098586748905301608487274463014685683718789508661288049944496306534048894477969561267305244831415566247359797396570201203266040329304214631427086153632085282499520314729675652361380959360205483955027586999062785423591534987525231922511716934496551088361374197407040416746349609531798858007246350902204660760787668130513696472407274108738748196422477581930774014072700338382111024865078134259661704470725408694538416994622971545604844486809109549542473143228748481272693736791124864089422013710015105439304826048675764974583829582455476714186077974659662195515624995353289911199406629797810714166049131693230231787135050175880451933757389563666132359938654033563251318352332592850014862679700447289455441822091819750126140960240726165280097203739563804098426446254586286049995521975570503086218106386018640390531158249533258703759107218623585994755589067087922004691177788542267053148400443314777587515019912393482514532092303348289919998834594377620575652080936477362276769704398854637140949001201245859910707026430706240601287455888888053361585884329971232814199951252501388299458149615986209670644068321705482266577498955007943188485049223829955284973895999006102586441500150037619067647562720370920593072148528926181296820401646208397900930402873942724051280764726137829652223557100445697022356106983028902189916880853752141414683470106942029283706565238947987043238385208038811989226430420817501046096475673095260914936148300444974141717094356784508731683287183326694945298743680754421824097947053404782493898173347424896657770249887705832147565395813252117338243552536234815475257017247728914262793338972645293343453740161237251861675209171562049584961946231131031566652196055291478205442686377485484085521664178969965574614274207208859343201704983826729503169885056865604465210277069387716734112990176081083508437077511813386940462361390310224709435849893664205386512403456465690855784142464428110459331539410568797352720406032720617225207503628646551294371552276323797440411817185549769677052922220601360759429064233041403587623965812697049372744237528612825674709867173025556981427290214868900875289833708319136109211917707255689425045571978109258535280858012461135455896843480218661013000901468975578569658383603645065029182802582251135005484391094431855757472821881312816335546
```

```
8955686328012579208234906928758421036385901817167737610283209156526167582120017424226539919381008691624214676338154580527560900257718344646522469894450826621813773550248919876205707015928769960299812947154147762015161475159092208768904233707914876235101480973008975171035534875386224239262133662034018689476861235472440117030031893631684316400281886414895991879295453581669378268369350031802354571643675894518090626456513484174838203106805545830608671228082398808128108348053523900772925156091382805012439745165866516547041727111196431256413311316711393049098300749565198045524016966693003374204110335796805438537617153490529051680590686948370701314814217846634812644125101279863731047100172158880666374274848840840812263118029658101930003075351143610870865507287492080274430349878007272517362925905503626811846709764300003364636942438601407022944923408225297511991839235450796387904862104823960167354531787110604060630499147424555297351420641866560262489455401529366979697723532407954972447576570988558817918488585538737331810647035394770126588013497970054861788972387240193315382256936536378652026211388897384542739785294836101688928109830764464125641473586561325694850537963797870100693790618669655222087521049427554661353483123817443143588956662290155417669327335088287665320833903436391927645172754110771965013156703721684896170111351572207166661034345833658547597995856447448858051644910536962256667057863401564603103435068287990600769337235705622645749426633094635985748500044251048780216489108945503000793393672017613556029153299789591178362046181200708688315865350046527379985394341048898760228970515631091731618918201443634174527982980723173759392751300646573573892090398841936808907534604901897588778017794234284914356979393931884162263707630888951300934727294788010475593766641770504047462524426233296689571548208693578333396036705290592920852104915004843551449668081019265548387406200091460780139011717187872657302838353411864571645996692973479444224986223748644441780591292985157820123750369047816682437270969637137761908699486118349862650751387743264556041167257579088133538362091223835294467700882121692826457492264324400704669135220561302271084666961497977457611664415554434902436489776227382123941709716497782078790484036254454537041062832227316965954678032555003874273731812320407075748183036561550804978897266334767981258290356950975511575972395448219033572300397992227431412859809984077196938291361091574592435214928796609376864340459787578207602256034490918602368988014718645316706705500206410662845023528453725915123961171923694303688786793094823907998503769519520251992774399830917234178427492937877754870450366207258173925554044566149405523441802417781271077047786845128273551485835407921198032913704482038624198238758852965177846519006439913076833951917185582506673542168049043557865070163951671900926165142465991582451528128400358201128987770160011764016694785286320756508871639967271885204012242547218847283969664640972321710676187601264181994572228815243583291948248031868869985133021663414529227651723091588892064486502427056918000808854941394946535516840966587486264464762752203751081739064930534359643051286881901684961490638342850861262943016376629078633672644770302172076924682976795818596972272266046856963810452098543609955040795759374790950062277290678205638717913259452200122461625631634457461348831131158730740180534387445796408664100710065823876747501979016493312735279033816823407186847872349707287388248396689986077818723632054829407210426097037072730409507156582691372461731112328773085175426049499321511099157152949889869665432445653787951494135086381990965341123863898793827217513791840117654396560112427132730617476904777042340554409638185759825344402460724625120566999307384159852380116316933115907425015639657598894204398577883904973791688567141283801012344458402326415414750710124930077264370554705437651844244425022596779999768549982872344764634376672921884244785255072685847300614235789059579653979457 81
```

```
5687860616972183333645732545551379890158902563658320758048692580 67
3572016991927191643199409118426643673049573912346998752156210652 97
7150030399729326256422176004621510809557819992724137060306040383 39
4092573023089166037987313806867580951191805064849162770490540506 97
3903748035145877377006788749736309235009170269779162300195374159 29
7564128053597419816840715699962602287894662875040865826886232821 52
1507262158508662960652838638661733707048697059050094494588673263 9
5697012915120391738721577475623868192252808227762868481773621806 75
4316808570508490509393893874901498679998586469186828807296079400 90
9994772237388560455012772678899406338724450046508703958028240541 32
1591513176304326525360378362788086775579334999213685527302874066 02
4022536003311110239298849022145456915509533050377398980639703177 56
7503750557763521863660069025143389817284622021089036907396366902 45
8727175577902161516674954551479501782393955388189180430267759370 562
1972238910943063056102174537510146920644948953144841979889428286 95
1624134270670154377292246056501086506957155031888607213774271840 40
1219770060357610152497031847062493706886460166953555915984018377 00
8180561014752343030906237421876690381094620457446198955123326020 31
2878114906438691250166478712184171860417104374208153257306621680 30
7742735209544882269128736081174559286962540699023428282182725059 39
5202533286861145669170454684329244665577430529345209792486810142 41
4975899660249588057368696921867766007785184601724612667853709609 3
6263733651185869854390174177193280059961759156736366726748114090 80
0980703474709238272071597529284189556545352508125231296733255748 95
0057538855124780075693536437757035571048482769301587449220389469 46
8197342419500500830568925363528982454676147095134860994743540092 12
7412011999706600910849973383929312471311472993835295474964498122 31
4481290485429793646051673362121817908982235707895914952677143271 88
9213980523549899219408973807655378159060680129715037173540388887 57
1846964627038814130397860000100393108386490345825353247331006788 99
0142739772922049764890721650719781697429439701276742051695284627 96
5647246533154210129678212732055480400422852517562954403866983368 8
5521375556417752442861592552070802426312588509210390546677034080 009
6232996852631337471785422860488388418595256551698091775347963889 33
1170907952254979001031375476868821917069604160158152180853901067 04
8232257956981278444478426749733139346210517898769762870253347064 78
7129338884609305183765827108922653124733060853537885580578735178 1
7669707270955365878977362959756928204452543388618213723301363068 01
1746035558883365655850655870405956202835390465790666065667213062 08
4889190710804888090483693416713011681287826774373111765198860572 67
7395625708030905456075119762668346691587200297119028020886732950 43
8689869333967816260915745547671955620737121324230971824434907236 00
2445485832230286358950989660408582519482238337742274756361563323 20
3447308612988722391934921068706265661948175674345865228571723335 88
2493980500799036257840680702719231318776244743598136794623749679 72
4430558332102044203336403929094862008542400129423347222798825925 74
0799066148402160927938082941706006706811034224996903354272632175 23
4100042274348183106684950205885277793974184248682719184690954168 60
6226966507439850158548546499649610586313066951093461541272500773 878
8825716466512409973628339496578743816623997266417379196717781061 94
4471130546638165753372983582838173687527749385281657193452952885 96
4720416682003666791085477115955673202293133701257914261437462199 68
9638898386411176621888754756973984406626761226600442603198558544 10
9140098025925859033836013550104925012070736490932744414506167627 20
8656282359616377339410993872211311157596703996611305906569417643 74
1357165725273632231106604334628872248545324425738182445440059675 46
2909064116092390859798299118346775925855053910584016870940285848 78
1388176155999002478522077323397819580375378136771055389862207907 54
6685423516396167218455489454730795375440698398740098343142043949 92
7094292499275604232847079055492132188746462567878070159141817556 86
```

```
194659020580246934553411240320605675429917014313727400728304305878
018317723543079113669411266883267771782428021973929123473840086001
485605856436107558438843499707982651165267489988211027970785738465
812690922308475552461371343024650903881823005296769636134533274552
201858633509872127265883595787581254042101236549359731511074022113
036440770728889185893829542935897503918635115378153102297056990859
199522220087187065557797682729294527133629379148310933302834872965
788595038950110478304519597957577382393436958724743357320404497625
567417769750507694787916664366280202290911848490207478566288032867
555130995885189272298811174882382848178202590451528084055150496246
517380124249482470457285694943049215267986089890690480481165784739 68
654499410912566999448685295252981485121451494023355811305690563332
052355205963278462624762021264323998106056409714363952715894673129
632726812707043857166149848221268375071109360839748397551817567592
129596582845734701143536112022987242353524782948657586860031921375
247021222779332033566971180774078294841458528229946683163560922882
478699049935255617778482161472320546339204056500006993299888638354
937303026256652248397339924692286413269415971989885375370726966513
095289401739896902550403523001745611147076653001669696874120402969
046135314190602692607269746396298721301077419341769928325036470760
185893617660006793213869159186368884178610491113607120060903228563
237633702367831828321925742041946481527439055399334751100921050367
621545906143188271357086557079717301724596183830929007191382999059
461368054298572390228244510639239969244339523133326706484797544207
764193725557995677519110552571410396731928782740562582981884985107
213093441278552491781431993329597138643171214730216900361673276912
660358638822843668287914774267478939638337212336438774782848818 62
535362370817529003207086223443912149186210116810714888224882739 46
492483857528951876553638166109476886555297179620079751150381950 65
471411871174753516724648003244020476555619456241785885922420319806
011945711865180366507165039449059234799759227389271437422600121337
739326804361297959667349405853137240387343896702232187862059857409
419731909054296375036984754974463934669538883094368692603304837758
550705160702449250735475148117117921211310266566391155507890793934
014245938604687643028234352469092256510732663809565796045219558644
445037185217339388195817201868146580174300521649237241077578877374
772458700992055360635200493253969090143354458581604961850166292839
345964793034834991526730922312770507644428089809606200401017335938
249027880769832901342355154534422006200591617782230712653531314974
198536071450714871777560084380404740997400976688961802903965683848
880895208663117496376222684361419548420840031764657119999254229181
239950508410524548376597032519416137426931884562182075780735497186 5
002776125443930665245222570815315053239040037944918366634699496628
801195119031092850282326599618003624583301942307649009600060452507
443058199868944536129056554854011756066301934984720407954895014512
091650441310594787472385449094226085874374499693554647606132970035
494930030006336980143060104979893499078952788185826715984807579505
612867543027679692770732779587204678455852348264022798325329642643
866207887155081147408141757874189336763518762236192240204582254133
244098141442384560765270080245062981039087109465244549468041922729
691721947169029851201960694238991434831198682987989275793146618177
806615813962005118867796736993826261371544639083852377109285552529
871300021912699633815109505427266919968186007916517857054117068487
916708535337331380454872380725005050693312742202394681131053447844
274444375015580051285281596148260455078383719802761785030923062108
537041548375358265173546591175496366225958969078489826262037407349
977257133192309673159380857425959953780992414115203249279305269670
370086733477824539301440844815986941383745397170519914949702557966
970846172266517050911494302018165341604241684299704830320893662 12
136396379770947055149680495093657532047296074606851554539974670 1588
```

```
37771154143500875780988887503849948157213189985469880002387926 6745
7658654430804497527300346696482952619749440950421950802192675 85054
15748788722287383556257834928702231429509323744853050905423565 0657
92138099872451775538928778934137375064280051636545976484479293 23370
48437196257853350281023434929047287858281016209516287398200272 0118
449811180045124448271793877342001082629588279585046981323491480908
361307176066199433451820566470947560160278646308423988482619418493
7363427001748318744082245779406436726327972688338955206938083 80009
982504390259790248888098183137319550350541149389044314812788127001
348806994739590300952291409988026613089217180990859344669951490941
594208150697073625928963270954062811901175393947697449985476398987
550054629753704320804785087309734446572388680148108035968512283160
842244674226494434167195835944522974641516099516996268188058295924
574999364820556137264591800358533274593719187578951174344717279 03
206417020713949494029308174984069248827126511314532885316460153561
184966264212649801881518995501988861641633098595856304573796202680
918400697601726141760698960366872383607693494250438279745778495884
914972375416123568086978787897809840196264073776691849427669421344
370977584174772740096975465937809500379800053186791024976770973017
390531995886671949029359000541013243398315004566766905880760425513
068611754941443679590909721241063392454359537875891716264706498827
652812065556662309557566225473798400553710450399207366842500403647
427024018036883906418764685842861909762112848291653350661176160788
160281948202337198669073378191687317581416602302400749292430956975
495574438521923874656132536000812027767558780238693823678441194 8539
441674783880933328886366632020622714713630871443966987733360961919
336184637037782862491869448788236163199981235837588349954181511597647
465514255998941828302267942970293510653244825885510143423173866145
685717910190926936704224961854161229064777325403662367123126001580
940528193668718595470726706388949938363028291516671594154034358381
952807089474809761974444046422067931448110644716317848896778274704
662883085132382131584658917383912525213450909753239242523182435954
544561906119817359340376919518561349845465482341062280116602933870
592464781751433623807604771251728852036958391064588175708428116320
067951993932337854960437043040672853605405677135707196418004328316
831638300946682887603994837775965271815028949475254617093826381539
130709646718344195362544633789782299866665652833812427270423096729
031923525658098733277583071812942787516833908867057245514799526324
041953012048843647149086070096756313211049888322420532033263808144
836137209181885544198720742028850783018064302128431736408718908876
990175457612548285144256599574846683395715482452029552045282553074
868830658310075557952103182091636642719725056876116686032701490 33
570403273450458149409520998205531610507289261428536131424231031424
636376716171955642183370345659152217574540808312778159564708177437
782292474620359008474007394402636961578192364240014026368115994148
534665743332619765362699758988051080810485756249759271993900000 0996
163264666683848149320937666783980061020747355417459694645114150267
967213147613509287788580193772462446468215997526583651948282985230
366493140589341525152401291708652418837168426052148436994261180495
830303185872472854533198920721285594747562217539159584527180667836
268628051268175882172325638965196946872952343682692685559419212147
340583065079911942310449766130569852897138191077714570815339381833
715537545299295279176111262751931450134084254851890314142119222658
138049293417674212747019201282245954695156088225036349234483206206
947635890849249125954491909219898413292439604938183885972476628998
697209158369134840621332210726946966417986855138519698652209056610
576060214618511323684977862315363795024426052656372405087692415 92
843360512804576357130146624655139460548579004789201872562370424415
623842140430981277088495307069113314467221935421610744001661435534
897167938694035464899636209355647988610619078376842252145823612667
```

```
288680834994869210589537143220334807991765344381892572529059148324
241248667407021841713007050988462260582436802368933162632788041746
962259853946526937990841019518782311402301046320549772574540665384
992311488888713436795231632578934833278035097252254703318780803627
572702003774132454244358298690171476421479280910341055649619925644
514787609374717135326051296625897702682839348769681490164475177722
239714918951098290307932843108775964097354701892412390539226915752
651998986191067899558889577112074565114204855717542840451083382187
382045689055834350110927167983479216262886381054572909411937552275
519929002734578454679702006160113429637666300542445163175087081100
051536442961217878215926446333088948691867737604540970334283356623
440390786790141170773672686567349958954640523696619449597102668610
271925277912120529517959466238500521940948210923102391523361249654
710793265773037481227186083500831233172336810779130227396183349216
700294299347335994361071157537998584741658700528599129438130116932
743289130260300417029267625886050065353575315318730759301828339395
704367722207557461547456284928894514039379654463258255806272541783
180527060000885515638283954692400721309972385758382693533880195363
613204761969869325128429575241146319661558253690709544905920374653
770974589953079508600695544438921585991293645284650445514569102522
647777210176602400435485326611487645789339971563983960469643324901
870929375059462847401491761052052503923732522470990050650931725452
730524750052609685198525348558170694804563358224556618315642483373 2
913099239064610541690442645240972554536226685468330431279675055582
233110066908703797294820047564840470819363416899045698387047129 41
032782266844631938944239326634419246642754175083825972650171812275
277911558116877225564835113590323769925848613853501583275417555150
873922836796948987151575881939223647218161431266457382710119228826
055773260323098256078942088139829199115292903755555696259991483334
516419524740757656789215139490598879609724229646922567968091657518
579374192551627112551198342990530629649417386255300220517321464565
188574932522942753216566558351618564049023282004555594135528909418
078559995555054990501266901121006376011248174583140978496681703606
041601496595447123073747468966947241396965052165862104988336828655
507813531551427778528796298274826979085645318330350425603177939722
079829736185209217842576219776792754919621107665533362710491065562
333930050101266236064086789407453741042799643162779419696140605372
078814795554799628029961549878226410661425324988284056318559967147
977453438075285334457925563652941264361234417200808655065711193364
968260084889524778468159770516427149356820335917928317716274807815
440154471992050204082987660242111582258095190501315343060101105365
509871575089531395354072725859167400667153339708308695555260213475
454909362263320745682613631450819415854273212949700674461636112970
976199601189219894094919647086541969224399749735954530619427913108
601429236903833480013799120514968068595282612546939388734364909293
047906246864998988493323503766723156005488940661236388908798891824
556253141892882330528425991872537282567637078603201569586093771736
369297635595212593814435311916092824182929041620117810398613608520
363169476578675997859289404265350507306077884454936897668391071 98
206731176117063787485125211021501283733233543624370648744563326870
140729421412614157858381965551412994065092167014462801234332872584
095975423697196895295411218756653220425874852609276183499256523678
129502030467026990401891299723806462044238448006229131464238 45148
960487282318045977418570058024178129492730324195249497404934156472
921834352848819672750811257678211973727605508444163693066720713225
322832939034866061222663900504560226668975661204561623396654722575
361625120067136583019777178344616730057404569665760591065710011416
473599323125793431091810123727995705943776159670989927685236536993
748165066442771751639159793366607337126269081184216937382069650 86
933268567278367422377736381997991985783464194126517421191221668 07
```

```
0941413284510522833805406708986962782679403178723190081905098325 67
0708508502281216413555738976284080744058368920926118375020348 36245
7327780716597778179461635994718234033664931031792424610254041 04498
0668449596283754307639569435874609863857456951585376565744178 62924
8695330416618217499535963366703641595216130542992119948031658 03345
2332115986056662735575887619771412113617281497999611953859206 22982
5197269046098299639507187401578557519326079233489391643870448 87445
1745842610546586451005430687087417270320838913658520423280753 71396
6164128094428370410812724554393107963225204515540241070742706 03294
7094273932172343179483151907035864431909962095508639098412726 46602
3867122467180134672971595500184514574032071767335761669041043 72338
7945509438207767808163131049108000252177457824652170744652787 910829
0733900239536340718801471220522140215399101631224689985086195 61004
8621493173721697245393532184374536368814497943581376506224108 15240
3315854713854365800572125347899241798496703729212532889231110 280
1806982687955966661932910616876134760127357109614183625134914 7677
2864112185859982763908016317795283809545587829491713068189428 04812
1622836910514556919627018366592350533516483307694755542077529 15040
9405190181035540191839178547813399790765698267611809420529364 72885
3714230435426496468281529287782397406316882122373237445686231 5139
8372583970689910936449988950374421668157449006566798089240525 59042
1706641935048816409596391628243168321436293614401089755855067 69190
4977026814968846369662836100299442637293757667957457287491391 91760
3147753217991110209376451019667941478693903830533989934932993 39166
4772590544173549043402872902989856101035947620348894598486817 77527
9745376761775739959655575307040492731673999678223259487874059 42966
9701814243003512541290216521273632109308233750646316264149232 2359
5455292065492774318872262491763738171334465184798088015151355 4863
3879143408620024253323592296896221237090088402610478255705895 2996
6101856028960633973289754720627437700132220797858821312600468 86764
5190073959373243314611377608574663891058377414738292226069341 136447
3443818014671080889255488742353442381448313743601248544016766 88462
1350526234765977665364690515679005115589257997632981115376287 50719
5884317938258173166921329843468872381579771241224505487909139 54744
9757281319471996537574727900424540311494660658017917689981142 22415
7233493323425225449144990337889840035378645344035468834488701 70917
9675190878066865656127703350754693255487739771782574987448152 64893
3300370309195490773379470251261916810538807456674678786128953 49206
8239904061221050925739381718440531615138398514137364038852808 64089
7509972555564565302536459282603124812026864254989319332304588 529318
8244858323439217007914797944908683859630751194477385606104015 45243
4366529960923664146046098275503075606371459115789251978626502 85973
9637236976329279913414971543547280289663777574646857956414270 671832
3934743284939194455249840529134697320553286614101083856366681 825473
5654463805742638551177262022623198214541563836516999406232004 11583
1743549208446095861554294717966158703406780715747398144178688 52356
1172253937476289392566269541532815266938674970317625157351116 70916
6054490731505423659796963723533859198417724007176593889415905 12602
2119557237348275540370158755337029227667485430491731121687674 21629
5387314828119329341036213997914193491696626712896650254023849 51651
0747201108039879798356315427365689432090869616915922085918541 88743
8734535219770064421677183490207074355899796488879710881529204 45877
0167924502103955916254223749002273377345402616929457615409985 64748
6090962790397865945877984874033577711002432301800598688066362 73607
0095660289159497530761097530365793633609839750890895951360115 651737
1310883063152093249069573527609470451357017011918924788882522 81265
2512649407264268109487626250804558246204741136262984374480745 93613
7886662184231723389731576491296235226328036498518808553286225 73415
1762135003757391584837486864158824405061562353307504470286006 68907
4375305737618642452167403000989960886603536489146093264618987 311126
```

```
6413966216836916488490935627175640273129072007706408700530974163 79
8281313106366511219024399911030088878270485483569423163174979072 67
4484949538582624173907717508695169762555838570518788001282864019 25
1588367022075810627771266049611797895276703259769970786428542079 78
8604232171387019842796648135064284835541030938760732070490815443 09
0742030121271590781333586640649439379493924437297708454928359705 53
5265406677085849323390705257245644907217484694427290042623438418 86
3346554983647670232383107294625625326515838774192520437234716103 27
9442144924304069082614819401145746521984035738697125400871634734 11
6458760772663464425312570154990383873157734291514453066167664928 47
4381950512194894888359544999816503666146509566939638328632834500 99
5164192802421464235570114008286897082439994764773535898946709185 9
3901815425688801356484197686184471931496425271003593076189260670 13
1041009780780963125446954956399752638999430505540759588788859468 3
1149394247403976745877056489412258558779849996048903032611720894 92
7983754059347413170386316671950983957244768920602820372863959421 01
3093696510557591526586873527488390618749593242374492626093780926 08
3109935999770434945950307973604425994222423277730492098184492005 4
5897836097581721604284925605666446173643082497583402956710814992 30
9620540075933798904722130803893928112732735466476141166233538913 780
8278668771960684647578488401622992150851636963940333557824281608 98
9531303442377747348711731634303911959770758122944800696425160029 91
9444365620770040792664657034817651562669216344991929163474235896 84
0842644806257287307200895596810192472280867687589481347456885927 02
0784590501716305293058728925909213784888447877987885892794729535 42
1793412030717989806486372549218153478860976993518089867785531080 91
7429201300910225072246147533661765808746819916168425851290963509 64
6084976089536754871010431829776992502431519429691257360914521341 33
0670749896293829658377977408480315734823152128537224019858573672 68
4745688207891382586132086028897419773931752265572955060346907205 4
6055956046244060661906225585256743927390967443175281033383956182 47
9783958323386223016676855135110482677685037056119532466688121621 90
4422766521603155047640455053452040741751357926249634203544896326 38
7875531661721664353699085922503130582482173077303866064746870015 91
0887074511766130474004315114390077867811939939240597970916012734 287
8146785864085071837075241936454744080624753128820549519583203395 09
1470619240924946977107163008534028369059175693701237962568944088 26
9756639525277771092482831965443397032018036945265410445175606575 76
4027226124710007458212486446273454778944080403613328250370357292 98
0033425185161409181374834743638314897434064670264001882582540826 74
9537589840470528822034499320891429556166893585563430244205905862 2
4602113047965225793164405414710744209766281654836125745420156262 22
9956168738187020870174254498991730275076076561220930373970748289 84
4587203444264420135117992720818805357024745585319141537462474990 50
0031060170133368907331019970564619750521748854795762069862087206 12
7010078378706576374824756291337773642585001359466717868980825439 53
9902566463963763513412006563140941023719314071544504457566525213 861
9348056687055589851487547868891191443613808752793435458917871388 49
3968005010359873591837370725316254904517211317709952660363074407 047
1465322055515276912558017232098299125645422608817953401873950940 45
5463292849216637057946153320286230170706963260301943478405564878 78
3958745500550862407044886741620670250209186137619977288941062192 2
6158928008867293875114114907228280401178881903245922826122718507 85
4105807484727441267981177730328819369408890670881671505683174756 96
7893688731536303873322818930336224792926842291463077857909510438 64
3185611725112536968429711115388234845119217033518807034662484920 807
3983633414921030978311415843471308812599520164529991171064180563 81
6167113184732898711533870574420225546995505939857352512041762629 46
3351519688065371271843620716465646990625301784288540415886043042 76
7129898259981232119266390246016504902433894005797066822755735542 72
```

```
004795062139444516594432044390242228722893510819776890683511658992 1
248837033787441133525293254268604160291856276550440926959811350018
960611962086377084923442197782053652080823058482351431019979685981
861537501028099709278649433833746217157026709517330373829513995787
115919255216243775968386952314359910293323669982373844906653161236
704901242643663097984301218395616886287806294101834117510859568176
075899846656108617512010546018792618154020569598004241559468135411
627263387721612432535328842000207998526956160450334359925464517242
751695299622636525860319642484817549318860687763760686390245689764
588703197481132299338897926482145777341963920936134451405284446906
214122913558397514574004199992935149757830482922745362272006022057
871574453067700208885327596667914598160202282511564660360570087192
748289147593685501815575119004878431461762318982424555660297412 39
055598103268203348692782209969701743045637369514647294994708560384
017916677975698260761230999791813953682842582226923074004068299272
129087763386122461908254903172883869030973607552560991478393285806
783580174932529488155954211167021668500998738687072708020315142396
770120056057146712085361797974687395457357788202645227932318826921
707201220248686029192267385042862454793379846398653877618763275393
362602122865288133588743980349974525217956792038772189689371710475
389118059194940746683503542006462308523975831629015580190865910420
028233040586040287444948482995571032329393219307147293724589666834
357999861257419741990314420832090605223875654862159289718265964941
128556699372160094945239349438967334087252077346205005725334728110
458787761805867069357181658912348364629759998100534363373106 73593
024673903722787909789331695205458894546489381845386622715491547249
232065469991825396871411021410771134095823094046452162526052219639
922247950245983986683261977582966323139202140711875296389495857 08
635284038990025284499285613884524375854567657317911874704193333823
909437600645450481127702514254903452141372669834488277499460901045
261698627963446438653814054422948054960832546435561837707032693637
536489703402041524440945710358880365246887545362347958558199670865
671354373182419300321410850461551760226141967948170892160333901162
100114661908828458235385656464872133926833651305281084863495691771
808119489008349946079330394415632575408915809234195779578355743500
347687589438707335629461355846346153677669639623512110583160714842
434405914120167155830593280700274102006176287213174175837974181654
680321778462030079221169009603451139799550544448892244298817839892
980754942915256501812386005940943120006907153046421806875422365106
401879879927311390793694416280982530474957292088590755525868457725
404772104141722452987512528491731138700517015249989176649726525353
361809623788799511879160774872012515751954759509696184113795830300
856373869364481128927499841243635038505944578004579945326279216877
374683010531319885374388395974802375599038597495494227753449528161
787683415449170094726069272194862906091315928670939467228737383354
050421426406881753363015501641691641062442904909320659090856696063
731066656275022470278732634216783580786257203314787288125635147790
984509919191690454444496028113562281361061920662235035607155270025
865186425652367594872143022036859135582062995219296072282512238190
082974809493244361191455451386149618842858749069701862451165088526
387991110271306243903531149774959657400164789060791340995446013337
586595189797303321671620241222974144440909485900250009487 74775894
653620519037212647004166258356057218892718703145759919839697234607
677893540020681271824654448793765402902159648358353462048165535852
358403125987425330248743411901123763693169121637040612839172070987
753879472158909340143523176924837466518949722211363271197900082666
149787110563142251829378248216576829836176283943615250441047948817
204198151771479542234616443621086398628597670463277033180658908461
813887688936862894202698438989539066608760884183620316155392883868
427496872563656603133706440811710869289191356996102392961357706322 4
```

```
3447029574181634766784422050863095224261518780421920734090496540 26
1991595039939299819487485724465978847559632550270344557765375150 48
0314227421698705646844850962335089000992309201669432418123346107 21
6358248154625885167762539333686526396495529366749271731348988600 53
7260470208611288988490165078446610270002027927610294917528277029 99
9398473167099027605696790400822171989582855753925696911098876648 97
4571071224767405684009029660208658125612789226766330841065784330 96
8105597009417212318005487060393687939365073675026609498178169858 60
8366650280821615604200503869795151955712733794910526341554719554 64
0855511709795853703145168146527818597737755526557262605701504825 02
1698057798562060736296979313287226407701889092800476175323124374 43
4038972915202376261972729955136333792255740238681427338107379785 65
4760040796872511782008703815949205402548892760408146627487858486 09
7624926329419828089194170318815361491565270751091643291558421972 06
7025964788846388722855393848756642545793430328584946456136330701 63
1736390343004679086405410795244236115260783573687018948848669612 42
6525159380651231214044034905377951794801024706302444361284592213 66
5013375314366116074749464120679479320097699281913225724726859826 88
4758655911535538042889565560876866329306411499691313825187546066 494
1326575368108517617933326763944021356772522550478266473886668090 40
9156616632032758575093748597681717382032585714520269941730820583 45
7481396633322087867916879511050937332687726757823059416818613053 20
5774762505118020878716841921707148625870719875488518687045071370 95
2392963868017851867290217078703796721502128514724532078848703577 37
1466186299366464079473467424980967127517110243353536572261046436 31
7338951789893884853140464857813037004699214704400113397588532792 71
7965222130248103118365420725309633959792588797943687736776160433 89
4172185045717138839920983471099289765468775834897835409958193018 98
7874562332153400686621162152761938926948944216221792515053686454 19
8980626757917725889570013559070831541749569729196418913872831174 09
1695889150410070812976770450587153959702156835884855948032583561 13
7174530990192099647286738590499390834960994075594883466808656434 71
9749261881390097388471252205412005525346421072806170256506363277 070
0734872631036307396500375593026042273597056398676817796602809334 49
7203618193967964971130688746517768188280330261129302225785390119 76
3598874527969541286844480463811607668005435802378861355227919516 47
9460292515319612754730311766368571938956754723255639640021870274 65
8532600092759743351774139968945107389650565172692438841319501912 81
5339537302591431318196463176596182258095618036378527076863714999 15
2201560165332089744395363978725099828461644255186891076735420192 12
5619885516271549967655448682568412682877655726816706262228951888 51
1034561986758831855341400844294493471139328430151992949952468235 29
7746169825395242808677376470455061713978353173743504186840360097 37
5575135676089497670181779253513055600686990584633164914675927244 559
4472917627832735953129867391142636342768882374926472663599907219 86
9643008187930218299848362086317283216009408002477296119591884919 61
0784407231294735475364414451705921060925264179401646471772045874 58
7374336121990127456143960852154614654407358912213455444260010361 521
9572641602952843814442839910866756539932062040121453413695139937 18
8385395620385474924539285349346096451661900699063103117690967615 02
0504010114765622039600054271804039314861869085472396944478715431 38
7705983293141944105969917878884378960090915399234727636724770766 86
0966598619128964656913427481555309117669484843185601983763152070 90
1214958320198073267500436264540852413691295457394044716785408092 85
5613024566412004614290261175081396791087839476640542708981369603 18
0140476310626323875787137649324190402285111913037502576543610464 53
6014583631307357675808521523363035653070518119323143761274902909 93
2395267248923336957015911157762298606122399085969753584856664506 855
2337675967356791464791603834042948386008223388538295784067720466 88
2709156483932221400869534456721473867711978688217223525023148238 77
```

```
01452031804538008666841361357366901562088915069458467444975242118 6
854036457970870702649468086936900822659422450728564559436620874244
203428592325602082591535471253852285973206407454477712384930385248
989382562042264376354995841980847855388416643615171332098129257804
883140578101217409497449578163615226772215510131952096986746045836
859251543033613069476427494628640929431550955210744757227054057217
548957043431329046751158182928388878930093754765216369900123963473
862593986019540389078453894036300956443233192307356777204186506364
092793993300274214833453724285597383776383128947706327272896023340
545146430613766882211945676110008517036750674891945422676882214593
461150817120673729278384250982729801581797314204164769852495367199
214995155529794745165399988438222038939140696496158295918288669 74
039308164830364653464910572416284560652107626944695342001343743063
652811537901131345508476219111419474112274045667138895106987583543
130956516393669745109846506171555568427805647007281747436135341268
021279436830685109309799295588500621420243227121790410833941415544
669066776295262498492950691922905745478174546321407572098 70809471
229114949377076100849922835542850719832945090127428209286344262333
468965030878702143487309210001771856491147310740519435367469297157
303313750029525876679599834048568012053165889437159046214107210179
767814234448712004948478920290480341689261300653274179340754030304 8
609294809344989922348407294745659314261495891618102513634364935320
990484199140186499302985119699173266704445151973748855689152366981
644214718882465973411136062270479011618359596119082693413504649923
573305157362187016353058311582654874959582161961505900024005260855
513989350911142895846722251860989748408484485354465517940501755746
588514566606393775583441102658042460059062981785830222530761854061
824174035147020648975246915979478202164843105286121748501507386476
274658513972420772870390403431014007874105368961675421137132448385
466092836191855857666995460001017661747414172732250244836530696862
629726467489673885284803597957151121837006476672783279872205027925
802642684326693787362892651775663277621665034021836381700911317518
703255106404929602183527357927298470763505296402991497232274853951
262487134590879733736858323813225332567427423454977735533875079378
226977665195828289300613390250934132975140588293522660909842944393
705099086588857172406743360962110284912671280153289872414072149497
895880243761107952607200633290875312243068668157722167520480239336
897731830064281203214978471831043719601070786239570727733577773371
092334339361018150098318854061112901737735996905516945733574055039
419421467066596794807383149785772977118219165538987413667496287222
815532164871502354265043425933813605130656870981999104706638785252
401565847373750796136673052954354503593328213698160931378401316116
719925261469025305847806112260697316893716334976095667239022376222
010985270037277975714235865701000810275939587042664878788156979835
890064586249715012521655723622491643460883359147430271640142354587
218403458651109079681824315335158365676752155701158658413600158471
064031437212262170613168733659003378703005626245876476538554759754
209236189630826295123183106479957759663116824246047307025743345754
480649004469363506260718042370398875087645521891322356560363176789
839769609267927969422718762199012776484332931421283387345368898977
129883856274612293780076053600121837674145196884139981297628712900
742783424894197409226880693299409277141919647724638799118353141911
932299661259594807015149976978894998010819580639387278356468989252
059088536824264100032701408992100639944055687565328160011575313 52
825966789229695578027355990665624198702580594961332882779250388169
056274408994602316456858039747696056884665695011672744942253403543
179744124711606810148528788785109848774548303522637342732079636021
220089959479629347104699481402084212549849053200501189079257229331
999716632135205859273856246073397153075007114601655592605129672252
391411268409518882343878812554489591720907166967018491381566199 6407
```

```
76368729275726215301994769605757376706620723858887128705849246648
48
24250540922644273424380728792254892253625718334287800883448269669
7
09728584019352958723646509053135062393483116058530184301818091468
90012680822433366330273394468053782464683611075379293627547767906
8
54975916089466341967734515904142478059469406550652471926839220434
4
67535773581765884635192321743612460324243274228857499184802963086
6
87594576792885720935545818164050620861223715829746005234120749785
3
90263536981399089162021982770311767799967274473899235643455272923
3
67689224833269489965057161590879869458450583172804438299673015891
1
98057886950666270435230082122407558932779904166339878267009108012
9
52428459012932544176412046517184461152862066183731939810885318849
2
36400842956408140652436476205206871161000539125652755783537714466
9
77275359233616243972652933351932253264130816179757021558625427888
6
33029824010648344080974032730950345071856393642966385863651060707
2
44659699322946148716469534452440445298899728655309862011645338996
8
37962780169886755505742926057561373905398535324578535909855385259
3
27759360600545542832773675389348908491323190278064796437517952618
6
50954260781626424896440288283298266615363970179037871448787012418
1
89738150365633521082221035168243155283588411983019524869897225884
2
66700799188557934505453987785109793609707150683834732182684280535
3
69309751828888446429634577188080816873813657798041835322309403134
8
59707627272130615330535011129616358081126432695935583509740047068
0
98741084083494226333343355172727727918624509189691776774287674474
1
48853869210641980679021385864558520990729680814736409284476412725
3
73899646705277440667053729129600398451262560816037417665962372148
6
73142118354523772875327469730062209537862155790834655403426849396
0
78555652931360588772000814013691028532112591903188562849594639484
9
98683256402919926859103874702507209007089779630397727382700054241
1
00629296516056102015970237782695926219167648701011343294348784385
1
48088428811559321719833310106533512000053481863171662625638057405
0
10577223833172016596959691537074166064988019258979671537087301800
0
90051388950842963700427077566599956175863233787976665200628134066
0
57230542653160701671380061749834421238167373214197280693871890019
2
61895520109246733850614615124094191009328904110609416007339013160
7
97703762755055834191157025338988317039726999419950550963514781596
8
48052602889356920219649599215129270899203657739585192949811399488
8
96021379194245396599576104462974526406377469822174203079930222341
9
93186971585411232555644731287084402954918674681830692850290445215
5
01960564000263415821434532606545301469437866080846829899518730796
0
14161578992440422541787253162145176462162252131426665827802954687
0
63775130223800194627198074688888588549492147745042360386070675607
4
54493512948491245989427116635304291739026398951631866663722377541
2
01288931910302535843861862640837034598012115663252093121394304445
2
70711713190142433234880520126604184328898644567750676971369470949
4
33571770544692772851478903589751862660470103329326092918546527901
7
36817888318191353767575646267851953877274050852186448072400352664
8
90555735213696164214913536956225363892932673161797253661687705655
3
22740408623848828804380071171584562455923227545374383576531293122
3
48567181298700487121559177300615015640049540423751697532043085357
2
84402702783996006356241412521866110548148228315480579706279980760
5
08162232966386280245697747586880096942164459965378341419040633256
0
47893225467116587547992844258073031647649362039085876536534979302
9
39849690545519957430636010228909812998027342490680690177900594149
6
69283329867132409763627604620559706227978934704812416672831568956
2
98839241685961869553876674972727598703998974758951798093575971608
4
11534273785026766760979826715669054069214640919722845956909083672
5
66999267501762978312586593528635018390791744200868143129352424954
9
10253004044243546025076456832206614333986728635477997840989886029
8
97851155048316622657176418778509685696916723949366157498418690157
1
24555411409389777894464952822368173924075016681824261318327349124
1
```

```
243278182581104483198793282164215213613870915833831624221761609564
406968584092923903403663044709672930155867583246482822105772422163
853502254512166349350343857862250295001017278727684893990079432918
317643020675654058970816829603969462587657802189974380236640495557
339845919334611895713616531236186535300340728987009639130811838008
255999804048344910321318862011908880965301631672755845145796279319
871659255522753590774088671582497911052269576283370778701640491783
593545509739621639226176852717932569163632014835171509086525989833
607544923110834964595087244687451128211642144220271852976726203107
818300000729250564437371551184484968541144855302464255791218421925
283826390732538720335895728226595495426501056373608189521930590962
089547974660189191287466488977618699712184107936037311085847885653
205271117909384439225204972886642685194674529737825163766209174480
261139478514172900282970371355403595948983652537675484555165589396
017911735658306047647605363314546623156234661630508626831816192452
196711131389800124408937214984714967795507606597376526677540277800
599156554754627865301317699167597781943029148167060285499914163010
429238843637049127488793995180133580361960510592522031218176456039
070674818800113023732673221701330454324607219439711287898483763016 8
003128660767750347464257251487177958573760627248881291288180644625
310281472672720121017258574332331728901072555858169951278618743211
679719368999727241839099066353864097424000423821320261074510996684
429881285258965355362185713027852351146373431000120593150477557843
630915825965743451169948985860831788153010990152627819802639173124
481926555606610119585263494452788267507217103308553823545855634780
883023723096509435942193496485951409228671785534944884394315260 33
756280031821962856054524668745446130392004635944154069171682047014
167395716624267266205893282822191389391330989060740956373003165502
092355363146090357239359775070152315655016754073008145301851985388
252395136048221938563999326076465239592621943603701424674060001879
476971534368103793729480800552521079851889155416933156832057891317
469269960861876787874023697001724620687294032514512749640316655532
444732772818097237412634417742030388629192439963149121649465895064
865486774628204319753790313645198882087128483371662703404640055343
526161350794666638366225241439167806150945738341035530000880543050
119175673278302001138413107131040837428763549994583869487445294810
751344198021599378287697800677163142099704585444686395458108555702
800020415374787659172644941998873042816816433294168405167035093549
997338443134094365627457385754241050571974817748807967714168843160
299998515608106828195164610622660608138929154791861498370376252182
547094624581009288640674250758481420478785130655964610104397278463
139985468259782669867365837513605567552455237742523969705019685794
617197919289551149322228481133297451521334268798384166490972142890
825845478865903224451776718223443365104797005821888369860189107517
339391334757785645537785621064077325625488968854072639502225100401
930245918806223130553845895447923401297317078506484001251405001135
354627058099368797323153652971975205183627052206410069176964647911
126329272791864806992924562821527549906358135829298468942257267881
104567531238453592187451898711851266890253215547958296139245295129
594088756903257332648960092770395786447155413811989977005302217537
812670623521790024080850750079087381517144609815263542813218283409
726362356182638647291459245880859023450801485912150907060027336923
385969088689199475541932183274750018677560825550790879236210552204
409367771143785178227512212400009172523222185964995800459611190235
249014214991642566717749572872642056931588990126378080396385697050
771329139203412620272814585376668640930514537495886269105656416230
176873020067726947673689101899810325782519865912374518621985849515
310429798626743658110374157510630434902370417708595714755706900431
449676162577136785547663015533943713254133236249896963885922564017
381594520091656020244524216644645953552946581011718689771895302183
```

```
76379881304951269137040779136911653871346992550041376448657214 2927
75747861802300768190723635209818030136582610925904226701138295 9277
14840415807127550146016432468253091850435841754188876838077727 9114
77927244656066807050237475773483804043382393464724675496039422 5292
90403089428798634283527893485464428237264463349333884041090197 651
09030404270947438940515739224211466999061186596814823850123078 9847
44040840013043757186585321533914333352271719374820240269239401 8106
21941371827745931835973075683299845786743704980306171140523417 9614
89981606792054819784542048949334606440751812407804970379789347 0788
10579026016286456141930688339604773852416584548630481805489975 8280
87094293855471822888368796305948480001391052837067578080715394 6634
76133686756021798481130529712336311053409634850129367693070743 4288
48468189322554652017020425186579585678541743196221530163998704 5850
69573265786991737263121696878168834986678799242493505552890062 5422
47317295890399733642696372212558191760249577621704383514638205 2186
59293546256057070107994447091172026580177748675155665920448906 3721
96369987038439551750408138022113322016861731870696919140182704 9043
48084589476770891755055536790561557074421162930758796049091512 8118
55396333830468929955231070660257561900855268817249968402617542 4165
85323441241615383763843947727503904461182156243985344063112490 4248
71357387466766045682783300349846788394640025475730755625416557 0202
08738343945977005278845451928209937930821469499111917043731495 8143
78370474085787835297914995965296132911464659568473372929875908 7817
61200395424124641183682698153451540063910354079365256867587835
69158586938338783963326579249445193676000603590379965343857759 26377
25606217591581846522945335651216226633949579614740193687736143 8812
56156103677703971048028267303891075696865263338665181831849363 8583
56521114738783377074957637576746650231325703000660353580918775 0278
68293209784452901883200077791788284029730520827921927337206878 0583
47048107636511800202710606748585543983689290980496900611460069 9760
89998659805947336989736026994351363233909462782669642846923095 6563
23274650129750672952295874147702237200636059148963965492552393 4859
62492094865517733224209773926169033352251633955740577618363346 3640
49100135232280965466228537067824917645479418064635763900626918 7033
37223272357418417608778173699668100681505095825878230594586813 1372
85457020513736554549309027528479337062650841570736008795557053 0181
94992850282204554910836700478013949504624217977182484660442197 5937
30297236859187939732108567657640102140167952393278157294408269 0782
09693405938458669942231746589786097726345414850876879130749987 7939
83386902741730466403052452466528320247764059842496662204235477 9418
16643775370351286449149258184832989787772034848961700573158297 1049
49733660931124637149098260893587293418415324243372769831079779 6145
98181328038888691746394354856432122511730902331473518590971963 73111
73536116117028637099422554530853590128523364897792413157325156 9688
72604341193018028313967259638320080198965623785761188241841667 6769
58878762729033170866987419165420751920966152924885119684810874 1523
01494503980942221092608124008855863620487409799912150329518483 7255
75658658732761401481836465894445360131951268607645621830149413 16324
35143689524748755461519028463061408009316033157380904519381026 3794
04047677066206516119058869938378606771044792914572274368156673 3573
51910994948567154235510353570435061230268526772160235857658337 1568
83426184471346615281431111709166701715729623138894181448247575 2109
60307961064476034175225360830964665031449860895021833271172733 2284
07542388008915831848336067834722325665877416334702983503174747 6800
47350931928202691826862233361573411215606316085983840383139768 3533
89446391467557427042933292577626834205123287575219082034668666 4718
32409653952761420828256951766147484062875793420776819705185978 1705
33826917213564553116305622056743892774621116357389062929746875 3857
02330343953397057772072139859552975519660286329896909002524080 50224
70967607867265376615273698598260627023970035006843406640044377 6990
```

```
0451195047846445227107104022326054523829559667683904192471938125 12
8887448957718371422098966785108755551669075446638501876404645557 3
9385270573800319104025963123138544744173288857728360608576987772 78
7755243310312766324440423356859483642143071806162674671189432582 43
5906450526246263221727249806484706780616369014208923397278986535 31
6045795852431040208599315239676780573958079372994044929970067743 97
8523233500455890205906628919668285867562799263620889774226980662 71
6994236590015925252144491434369553129323498401228258600472378712 61
1564927997558433565013208317601439825235160427519749672916963622 42
5673578494195460430008685828013324633421375326168047492604791764 11
3544918875027542445592013102874027762185948933036544418024158490 42
2587015184711498445234615844614128067976560049284352447023668143 55
8537723588464021317668475958836662984297905949439423744059292615 67
8048492654231974695786565845743068892272043429078648840796138277 30
7263390594628562986054623320453384461750649289204791688400401110 1
8423434868088422356591178438549108023877694729170273542191569426 69
8563254653736371530396434611877498157557391767849676263486940404 9
6304617031412441040220240037028892307057228977506462775217463847 59
0959285625221300657604072467676869962990777184334686319459276230 64
5230965955872385569188875940952820138316929789723526256621022603 21
1753643309102918959660959837428876323358735072890465394705973320 80
8813856204329993092676708246179454318609014514914079840733067473 82
4973503175263956044266017867824858088483446542985630723944956287 41
9572498536507741499915942457915514392362560383609008861047302163
5533487610697201254582381890404647722139986139393734481607713989 97
6863808511177614555824668035387584006468904410760852585707900581 42
1832514386663268815882994605932271428346446773008278980385136424 42
8281254847314255686497416154692310230570747570191464608201858469 30
4644199897213455644401380756206949177072501472984336849215029812 35
9253768613345326724179735453980336483440699498367792675721566735 33
6650918048165498077499075280872902118769409805035667528457907625 58
2223050004770524079921730346086074093533388772079656921561535176 201
0683272800052607398589916403758255217102527342418718600211567010 66
6473858216634472317319379564518567789856706607497209738882354672 42
7871737107809165066097091840406481274860374296805116236494732317 76
9475701087261820492784717866338407888982832854186200484618984966 99
8013397841637621187477718147913974876679750091540701448832094487 22
1908661203516980200379683386200022336714342044837067298359300387 98
9555999881241001034926268542731811422190616134290982449820595445 48
5614784448543258078471733492040185442036887445495922558862019690 825
5504874320503247491937687450417244361411162798800397896115081557 47
3992101876441761953272324347925708107648028299054087381294477309 14
0125504364021351709963181062059001842898924621350056868526025324 75
9520104694228770078039960996805641017786921740630380135077058048 89
3744101712205173820998675827310564392974536762645104745654625109 70
8842870239660751779909797526205113445300402972054928229633982879 09
5349027526428662066553249633678405152535209155014264367624927464 84
2554096791461962858261954308594573909452825929040362127808712723 42
4447277130254551998870370815819480783367007680624590220683818219 30
3735343310701933795564067911796363440532850993286893467781803536 26
9996581059334448735330618324719497335588343862121653952175919654 22
8038252760492657002692000052877931182827532722238350965913776273 91
9490241054230931198760471735769009363798892994922283155744622998 57
6934197286282793342601271285885052956832378274835210529111938586 39
1213752872009058138994089390542488562028006786979163119955098992 70
4397020524617734867258429635937514010872026043674604019154123416 01
4013292079970234704063258803431474118618656804490901631152568540 02
2059978891449187463810151935340996004942829685229974033338014332 94
2837003065371879906684060890510684629723692529681036367914154834 392
4308721493417710294022733536037838139747327742933992242447896508 04
```

```
9581343657272771507559441079320559862911971144647394718116288869 10
1855484536264950071344417737368283646425601496506140199240775092 62
3100223366998546982054170215216011921691591980322412096572538749 33
9009453516032575566106095824691063054389242336169581390257238919 57
7780564416730818550061380082427349691820410139771325022746123016 08
7383287353763729020786469323125224800433011152253644830054333845 41
1881252646239900141782613393281418183480355944415087499301138973 42
1038766960863413606687587754147905435777381467852982937094912243 32
5322343977460271442576675750096989426193399596177577224452055594 97
6403717868668489890191216156674186439284371279703896032157424046 10
6696145029919933126073003356096881585266598297290849339499438287 0
7744148300009119041431035894247101194524311489880184620159338029 80
9108721318082353691431474409053122011481432985809168872364275305 53
1916953826956847051646360522456802331826269911562582896140854655 95
6340593695790664499383340730657778708264501932654315888524747765 5
7283557111436752896650355765767931200175345308387496865859128622 53
5326712840358187342127141852138711840021178135967213884327980842 8
7713517349501582583400148367959743190660493717800407094400463626 91
1577317317088646570229761165202615519192689089167177454252146061 58
9586183402470346366524330474861074817469167822830066525867007767 03
4559466028227784062899712413479046779595762227403284732876697522 9
5411387818855734855598823981661466796047953624133103122997850243 05
8022261926619620020636159354612355995196814960127604930874966509 35
5542835383612530254764362357417484411106380645031201000189472005 57
1387397457423742736676477486303390628861169685821378948257394781
4849759344785595612703273526242665965652434204076877558149000751 225
7123531598327499003512233035109893336144107367169191840175058104 96
7027136975975666321121419719838105632454627446029569263483143470 45
2562388592211262285585141735421775746101963558829124365287430822 92
6291367638377133446741927948442844286557266500642010191692880810 94
8284725304207845440481680209678753340102687465036667583248906442 70
1536976523842387808183754927462742235802360754271968799709096690 50
4400828932013403243385160818447076536205664241695354141284199660 63
5215707777966495689477377020818219857118769361177576671086982131 49
3539782449778505992643426132998741199549436968225614974557300538 30
6329295729441988811917773619512160507157793985716611744321988048 71
6960209245711406991064415660072419211481463691130964169484828449 35
4241624023321469186632761480378485469302814569886488568238803755 23
8228827790062852230666469049640037231579550509368760170281226773 92
3199554218321299303262945195532236935969964679267632148314820403 10
2109199215870807192355498254561191069635646084131231511159581986 56
4872105827287677718115274929268046052280699952733476897058209265 89
7292345003717370843497321472928410189197524721220172362047253168 69
7942555259248429211102220330672011867259936319551828445744883630 12
6124556041238914976306944073738851801964599356770579238873144130 16
6792105257341975133521814795673220317091335581873756396309726026 08
9590334392532161788095458875725254110507652011915893975754199417 24
3306494741279996467271655000396157430455038467478137020663464417 98
7369500283147129821011537322098409339816041526328926946377713770 76
6278891998466286340172392608869664719648825039910502669605321184 15
2908950147617319117406559604815183048856869813104522010928818526 14
0889188542384095939789514711056615552945142921992313814508653523 07
1475557870699161002619239944872494480297773768209087781442711774 12
0749660119237223990933813619858315412221784792685567360499510339 5
8909689878457150048520216445511176392932430709879136791240427925 56
3934312986498779435887407412654272130002738729827663052879302604 23
8798486918936090648430023171196695813339877305393788287799405006 27
7571437422464845857960364768678787334052380530982686436266547945 07
5590275256578510617923865444193547421258262868028835528245871725 79
6106498381772761621860191374074065530954128249552494706059172414 09
```

```
61956576374155479558713756022306335514949094665412160561271251 7730
52657965111624350189355132814326155201348320058221134128131150 5550
48347455494244458426701362160028413942926141499435256277887551 393
32181458464540403905527284770044004350197606169102888971406024 7007
63612563333968348508317464958068373114563715964522441296191915 38321
15450308984205130366083877720875215042156039748667232307791989 8875
15906042423324981658043294474536455799233883273315637406794051 8764
68210626034150831337951820176956127879941851929466475342365265 2156
54815106936222133281312080246812605887626983602366156798285980 858
83808551479406596330084551081580768498203676773474170616633877 3148
31470904621299422401296881614547017722782002041176834778910632 8381
90606123322526651355958037835571503870835891196261260090781963 8699
31484748313199689852312573912733770417207314111030105278168399 8235
01983385999154809721623673582284848892314018089469221904601046 8727
05913564207158857532909067565479863435465924117925728300256964 5336
65246875456254249120676030899277727460748919684430478657233219 0854
98712128538169747218225020620591280343100932332625776107474140 6490
20003395600689280584002259064861436060161939053620865743253048 71568
33760340829011545034841068634755776040766417820413543438884757 1669
48625982965626456469691738966946543403816782645658704070256781 4734
33358756045757797131109448152710681417165768733001925780199170 9718
46685956734821266886097979635221905638097499525897138226884879 1033
87730419988206597109688927523038762589843988772166565119854439 6340
38603215265110429896333557020229397249297507349855162692276026 105
04300413598846348785103117563765751916205096246322617880005150 2557
82253650508000395646412217571284616615639187668195373821457370 4447
77183861174324111759134737263954150525049289077126335415304350 1490
81366836036571722116344656318068460480419557592036834950128562 3402
87464393884339552607633630545875119697426474493593768253278590 3081
46013597701076596256639918833616184586775469089543760551424600 5811
44595332550064107531206735658004972973257824303788447813315408 3125
67166376212995215550452304457438439297592665117066860426577353 57875
15304874740168364275851275724452560095314179207114281968574068 2247
39320534891705858465983787262068829801623840332486445985805213 8191
88647758915433625566828951321468058258544439046996477472265636 0598
78431727121669950129784662274986033849378322309122614562533498 936
54091776314486444785467927329839437334211227323209783164267422 6560
76231248442514319056275283405356554997722460607316591827846698 6788
15140030926681574095600874378009218665932662734808137083536089 9378
12822139711211705361568618659182639391576208149993864214382430 2765
41625169841651563781642111593842869773331226086609057901477192 4663
30055531867004002679806795932356658028309288344627542558555169 7871
88787649030865842023516193294765165603563930140020485928184168 8779
78018593233763799582468738796134151448663463878831310432122426 8953
19848019880592384294686382631762355049980456798643079729644512 1935
28661455983361134467938720749814578919901244026097460904156052 8073
62184982837410558364467507611723786859322998892306848652780532 7616
40088125162458488806834444917131285162004354925506670807708404 4886
64544007436383880240241484473482595383858590225908247516385535 7044
77825271397003240022738923228842625613399443552544102620522946 7127
15384588852816103359736834058306164891959902207095258867683792 9176
10615512663072798076067377376309152694736352505079971087284932 2633
95318230068713466143763735877210764149232028179480379924262134 3379
48203343191146388486149819166012130302663371911966814616871096 2704
92583371047263774273909987356081598794248202446321215537507992 7124
55359413158238682204541367345047861461738541219771593085239757 6644
44959247281790899167531940056226721157732674366869977229352480 8629
18933214919953300691158458662285974505632995560086763533138252 8854
18579051317314584772796909179646684037400899711493317594231623 9545
74754445434877475115735559655017848623098521956328825159322630 298415
```

```
29125650228524007929364638019825388597466853407935496390527672108
9
60240722690216598695770109707753145210471094066418875528856976217
2
94416158589036717664925067933674039951284172407200686414023664798
7
68112524117517702840593470301717067343810185185547219617153752137
41833511112634443088789589844547406377831835863522451536234014347
1
44907937330219518736399734604026816975850890723119174504043278586
8
71975258809453759273101288493428595964568720864018720728457743839
0
76390123886723218719075840952484629178762468470810868167616847808
4
18841328658390670000000432945495240611738793555114591145741370885
0
93527165492311706894790903853646846032949561753537723232937859796
6
59132170259500832027537273249588393997471195361862148367283162266
3
58871292976170228437562373815075620641993177656735059414889063564
0
87428181514187358546925223957003761365428755396282582996375939904
61696724389898940838072478978621870015497348309707962945918917926
0
80571282286890284456968701311416598913886312778159377472351868296
3
09105880601095529017555668067572254968349720895906895488996544568
1
49174932503061404950782618789553912635624858651335157098999445786
8
36042749019596880158838568613482654367950336325246646421095892151
9
62633782044580434647588450202341684391178717116779371776792404795
1
26102941822464086218294683878442580970747601966277876517254179451
0
90732774938634894437289710979113943811306811881173556590443816589
7
91527007008054871459137678941641546903043277405058454158421749442
9
34075533374811014872703312602716369190209141743411461845845420015
3
03487511456649443170073322928098406416266674079739333100092890000000
0
```

82099969101770717192303601100704331158878388323662969665402226484048547764828345756706689055719050505110521854154351546938839355819737378138913968113261372724346005931631458890311254293622905615592792501747619619910195167789816900207923762253210088558477068718385012769327729528040165853397238089710497459473810934041091197558135961581559389688019805399375463461872776231212179233187692538078343029836896388458987200498459081033244747868863381950536877272419267534868383559434108549800543903363556858206384964264234653550606668157315771705647286943524697791931805605449457074120911949678195669619017236594583846472994019056013387918683672744493434259743126197780478441599321756578481011073143501105785760831582855970686785235841705012783070222803804592920817664961066409274997298898052681111839164788607559878234263052096151044132081507521424043245091096716504212401711041062533262157449243613330923534315846169830245868574986064524304733737315145183657493400060728264176165617058323849776432865213702932873548624904772042692090041485147895220568562854112859065337345346962963857339683232811444107849604819892555007213387592336805188689375263048910179867776363736891494769885842638756824112776167987296526124264133885863924090284031728824496655022800162806961287973754602491331216130681412703629449466684817775945662784049819241359497309677514864486430892988720592364760463287458534882322545867030494809787920232416573790119133641491107295498809053908396302148301180237004742420920777008267185479760247522907545993257147919511374924890091687684446200286999152392713417117055769687891561782367956831021581964620217060526806284124709017696922242925060323987321868397079371515046941603436673186344220750700256632057891475568874755506471399800981139261314848373326458515230229842346444800882973494215199445057253649665656310648246363116755307511897450201413014321986183953224033495923969974764227550354851301994236117657651587269976895782157905919022048349187765244083306349266436849441229135123370126800299427994924158198854459872081715021925057275453637483966859686973857571246556932990519250543008290693910041421988650860339813230064596621813773076080882930242814716546785669902950498843393881700458294020938683907753691574637744665299901040175526716729777435811317634642655437300463814890756336104809214688843063736751715890526297357477156095962726230215858555711737464232098875253268603066118054794209091412136001419562237883723852853372872382972863488207678121884460163860534327570128267154138339805376207368039527170012437539655311557020162766969778168031964172156205221036294629222128301610470109740744705766291014268713515508496715003222094381109889611018490785275267739201063605907223788016686867416742585209526204037719047354128986753795221665894381547838187387486082013790472713714997198893922248235464819309953545151255389034845994570933509846277002061218087901747444163067588391340038598741265963874837286270125277211191440373012685659852571701959467226047154042731086738050679363379825662084333142894277583069256251376006214247468165814483471051054650387189623887138524198700646338066310754789403131642406415480453540992452396135262247968988263638175771497098159319582507861035392106027775230521651067733984327531426645067401048428267523433731495458009451525603298395629302254765233678082735923290746184184347689103016134483334290175522624306351131969879205084912697194727101824336442858787553529328426972982528248435948278782183350605217795138560210804895840208012882953878379794205782240084968562204851923721711754355361702513262435492722236295273580070977292829947516658868924835799904059894584612784675689609599515570905931116137854160191251552110051005274977130047879292018260377033457938125164375482823725720732782957021042057410983675109220887927439748017521880988408409256253245834325973868908759842113000899455718586597243939910125087051614673424448737060797383434948490516838068406094807183830676444988760160885932913174404856200164455201

735532135704743305668668610294653297874361677647050188409474234148
276129512519742082498136631844413699644642310826179297998539899803
796066752198088011262677173926816446599365877913266104671182315599
977066885832935173540350371888525005894772863865357297949038055292
468112895685456575645972578974981946146990323669483573921958157003
074074634414172537592368753624956027440727966799287719790568 36806
567575800251744645250899348532274467203630882154451784535958804186
665518640683052618702108751121812619136168336846624509864501367906
427706084227412724740368090914323500535305900384569644154743958816
789568269944518242080657304350122307995462351115892331466593876381
746591543714021067886373441050389565579804077089240867158031464114
943939437623537259993149415225079921730391232967886763418127110664
103313030796647830084208585063777476392156847246697969426481628860
878748008689613775969900754628118129577358030370624261307811923644
837997395470336228493556734478222739658444552113782752045137331783
285064219963367564717924651166131143573751249499769325215048323344
045223784981868634208718438720251400074114157185639735114473445880
398169968902345778991917490420832437760367356497386118406297863196
463025571847048076351844452370408089797936615200142297942498256 93024
049602272034015149268019708962994843578435612480184296214076190669
288264558412371014111131297927147218627673354278675877365414234391
340903853549416903703376579266601581116054751232768748621735110542
638966501384547233109710764464196595134724568959028105494278517059
590872122794972533279552453802146221398253295900327023130990024956
872901011062056829175414806738128466254712819544151110807691164588
781270498721432952166962118341313964582410857305819211212747544805
991860409819857550597862263700997495588476632568180223897898165934
385941785354489394269489476672314224992076343636565207256928365969
222436697421420958235538747604724915389947673686190096091202709337
536220039069929693103477112979396976489631718391532374435265215107
263244143792982145842878396709303285201882289327130673461189504428
540477018727813777078484863449672829823358070482408531422866612055
947751606243795044628670702970405535383912873321647855878474639256
473598022882823974649395950331837683392751176738580376773206165263
413563236799521583254046582298027058561227358491596435760050610521
600316345514105468794537147432734226571839918992769530890655617066
587936818566891211573331023120530819071546564847538938688857321524
853524669721022831623678920562143298182328387902750939084395316028
091171825357745384851241663769865846747487951405628007746831168997
967662397974702147894428139832023120113138952630952724396197672191
556641628777850543779825910070413520742994798194196041327082246594
211994727994050247512065238422998892781022715914854837569244042325
944668094935020502163673751205046373600821224998714720544824291226
939103013791673314956176449623284982899337413787176590742314052174
832901146043528651687669790103146978073849605272936918589322840748
675008554264508203662981442131839542180212076038583210000953417184
237052902202692703410654384892775275332579632425738246407532542841
528174803873345947319749016962201876895421500154027220602910204107
253270902619757656704584975011721480412519616407601867447793493704
546073225859089096041672412704171658087305877230179409992764393962
911626485949134768591630674368809901519453885273491484510674668993
386639978878351844493769868522785151729698344071138553266161451091
433346114535039180806923466624496914709696919530795316219884442480
784864058911479715245689161353457782757581602058382748491165323281
948354453474164031568449713652627836353807673776288508700458737891
194810040254663595383996043334257991323784132119421592376479 0729827
072535594881288270652357140477093047023246336115426757566003468832
187555430066867084412960050560442710326109588094507798357131487619
015080661905204536141672087418300464877109756240885367181651105240
436530333241334882464692384685703338174486332715566171894486648362 0

```
78870911202300046756216218890757430209952192380119782905698722132 2
4636530887737751573399956660317687279848319662657376895230385324119
73607070652153248371472992644395599241613091372550089961038384 1689
04829047941263105387904414331888560076700528288123837812871222 9161
67175212200520703525511061342533548900316811829270828684586402 9211
86231179891970851012537665232685815226901928133320613873902629 8882
24587557662389615851428833308782175987303725102538512399685110 0843
65317182618413723632299690967270429683531102794323851869928140 0760
13823405943190615753631546125594901406435363821508867927659690 4332
79255772047864721494205730632308427805533119740769141953523712 2419
98306956456257721958970823835145794015363835548724227224918041 8095
57057440853557032177846039511880154816592335884269678072690871 6692
33387933041468988740739159426492649191039267780567328580307225 539
48986285623147882533830092037125741592877139255173564713320450 3384
07889033653675611846208154923055020535144202425477384750223060 1402
24411378050244309738956733389693237730486743205665480336862103 596
60550318541126528787691422562214512734749783003436322734794937 6360
28132797334217362330562333045739030054906772789717630389371532 2525
07532514233159255122076184159592456967648143737388223626844997 5004
71016275824645455109878875114951361857369616714969075047348370 2410
13892593150550206711128313727977254840546627020826180877352014 1128
35415524586117263453189169003664631019488077358683046477622763 9868
50938812046034250068680464716887177860255804405531979334303785 8216
28691208175326862140948470221466032299635816638520862816554564 5102
21042295520331324996112205730208575299900475739902247085929323 1830
09436457357470397320023062509135723477003235608520995342192227 9300
08688495306267743204670363231564837965808467698026101479904744 4134
16644628381749551378896463498033669069703091091184697297242227 8957
91523615127434530952552339471505803361424243483234181668736784 5116
29254763215730457338209165808871077672689668261139075450539945 5683
78518830886998744244098700328985928491153497087753060491133149 9782
60370172292864823187733830889579015029851719570963574100349052 4692
87317457584358136792476018826482382725408270154053589413747244 6602
33167341611091743337855766612802284946539943048204986660934264 6982
94676939907910403835816743469082476414551090307025424130726247 7668
43764511560068897690308326729490187882965588276980521008865542 5976
12440051905975144956128645078518761239402580873629822924893569 8519
80437100142351564230616300206252736962035318339776209575465726 3619
25334387863836316749406233493237096773421234593371465723351300 8014
08382479519576115630149193323319559188830529152622495591946082 7280
40327590614442171594529262462608191998690352885660745115749504 8427
56262860680817265694191409854694632358184430645499116393766884 2444
27139215835235059739585258209712481608821717998519417591249400 42364
78995615143030551558276748608878956930719153413489017881155462 4198
11974262395840658198600663625727306864309275519724456292194153 8531
31153834867261792735798869558798357626356043056977595609433122 0056
79440014704754803422250187570591116859437557110868480477955284 5544
25033466257038840389402775597591915022354280168454013341080295 9428
10291611216146899722316191228879522216215905378845660648013497 5713
95082149773805468683084409450438456068007461483096266235167296 3690
59524180267427991188721540930316888790620677124785898729631618 7919
62901805604757507371753279758805219887548419673259794245120190 5270
66287047949006796953008882344182376605109677483033805484721124 8619
77858012349649494443796738745014777694601176515266600955185146 807
72772688597474892300094327529337193596631621446860915316286894 4231
36852704457131659803813092633975290887854673238769145273622360 8116
24936464693519279383235388393779682635741285295665822864745116 2224
75472711800781972405682616279411990526676927407057965621576415 5642
81136071045403302462172844347311744560088996231450766734188395 5457
09396722195283436649305833049292150306610929835533387664642729 1264
```

```
584281752768896259550620254204857523962627003932183499576842611491
197777514207678894142983100640011838255139109366693776775871352703
972320175142155302812561285704137152363961687583147128578707017167
335675597956570821044933014911495323432243539926751164379164978845
186686310300636014001685724953216647501741097447818368275459029957
137108386033900391172890100185316839043286584836833880656791995794
585425262349940891885783942721710708333995032880038636858757615370
036263803402974530935670974243112389709367502759588577417403903581
535724387061406851805718010980813152725063352026267948968693830227
360917348711416822194254326349957566853799963987156086621855132374
690956135128632534594066790328595896416532231542448403506905710288
350305297249364133355219517929100715666033821703938487074244305516
921873482171320654968673420316011426324972802979335653032643558238
476079234853767103340721454866591612010309793002278271554668865929
202952065977335229636339452485972445038740951602858543243965256444
610009632820457430265066079403920339688253988307745239296559995609
408822437044673734782908175552175822298366909531478479972485486131
607140048857566982852099867471364990905043437646212311077419227186
639916151536182870285038663408496476952215476799548211581325286466
377327825527872815677877936264773669451270285789618429171832914083
346938222057704613324123480722688895055971061984098456124940462373
447948547411356177782419304015724628386236174384960107078178089093
632534450658460385683744364184828221452276905818694447723028977707
312462854730542924132616256457199235807925111969150295132171340344
547548407067144196311060284267556641500461788405694181145770321667
406647139437050123837005545973804030898578063048272753024955474479
653626908865531776363010652242695708469743140891529358129641015448
737186035661955206519320789978599183323267156173000120156211375828
325027066871444187713710607740342977026859063889168920207691592900
528536334805249228409835362263133417254392594312021345759485497560
618285044399500565016095975134770758713854475429893935750581805050
286892425533709118055700562227736304793089629161053399212609542681
217449081968572024942547985089040104130364115242947492057446502902 0
449684270585592489374645334565909785234244231821263234119509 21373
746651942929419631223160265254644474182956201892966880738006930869
482518414203050143414112920722489818295476229759682663248045627177
971891184763565237762263820036022881502608870251340317839739872485
317959319854566732256566897727833994852374500378637344288068490983
039802098757581286296368591265494226233320327140435535799329145992
975552620096443979509818286087477783490511341371712986858625678687
094648905313545052641339127717798189927319874697858047310872568757
300777750895584773607312133502206051484640061570201354769512721 88
793259408315284359994865796178514794181592226484543009425878371929
767686745910079650473847692524061121829837479533289111366479926341
411408738468384169545811190041311660598965113951878441950772021679
405154729374931333683648091588481657159276512235986854487656797976
167837945873951572999025627683590022525372485622736433411728896630
598687917904310011504695856064354303399599691731589180939993861849
798608575118838931833782130713005588036151633259593306841151954211
926737311148423067391721043696293754924427572466227193000433399225
581309183083127692917900426565198818514179648288974459335853167528
762724329687943491596093329507886912934485286734135672174002853198
604155609017889087421177888097382379186756953928807137456382596940
674268573141545302416823525906011725540908677554479927815597556534
384865881169771642584781610448318684476957021174158128536036653434
549998202176254920122407301479898872165446540839683086918935342706
854647077627915157302938466746291149397889288459658555446674578011
430999730889128086924703078299038050986889994571758504449123036450
724192558363480359147776191140621953047208135845143183636244562974
126389098924189970534169914979437377668562775345769066631072121442
```

```
64681534975188921999495196856192602457442567377918070886615224 49386
136782502001654481452975152367757652252631888481990797923548629565
30809169429946843292710911084456813445610672549861128507224084 8593
06072591804146909551582151306891200075621289537698323030607064 5735
27300829395329112222894445671528901084040337781879385753832943 8097
24387629763571273212595055751571940150975377375274406400532368 5974
79275179942619299917947344436238492501921145816814992968616792 1532
45161352256037969854178943480227456221819320708129456314502962 2296
63224128102049705434634886060304416240173974165755648684964905 4983
08001801663380495100939840373891380303244298379268560053724814 7122
05533051885790878390072755312158443349047073368082943815645116 6875
32572416946773497361864212989400447465053289333651931241010912 4226
10723700778548244919245494994386997302455988635819031698535566 2902
51960593055573400219721870708506511774271723391758475327775709 4950
83933359355273196498399145021907182948818509177139867181607418 6187
81711652394446030973181994299052106511967902398630362873225601 3932
68970241664514726594683069914703534843134491713498285091407911 4883
24546057140575721240033622432265100300917920395572457690233611 6156
60299520650651414908698465606421440261195883828363907316920610 9056
90983496136918566602380803328471810370212444015338938857207752 9411
36565540208102392430182497920812183290898650204167254252095585 5322
83566845895804354426948758249952437797043943055994067941105359 5052
19785197706013132409485515368808123017180410538823395196468323 2111
12057805771329196538584742054892959851133969642270578393919589 4320
55165814827932930272186227891769261873160748931553673020575373 8514
66596296035917465519126989239816912956853424338710593709867613 0938
55506620747469317586655479647069533572979479916776037966434932 7037
86288113675539389725502510368204796909481716494729866579505796 3459
58364609604709919944281804584735952790083402132210565658799526 6232
20358774915903914081391781893484817749458930138248405112259102 5012
05656914577124576115794499294401438612545856904723436010389407 7784
38001918787317457654195713567957294809011391582395552166715470 2521
99225665611140917423150611367855795430966352877778102494807981 7164
53740819738403952202645708279612446683146114224170777664599011 4338
66271600007231677601913301329407727999780250002063410863136517 6305
50558432415501259457543603784171536966445827561627664496805142 4502
46005519200686140812125729846046505312661947486827631047587848 3428
65545751671474712599358020288735665126490938439348532633106947 2732
18284736199321173724443480124577135772674620129359086208433586 4882
74157044896272751613219755232040809935646878175910203122029942 0244
77536444295043801503651621788119898463426256141306457144007005 0241
25833280845689763629556720955464610188208325272709777618316313 9827
92762491018199054686832855957672030804635632284171354519794316 1206
38499840597583636724590549621326339510654132986231423738660705 8493
06296382775604205200239786309521070939267511710176598222602551 6043
56579865219635817551838218682418600091990284134605988849123813 8952
12319829508276225311241211960403265301979247307825803980314068 1143
86419523444171265646705648326064224974020343273442951524020998 531
41407463951819993738603510125512045942290669504562837030747614 7631
12809581232613895421222934761588064011256211934165718622351637 2938
10485239081213588330904372253555718174538356464448062599653132 050
90006029460366504348649369923449364109163984106508041614844389 1430
31815599739414161971510227406087173087220460964712064334314717 3862
81401136870201032341717192143474829978087603608193343724806248 9130
98745213422376094644715229022596215591636979471501296514626459 0310
91408656115433675850221124684974387846927739764403173493383550 2116
05874075125972721291924703934362092630600084057565530856900024 7263
19343510578607513223352057633706046144698492556943091491449813 0796
51579723982059084445057920494701168074746461697595638630816695 3669
83846829810692320825745488374323920710186720612233815280611661 2064
```

2884199672886324428598879763421219683285274352024840383689836815960
5801851296716081275916515273748241307339411143007955704572668325480
2852892866285141890026550178535789720027423922392555937119477191130
5760209317196329984420272422306157108355931971126904205933909705840
4797205227600167221438783070169139833983312875603294787969347059970
7192475515958615102891463443788530767746174680098811225907754464550
5660873667070562955640104100116139263663699300887797047295047731300
3735930189493273729609006890135551669803726287372080227178300684510
8009763606250810591590641980001142846552364373896605418619534111950
8571467750352245281829574282372367306445897788046358689504817759360
3745414354519393999218681646587685744162026337449800922490135577730
2601911357700862599169118551162945167106259209824129823588659390680
3176904432861526478099090894053992592456299228818927368050762403791
2014677707046168833690345948433075280508338575048608376210781767980
8607640805136851140068761871489552838895075565621491998851928015860
3755709546340558823004455650489577977381199756175997675020632757420
5204845374838071886107917010694220218277401559004386142267299798590
3884121545892856287518482567145028100628795122830645939315080853170
3999925756411418507674543470208084088692145620582667375049502713040
1522789749812041863405833368024063334939882089664780825587898746 5
7465179933553414305860085449275788522609483375362596562497
9397245111337275092089168401824882701921132668303949312291434131610
6388803051992055714885079345780452654202001845365306595995538031240
8456246000655926357965926089236117147066312984207401770808493846115
7024656690601818602632362796165699487603519368288776106122399547880
5327323486730884781062237317389735506862783071453766853352247928530
4635558360480765505090209399941099819622540197022452347314348200964
6986373074426328419225719277682328344980741958841539828421249053660
6746526688353149997345231299475828412700805359649184652223241218490
7880771255632922708505922550982249878011084710990199327132302258010
3140753376013728890678242094211756100778010035492471418078734354830
7332835790933009899439522524894860613422026526780629853980899718470
3930284953676497167657241682177888905106857715926658386766632419190
5546076815297407362037096345979868978534311259849648752044957587040
0106871825123991316301514061733624566327203899773940706041938254450
0917083123574125443268221471975350517217213983564284740103540366190
8011170483960147876872719392629965632845127427912266960618039045890
2773782209529736197667320024447776753513454054437118826672896096690
8348697611945731977957952869855360034592912568121957142302578592580
0425526907023673568775514886516539135924595860291153732192184111630
2125056243060149876511178126088014744673499237182440559860570740400
4511201909780434376631494333648944848835998803124127060538611519250
4198406477335247375658728098874343333653566287230036917953945731644
7278732982909310897824467651749905971264114485151839367692473209850
5900082717942514698559452025465970282096520133283046514081714639060
0107897681665711509929670433646654827645653270093532924003816333840
7213928947931876556545324636354862063510383470686195316309864835060
5496781308127984091982071525932950487560795978133941952220297886880
4939041813727854095077272768648313251640311829006119933127504417500
2459431750816926596650723994242522063174445255440021361797923373990
5739178923632773787828514491222400061042129312539647507447835475380
0931516606487005466575286854854904745486269703145908214195562775 2
5937224257519113728141202197400444056583220224478966026655163922550
0169114062047528421785182083062338389872726806838405868344957023020
7879203741930171645710381091244979519163422581479402653153774810 0
1692172192771461450224463265430330064907454122403581830851834851440
0535140351939710790679822464935640913025446986283955333242275948200
9843094186470882956700692140178060262907381130440315627058123463160
9010109221342486294949304024358972806565660535631312376851134214380
4341513890883149191522727524803327578580237487834098209031986989510

036390974724180442353339580897531307030201024327204510295280632 6618
377179801991594177925279929413191683956366887293882542375426879869
994088763398278664055358660929846313122803290199728204262951574981
629121314901517659326968509934498611464834922694934765489215871106
126831219093124011598416254331462804678165806809986920819505159802
109898252822704173912569320931408273556556435873412680349569770070
230516187201149876161343272808788221490973811575767461956756463782
055005023430440333994139635130416016667849446205374007740541848097
766622169791391044503837868648299420337259376674383256436914506335
699907494254754649616240499997640533124121652028803079245215508687
307034528676850563888021438595062088534564480038854551929255687409
349529366208319737579675456405618531159624403993036785952853936259
646669498291305221706790487239385585860286440426469378192291664141
412724364311023496415010032391205108157977615236477657734689198654
313767092951781271861010226827712534244874833664425774120604326619
177469791330886581122650076770584228082411248050298741986057473571
959163150333998390205833063062874906887933506876879280098081155287
954381207168543589906071040246759430126817336476240760962654847955
179751207093223799671447389038868284970748553689206108382856265066
745422553002528911261369607771140629854160243157238157271647775597
200438741847953462719086352643790116388768808512285341514112546302
345171157577956863364385195133215620669324846544291416256629114066
002892300517394635385646427643400706629339876650187254363235773665
568258136455684929237399126547884570998620973568870549679485917537
086837944385954447443944161689684157568781522739396598666055825295
896912884366453646033042452394225436991630883004414483649407787638
680064184044224535733401114469520823777521593286439841425112222024
562960364070496734791549725125494205249007431133957336227961236928
086384085304612596656533990878558679978375963953953205862992183746
459559026762253778726229144617760203585688648415019975922225452421
398173402840678211796896211020344743708275791906348375005135799132
680538174831391266233737509208648806299213188325393956459935567526
670322406958565168150117584717567501087107228627153602301969559981
877667576933932181141062097958336262714422335337854538894474808201
141251722054786956176201936826797761215575951673570313483365981481
947780585930318621777805442882640608284600900433758210041876953810
085509671444888955589401218054331046930034785856781206645222845475
504244613125261579556955756734296443744545093689517386710854166250
403187963786128401507120295618620914051016242168009039515916101 09
599177237741591722967264629547400376222458993468841189155189843200
966258928516171619984133170468964394345022126503074133651030409447
190241865698258590315156338633170993683759592138723713201894656315
471224028221900331127895056475982822081462142715150867750122561332
963523764000067162476711302163443942032051565327346022693904198423
306394286002637997292169438332627436602900275150890837424455701840
673859254869774517718387847222412606699476356257062599492514812175
730140519058592218381110408362596415280785009890429542280676790296
530780798954065889305808392045624413024638903467849657177838436370
249059364732459885166919803804533806259831554452718809220133177606
302420859165885910539571118582482520443339109150701514450559178163
565816119708795101872718562437906098358169977209867612600753342963
614773579298894790725412579158082356994539209246860417573774347206
325525697471974751472552438664397969108062376939953364990580931572
692058244619319573983154828080202214031384374855781000669777614 3200
955044042297204642533875120875836155637716437578016587867503766862
171997277782509654797442731091236378219600820733362057803897287876
299080933013066579494460878476665556020494676854930182477385241 96
318004325281467011973402657562255296602507316010859013787624298509
024983526493974533115204922040484449693668170039181737081869927 8067
320210467522958322799709229130783014659225165547600328625512191692

```
1431583881955836452408626051057764070847471941777181870973784297147
1576341893825448751695747125735822553125086118190607722729432790064
0685540959802672428956258321418502088788637885494128289102083275339
0484463517373332465369208482382349776860851802190895493363001103825
7598365118308801878717051349364248541411394691775072023272269470035
9163756376787343222910601423245899769574110272082501355467067881871
4486810950386648492378113381141609774425286461325407820803646495338
8341962660420845177387022687991044995908975825062522454344391953967
7703243468432464337732781369945985069198298545040328511255862299726
8891841997978307299650776997938274528903895796564962823025273215613
1632409575213056060798667849219145747474554254942054572819859306837
1750123380477889368303102829984667930648804962967105485473827024897
4954211296561367553522001414481285719631284529555336949483321297536
8297352333228015001768553000439464772036370627848562154192052988302
2494096715781674754430590711131058866271807892545605241577881688848
8771874527147223222580217485631529044544310929726269554934056150433
6550642097993572704772698482029660003493307717293173265317777794922
7791530149525129001378538387544103327242881734376413003093593684899
0249056071038999686830496554369363614668437767111363495541209649713
2561286521520759844651267798175688212970577538745124226357707710455
3028521032244139809554354670621821779314526732087523206054819806255
5231353510076699649085880255791833388951347407096941585300725026751
4057450677123577968184460987917526945463713152083452471650135431593
3430194931335209510513094214684701461131557402290250927740376366484
1685257140951102795850267468206790380739112674430141398111509627054
8596541251108150855267884094650594735799114349622970883187052912752
5019048625270244352765868464990774583116609792333107994046431344059
8338990400028303421981044957808286642031137809222660479261940443554
8694975353437274974153358215450126380529346781788402788693682095290
1758649869872893890221010378805515254357441008585444293316959874387
3490899686423883587872244967054980111721235491041522090718055314449
7649046795680097736706578825819802642522395608368661559880823740099
5766768678884212206407059297160114726682628189295286197056481883398
5853422302199512790237667925264406639591654850707006475580049091408
2355164443935730821063491251644897797254319733110177810182483178211
6774927741074227430968677645456066481935629475932635629739043331468
1927031876550074391829903226833240523902861895330702879266947518675
5156937052532938209780012262676376033119662824747621735334566200900
0399533352252815816379728701224209039727339210614157326644094918073
1774642665347803794714392102563555842371579078722874896655956854738
3942232266077494331455840059118575642070354944121390930242670047345
7950017465154245718044680485938046569434669096896599906674925161303
1575961756795780997934882803438683752679397350955817285580749363775
6630114650159546921122724366396154964419682413945760459565756692154
1582002424162859722801827278126071047385079269870073967703376985474
1379545924565485795955317509543585957724512915131866272845657553918
3544299403997849130529351369409896321822979077417399564212337971923
2647523679249134972584583740248668777780231058228590736239246214604
2524447162290473474004735696047914651146233593055037500415468346616
0511829101394546085033592943802850547970334218449281781785985163324
8380769719104045662120998023195730715656032358818151187953359461038
8067237512101630937454975236383434980564807472852667379698641818393
9291821358427935971103592204349266105279632062639705282010705536222
8885615788784339912369688159364437829645228716456476767107235053225
2083132114441867648986152713271814414732518854718353002944541690509
7618525860869876063593110013786116006330099789276354262160744808882
8212712613797715530969901531592186679410298393171089697372283402434
1628240010854567982454131995101374730278316576133600960944042040141
4763559337077162172976485767730211292914949100787431547067721356618
2593721
```

5229983627955767793563199133662685946080945209157436976185890 28765
8964092693186314310020485058378966804791221889480483039765246 00647
9743733996706129388785431982593885535093279794452979629060327 57826
7575782164562765899548742444537320523453378820179191894405026 9212
0129557162433575693840844658014242449606919053864743515718796 59861
4492041893919151179126190910805598106148355993528669009528619 458
8558647002654880854774762984796160202061841493412306336474041 10665
7901785716782153926199452905276410683068596269347592605253427 17931
7072126919194674831165731971567070249253754338960228541782028 20417
0090735316476652634748268514416869145484739153213369028357954 56128
7468629318610279072987575732529907054527814518974558607570859 60056
7210146941128646611767309014222424227722794705832176665806201 08084
6168027528705438927212545002142460655471144064124121246540744 63772
6802645885589635714086112656572194377520640138517310158412554 27645
3243662847761149274157656336028164567002823004926812484775003 91161
8349986712086397968986076273107259718504042428605293302611053 03313
7600714723235705349467578859836543149607771637310914609709201 49066
5729290914983223250813269471009383627421202265398955351444483 60226
1628547364772438182481027630284482139245561827125732564998490 23194
7884639710014867835594024152203135334286531534814237648138868 69247
2061947573773206155785165096102435266184473697999404406823564 33010
1137006923161703093629979315605333631927011076123698498064544 109433
5752732250151714848235578108008684820737540109945286005607994 51431
8608282148768843594108014281306856764659496716995258512137086 57044
3183507877472844507843542699927213573939783103039433248273127 38893
0913780293594573039470033407746426433744377840269129693845015 73900
1376535961547419789259826587598279262338426361776221039259900 92985
5535950662969717903821244861762344714791761056885202621176010 50600
0883814631731308759581954129909300524214747056290458867113244 34699
2368171875980980293828889566180633964638979169653827944064275 21650
8852420441770437626866606293303204613538703043801203581538813 47378
7334977482222167879060251860725925670733484217721330924325876 04155
7096877732758418954637690129631343322844738263750622980984042 49600
1570870013924066572211926036100033425181351656225087520927789 95254
5202103420649426979944831801997452502714557010004233751089563 0892
4289667546668977003011583255476623280962855298155173627856555 285137
8213749332496000434904520024840963987199516443530485558892160 65572
0329599794026475794715281963154699826097671509758524630738830 09255
9190659906184939639671214742793097870430870534843289920881840 90986
3639807751611877251458330268641139318689836075542378059548111 06944
6265598453404005738145053556629469233636638434264490657144671 77158
6156627096510342618864152837302194555896620293917386201566268 17979
8094201029163962878596589331026287923855505838758136693019694 87534
7471501042157077908806687921748975929221224382676230024065565 14737
1793089162926898317341492873052458774157721552735980526823070 09965
9214442500501470160940240568743893036583217942610485289071701 88079
9877136078015801004040501775228672135046625726422563973134056 2195
1826770817533116536599456917614615056829966598775415736562712 00344
5293313246716256665532590253250980635512652059805476196977598 98171
8021711064554013929499681531189078503617082726141079839356158 2994
6906453776608268565421548274278211849022127060629204136340971 69912
5620296374726240559371945146197217732756916898233566537213836 19250
8415071365223418891023698387020881177838349047930256285455585 98951
2473223388525791415815851761867783159061141122159287805100771 25062
9064252577466654525374178041125969982956518666182069166511667 448054
2365499292897660554082991531916365753744884791663425723308897 605994
0326441151848502243269337568284615736757532226884466254319672 802
7027437776372432719775399004654679287608577118080139703074072 71190
4189936541729866180358070741508792724286242191598426772890932 98993
7922013858244608021388589184687119142363802114290362023947801 03644

```
148019963119736987832713915919931774419870381721401679155221589181
623215154280678586955706517488473651675556759135965418254405480731
851324317211776922896875465413503543267573352558113061015291913827
118601513085223719002884141649145306731392959238090967064520232356
066793587604011675755062657024889116967375831762720179004604002676
766927496278621108998902463580203760714919435576636536822709946028
154126262879328529800288559749721794999346452971230881052580397660
471322470096447193654906843987350485087816301674168962669160726477
011672776551560400668347741381607878180123892126725194523637245378
083184688314338403082516719814061896808905248332056668746190148 32
683677588187642278052456856579825134981895692345973403594569401132
271506381342976941040570729902623822971901278968452055858071579447
810590716939476396264189644359263506729821528791549112750476989551
878625808016396035739787347033718355353548818150332587009552289138
886387508994821107298691295023441319225784610547353538370151425405
237384778635290332416834781961787595917999446009873498441954951158
799930579878992903421577952896516260631742498725530958479132293262
504664081334848891179035032424483945467160491698446320463883822743
275380426923034651540705216879811119427581308554330082587684711634
809137564866412504872280721898537008630499312876689656668226541 2880
531388800434477408246784768102107267062839933250155044626686071734
008421166702046201529154254278411310691744575169512390098578235282
274420754499292819650004767892010781533101041834094440966632980717
740110857974649098995481315159606434800760550664209849377084650 51
694088393412891983399885651999118005614729314266293063843863158893
862591858848287384971208691414261061143664179854823380323562875484
799946191347001989893405913891838493692447230234210046047376702332
952959359446530787206037165444238556509190590945095017451531292942
242149934796440032560191688143058947407526565053128469871744703334
757811783396928108350588539701951172308336620344907642257491172661
128529478933141844415554985700290931459975911187296313182247970335
266682887824680904168288213369038916899632210467438729539740195313
578071689878112695575558657587617169205868390457677602930131102446
022259372033564304629223590449458859046499582673604667133500661543
934669764348126667093512109116083337939337267991603185348005631 99
298023527418703212225305158573303396278164155888388894940077054278
363529287047932933246477106445403736823559095913641872208579260840
797309553704461075676881640873789492880533877763661736602133549316
169092570617682630757128200108271532845593567781694955315775642386
521345637026956855822045326332889120440962728593476081780796662 98
420883646929153989017034083554488470793592920127474314705416997057
172941287998748276240678067986053868029198671660581472242218383436
005139281646183163777040028591283667611363872888766686245102195 96
398052949907526465383587004104155700553249172606069818386906091 51
717688332025209367219649323031369090306135294572865839736803315698
725277875823977229701167027697798588324676848337304710308117839597
932042860871679305456470857044761882703297122614019264048613646095
430746814788366994977374152586060341346856086426836451894548806585
099541195454614193842715203157921617371415216283586705338884137039
063287846683679989733658076598059373623241066723488845984722111991
595002836170240450133587282708620331029467243772135100726030300793
931172448545088685826104819377899219937892037960792456877929314766
905590858178973274259265121561072615470988929983177479801465034439
187906812494962739681873531116657313732068220464047678707894840494
454489956484665160976060454735447005221682392004828431199239062654
860265044867152208276908294858949379157598937906854918798722873848
094801168303824328316268337650470011156743560423905239609337159594
607337407887090978619105863099970988625801211322685292755042921321
880861192952747633680665847779147545640478873215588800396666447800
324461521211794672545547634217045598643376565672916520322698526049
```

```
95069093945808064319472466099952814732360593092385050283193696807
92893659095793586275643770111584101792609777326619286810591572594
58413631062803307600269660571080201402332697163152360571014321270
59162600297085146157740317346093864248612841320157490955996541107
99841344473456506977589972475668508043400760143891592571837675018
58402875253239528073215581515188128090794827140388330249784369843
27326321852294750226769345340253066295664681539985113208060778593
85573740046526497968599283595347672778108634671617814736775044871
40972745314538196221188756376223899974415331939817256539049069980
24149799332003571729050798944076337656115780628222004394937835926
15575450269251321321942976904476360154334321509829191313914356015
92520200090431345564538536675301838456110459031577437479508075376
98551914434029691439814704175162697345808443002702166951445618297
11195131877208277943215204897613708847815952573520409396553768354
21742976457160259131551369266975336791974753577134977057924016224
99465418512637188235156058303147140764157270169727483401077839880
98181062124980860359246435586873716212435118452521392490378215110
86371947314334089752167385424196049094655043794928376746043044732
49090672284648588016291376971106552714916032406136840957968082571
71964730912588056987024724558487984339917337806031127768512226315
79252260962252205615432640644400786583495324849121394363252522274
16544604048719165237760800957311104825056328918577153531539181735100
05961956742990348075618490234853141177979534123211237631988310250
51638631416228631474410893055254742372128684509068980470573087940
57756479653498934658116715951843128621818466681685993074638612579
81888403043748932724323336286569813378193088220877461054488312066
62019097581638816971278283182654844215681727224409900463320723609
58821638175488648772040165387124547074044526505855268002244484343
48972198294378443881562586461695948468849731810803658564117437808
11839994940906486095700073837055518893087703694550623650637903749
63182498637383668966794147577895103832647159088559964673783466980
52782686729019573270674454811853590239081800004829428356835934169
73703319373345371152156124832739946513968070000047839007510628899
56601827011398516159698252896465540647508996989848025287502301830
48866555560009514988520278733056778747743228907822834378838575069
00558359997012343408584938159691515260848950810899952518871066772
17316923444956238028760001505184291125109288974165231962214203059
69666098203180571500959853992614074575876562174652157411186884479
83096395922918562932676539250145606162987816632621410659740434840
74460075637525179765047748330918484590549320535828106114033675175
62708060052185165974450821866059427642543532078599488228553661721
05945980516263550451153778110992123837565134592472196685047694102
36767211938080945365245559762814012131183798470928613640131204051
38436609648194020754951549407307285472922938773789430676853056482
16871250442893012496746686216170534387699032348680886894079227921
74266879140695160278261722359973143882148490140807158579419139196
73572900790405545442849829901362970310050170186201350782944143139
16031192838370690006548340990145948373827817621820339233458476713
16544016953064033974601679449379507243621941040579088036602959053
49529774433375019391297133941233793278550325563485042044466626841
08356652245163280915618068131007466904072058317236047606022818728
52241391312725048890992088080460592114817573119286117520734638097
37673918153611812621061453437081260429042473284123758460076106044
39492497880178988794737127677318665116126232510283055400359061112
32219645805300344854408815306715466059081666076707726150787357634
55959938568488431820758641462758688830155336517604376962446871601
78310938971792555990807743554523024819713342436855991105688794694
23973239816799483611892035509104312596035603646949699325612226120
63742291054527683460668636519323760588827098726516449455072837106
62133616667721404179426113952203796524171624731205932259919501139
```

535845518877175719382694300701509480143303103664261928135603290121
465821962171272264366174293994893678094806922221819713655472101366
334424295133797071037941063342103271463169484284088692455955324758
113375160939375156088396917584527303259494589656469494880216947178
151013716782023422473281144677022654389369966736980832853832623227
733255415956366421470627963106784251279562819190985530120611587113
432645766815238245780656405035008880365998736967406780185443326426
216162820250308773264830931334130190665231669252831965493430327661
806514004045488810242928609389322174671387270196458655901941753991
621164946965818765366330316039189475064403343829841662203048840317
377394052257202938389273565433468631625163007724941320129834853991
766460576056682790449653546315236614730419300599862017523430805859
644431014842976268128442894609216047062946926659546597197803295097
705456963252408117167250627160308417398789316372703183833273203958
647813040488258019965699107385204701626081533057050379473728941188
428603751257351672215424744527307658778248738149532957895863016846
392649948029931809301508647439039597738694101807116514246786566392
133031984860143497024483877858929377027570478571924108907531885246
051424848173952956939829902212357160846355380856377573582547208379
006440437659951825147922262081771020763230658269270868743713618189
634759618235227259459302545590575055584876752960422540447238379797
953029476104962940664899632758231612228940203046368226649321005633
221086358879876472003872010074224935234941410035856202617451349185
427657944906749865507099669924433160629610461886591168403989140320
896127413961601370678335217142255986935921057005567181624621757161
715986543927419682635866129844288915304063926846644609284134269828
368522242985941316566816243272394012900795674018120124518714350001
792840711181834125755099824040140091423278917959539335373665962757
784112211164254608795083900433623989613182135054349385573501855794
315629975595349634789528259616261539588994907003300844012081545395
816885007542268375673078000013186113571914723146390084845329708852
145850570731823931205573300589129839766238970475590432120592631334
894985103462134120335751146735481705850280310966420252508712907655
924750499953226620106629916144978396279128849471750994893154192116
815094771918741127647820673102948422212862137649387365874402596874
370070080773208368583929617208477758093769718051426830409157871009
526204083579907636912032233822558947523299305100007255810548811084
638668962198237212750552590786453052618848767775668529029647332587
237020651796111437400386249330251460280678281673275872325018086231
769722196396515510054177028430520710754244633204559783957158335995
776462207484915109166490088450784620676998485523762880178658398262
985967104739676333354005587760466293822831479408640163613239583256
934856708573453216317374082555355420644120625085409758696478656852
234124565152108421803248583157926849976642987098557077690613415503
939868364726778059476585976839329523271827315724135473678605273364
330963686993914149321371178687118399236694764536457558965456751176
465225878734816604251892127710705464702578423753906655909765861718
670849335646036264492194577900576361954495929317791911721237704228
952656527144182878051962175507984378035230392136569236609338833797
100396974324954765258838359227093947131963173191894102117299673490
629214968426503551972853360725520286563210049589244100233453597174
421681320072462062742928898621807869899736836063008175137260213418
030987407964297009770988317861290098848426448754701533917488338156
997208675484665567497402580591609931912029828523799806539658803896
291531314886224458260294559745486250776157008659187252501078647939
948777937407292829225183081087099453928042730755945547793641021 67
599783990403412974625897229362954969999744435849764079530719603302
240143627744645085106555109999875211216665390228003289111354112794
715221370828050460567167019824521451325228073925322537721240158353
292986309670687341068490531499763920047815086683384887874090578709

```
41165459360141287385561360984297995062272871357324890605106428 9044
57300275098944186983412196712987807982265353179142405728998507 7177
15824559026597184890572347768245688247264755620717562149607685 5361
64780989460555601051751546632260205057709700192460042546620443 5667
86514570841102593374087336321931856277813556826708475041950138 6098
62768107202776667132542204290620393559204714075357340711330873 6246
17681072427934427314478698830748088259099181417944936107576013 0874
78038192904756968721265074063635926855669868181803086267010317 6482
87830132292463272731451758049575705006786289481462300654112168 1644
64325219220198075418352567811594346757219877969826952958263772 5386
09827426235015594071747768141395449897563712808904360764656449 9633
13829306494305350623839586303442910803662377903625668924853358 2783
42070330502153593469189675181815015033718261257745644182295302 3028
85024050639293828421895032953042185779421030216875335663713677 3030
86062419124547399007968433094632704492031648746284131487198849 2237
25191834032279485476919655172214877081782051849628536881569107 4313
66640887238703888520267167314769201370980078542923028026947716 0793
28562514375098658000210307820845436286735527276973827609713038 09401
70371040986436836157835080461068320096123402849805494106668083 9886
27079058863939737263454787968092207961954648411666060152345274 9438
13399254536000039487585718618134086862795295589291124637833098 4841
72538752504717134660544602651877452309904714383654960814115575 7
86433346201115987375701728457829398907484204633940627412439485 7030
09195643231424079152625914405753276271258167833775334829011389 3264
99866754407234702007441136534857020134016688875185496460523471 6649
19042629134269394710594370062084523142949584738139450583010754 1482
60721171903599341842845573198438945462699125794855107014632712 4594
68419180391810330776920465270588043115918264437767845825048313 8230
88851095254153507074534868882452879469600409405807636262034286 9199
93444184447060001236949678290382258044165036708857724378893713 967
89598153196917688259570809706160166246712372977960054829298780 8469
60868235533257786629958481751678370545911426727872865390835330 9835
29267374782567636813047374312555554320405294434848175206332736 8461
84182362170552253666507705538196162286891724682235949432353051 6884
89797965435228138128382513012784087627749436871571062524309420 5256
98125198123953061434440034526738995473315508450655715397949387 3782
57693176096058036640157964035679503839576740858259846505529339 9034
25435657331567727630190792073553507200230839930295733894941815 1335
51565702092465129623360422083845549403785825402153729986403607 0851
95621262904710235313260593972109443513956650444108564796673437 5139
18658144666485956606420002230450053754974958248868486937889223 0513
50234938768877525332748175779166942226113816847358956908951321 6810
77028065203313728393968878260073358531459773044345138979704440 2487
66227716664612631636494503844046146557572624458884914136481048 7333
37244383217093422248374941410562095123879004152860703408334247 2258
82037306854261805677908813624064018621076083318063786125044048 8106
82017827415892266085180981703135902480836772580117965899310795 5735
92927672344243537866872863639971305997766266303814750160913414 4807
09500760273642226305228263508374908679207026619061633370333146 6608
62742771473617962880760891254858505028613045690792718364618356 342
13769478688183874170833073000343213005606912139769282455970345 5319
42263192145939653941347035137009952434565128508307682920628933 4256
39756595374679632819865537967055839862966073491088128607209362 8431
92251220970338165291953573009692953728603703279918706706966904 3778
58286150118691489762063609102977507381503427670219547228734865 8049
96306002364679358509588631075741552993689878809864451620880573 6300
66628840754040074881714620670199218883164310368367392499953520 5202
81421430922498853177809036957255714690995238279886093363145045 7792
40455254013668204000669584989344621942626011549448932428316515 3418
48547946219065097667201642409554240701048960226753286139541326 1524
```

```
8644140905574684352667455998878178990766022677436217020596459132 44
9801834363182591712636586447173857919991551446971450763926603035 45
1149017884383081786995094517182028834047566097357493870778973084 63
8127326819397488221441679967619969387407602490286397879431629905 46
1362674212695206055120268809272503755318045667300836710163786106 08
6391752911212613396093234655661023281136324443590355684406890552 79
9088141482467888449470228705432393414941180078306680148536573803 11
3108601310361977157403395878147076028957759203010825660054633124 01
2531013717789697076576002153041763139169637195664072250152533957 63
0457381140823987507308106063794398243313690818913986431225173516 97
7464415535058422447415070453317448732550720704367664848002664748 75
5532672548968859176187224355783181788282985787928900496580122056 16
6393213880298901492692731880087507341685950131315966351666477175 79
2731698496583259360607146768980916928080438571398408706223317438 61
9571936280001099678381559220605000432792825455948428085628234311 585
4522880345308322817997004912084717729579188082877088044366900007 17
2842496830932848538041611774536656356729727181299495556007164727 29
5039582854852234087768267490198521235656949625361468009491849524 38
6694342499172392659831533029641268257341708214498309714528837546 03
4670247948912570867152204398775370519958772126531083713275985224 94
7705124578726552503903116603947413127949138938171391338537546111 8
7184612824265101565939440769893166290187410952485797268609389459 568
0953581139170747938610105112213264172773156509812031918817513351 75
9751929625390822905370889773986859248718225276987514314140461405 5
8806154047927592491079892165415429140406026779075530384007856103 40
9835226585128382002470593963704056375601142894057539388649820223 51
3128535895874124724891363128150815776650765232642871151394793454 3
1415351518277847958062370656935326144616335606902108440005625734 87
7016175648623181565850783240180521678694109769763553378126661305 71
7317432778000998592854434377702069618032957467220389558895820848 36
2800737833549180248974785617152260592623966670051714554601640184 61
7369031286586108540659935534116421874880434069850235743270834879 42
3755690628714257348600582668879615982774046873304194928719631405 34
2223457921204643974974143115498183218393005644895982474401440674 74
4771521156825695785468102727365388196219068160458002931614489820 99
0698805413636774408925445774038946996463043653350533816252375221 59
6039722327827444807881210857378933440779017385384421671112812554 60
6779341934543588645899836884514048990698500867850807179682889438 72
4679007095464087416990635738100902740813426043167640776075863817 46
4482632785549757842328062454199651509363561481140737696230265285 29
4307176704302539023952479566657345054889882214054241409619771717 84
0992258497783772750166122671384270738570338700754086507464615341 89
3651847055855501523187541574948561204937985834343784595536554962 94
5675889007969280621290276990781556992525656639395433606226703854 72
2371980071820631540235749190782032495407857808626147995827321394 73
7715843307575919000819052198192963929523195103782118225842102580 817
8184929258188628611530924647959543296611122186665659142131386957 91
7136391041844172524567228004907689517769322319209505663835226039 91
5746123154237143796995115089664665116271157215949354905430481645 58
0731219331412403338831705500691929101642484533104237523531370761 42
1293408855601467303201084834771830397998626727652960820267272295 47
4597816482280002934158235680167930902907288631624257514187978982 45
5900230827823530443980739842410913577977535762026404881911538929 65
6671535966866119324983805180239137555202190939782857275953511768 39
3976895210718166032308542157463075757418046111982133083479615235 27
2672412335720178167855927454640251175574735376495029380003230454 55
7918155275354194787635441963569508796625937217127482785502023405 60
3370439608555003458522663662536626186095061849824762519913313297 72
5342401163385066164402230975277957130615100794156659051461930755 81
9261900217815047277020390497762114346250437834036943226200874973 82
```

155726871005719433213139576292221447781044112129367880734683585164
783857429296792000215242942972601814780484380719640477496609906450
781422536184597466310196727362088186973414384327990670599680574496
231963000325531249050404199435696323722454285918755346884390810797
948666236112001975898165677780674123303128306775508918548950767582
580112989797906321705603755899795330800969682184640547217010300007
732461791783125839048101764408743570791664345256983309984998386466
676127393466834442006547672525394669334983111678097394975729682087
465947896779634382108713516018393395451067845956513188200669533207
910277126443743335032403781576125851478465486484383504881795243 5502
509675372471978292024750903869791502743934743835656634449081720528
457353940636428490432008445370216763889144393171227844919645555801
890604876708832721616476494605242911180761745275231285841861521357
263161776965669233086978637979985464131700396105045850452688976063
193979501529241512056201651947937688365961114210857142015283698391
636694535903614673116616853059755457453849126011357086455533408297
850688220415175774955918899944396049593876168588742256052678136865
440051951973379144391309561057128405357624617833475178709402537198
071171912233209466223301262853316220543596148582319912343115594568
030891761387943553175496614239376669920595382514933546847476597490
594240921868564403009116602112437736698330120676110698260908937620
989957941998017990682665583432475017914118912124593980115961 77694
485696452218321258066991487248993365188734673904472714751701 23678
274051999731562434554584763284902241316944960638912465802432282405
289676916418316475943672708270443319071939742812465558529936008798
174395277137851493998963404625719356070189951440113586747334256118
224975031677903749980156921428084467522520100511325824086451424204
998432791608149497647802464084728012389535146087635295228779385 72
890857856985347294811030985165020537099493209080674043820599802027
038455047175106913653834854200107993767960645367394810980538772075
721228682583184388742623293598007971053295185514536884351990489672
081955209948785487255910360285364308946102027009511507192537188328
064200011957546502432849858640793235462955080980281059720243608038
334441025250824053734750857925488257828130719723099164119826869338
473146687339028608637052754404176168119787458473226735478573019494
104788484386269304039773202150807000430460316774685718546768797026
773827893333638727718751741174362913625820416134115936281337865226
155386282806038222322440998237464385988328330364690930448809211606
880181569072230021764846357876783193492974347752876567923556840443
700818692843661647327870168498654888327085924377642559587141485768
479430386555923406880657236922356042369663821299617039919194833491
798900441348331324331939323674197717186505219617516901214701572663
171259608228911117856987155462714631177067738646225098950646095305 6
889130488030384683197884094513375630973323935911155208547101736 92
516339074767149014177846377918413747449798211203234026043802081949
693211377729124931820728068223263530901418509738280022483408516461
298452981397689688165001745890210353185844071098767238752840075785
715419554453955204805409427071702212783770000304755747014989518962
649568438412356391116750330509300321539362893553029628140000487130
375900252260545488769771515632124187948105843827419800043685716797
904488157840822171208029351587828498555486720695303148539135590760
678869828323025976491933692832597849038324158772153378304814141851
598480940144807213351015547465606380879525895087118562464994688253
100588943389629815808701009264980710654775639973185554595047090510
163682977344436600109260370148790117727932709591062744084604795 08
630612341361000455907882854261633817505948669180774790586834965595
474865898847182119540317335414167905798954949377199016158001 51662
585501389150178218904275209287100056754786442316286693003466297035
917251889504108499073530776272425065731745147616402470454826549633
455122502335201426257785901426637197365898693106174413859074484947

```
4533765119165161242994594857805331328180155211000696693789548751688
8566490600537905682999217467527632177529704065014789759806514326914
3252140694443731007325518623190191823333179585907727492499166488882
7226588562269422902310063323839404378959121250539249662206990237490
5024718212208028104763887201622955395788946709489891962333637466540
0188948027915416643298290091880114493343730349876135918519565036490
8515976982237202249454378033879841299638315414401054952377428750638
8320651576418603210537438813942372770019329684339811234763501455960
9222926175114680150238847316433838643167307307112054909783671239530
2944573379860728018129090764199737183019755201457619473418717897477
0335574540105102918027999327645645742777598433615388952949918704140
2378147688520831462768795003867039806405353001936799198136601886289
9639215904185602031331547575103840918829499829122493612463649682280
7843993507228180054592742182853848517015501419813532590118506538860
4477549458542375651549551363044171587946532190868635731860282135730
4980900664838163906973290081552323141826649354383957690400911605300
9735807966330140198870255078053794034107823322413265140230409171740
6332472032772431346990666434125461962101534678535013210469080537540
7407665720893055253904805764501768342450824334642469533850333435220
4492172439828596591175174456142275144011851539696729512768703088130
3623842627766588957022450085530175899379962002676247023104582442030
6661263937734564421814676429975143026225715827190325962617158130160
5334700133954444793007017635515951224629600173347136647797280805880
5068949740016746893449355508397190724482247601462327357535102227800
4938652272072512223537146539655720047248654352767891084529375853080
2008374415342966539084340864920763456508176905278007222834257305540
6337940736049615020276689935235413821715986500130544837690499629340
4405008885022580341644488948672500790602944744676902294170361772230
9948097667214778772354331781733604282243916293897414315247465377860
9576057164377813744146591213895789779490898209406214858651358941180
5195604913982186101678199765750287063499561275065956279086221483950
2161154838116864535935322863746298705803048220518507123516446136 (?)
0616268325998457063487780771106119568715232236195857980745427053550
4141308439116839788352626381733069593640440001215744670780633484340
4280266824335689928959442767694617023850168675649575188398443767990
6961400283200165526514548019510583064519419042867757831838388578390
6979358707548916237268159601399906486520512209802980995219808384250
7559888879526226824172646680902346809341093556771617177297277938599
5369916933462524362323539854504852258156518371349929014831264616740
1735779696584267666632414698099381905931676841730021346905706014160
3663435769953202131578211635156661199290457919325811836887309454180
5699710877023398289830389924845341254312942548173715415789795223610
2633900853801111099271286217980604598739845211117190534664234376210
8220406483368835344629234602528519069094255258548909925137275339850
3988784497437108747404395341863068806882457623918200804952389033100
2513305639304711204665037413854793555672312883667296679922949611840
7516205055137764950530174841225123222440249061027249735677616899160
2088767287141005344914770983415372814618453645261434087118384271460
7068469961127601150163392070939361887652309552685859907378585253640
9133849097994363002274481609627944802785314602438905934861033100310
6077765533312021413692456962276552223627648110075461114892099523210
3977133678768243107413188104031757198701388099296639387286604925220
7708305634424665595192369649035173838550012677733575139948092226610
0836237170510265686876712988698950225254472883272809782448835873680
4247917500779759704249312759228515781371055479793866689752407313270
6401075473192725890363716097973319786219416944124206383302793725730
5667176327722547055824808181153009561179863304300461781707317473880
3886899303106965650466528276396493041616020041405122930226354360240
5513660533573807623457639036318780256002845269500890302955212798060
5749593284527762057342147562968247909858152818573854236165145122850
```

5707104467331454613197372107140846694600394953548369325797546510860
0931513118385726579303235129966064430057158057408689445288558167953
3335305387021908823387586556659030137115381161561576610811823632790
2428386606892353884885249215775066270553180073658365098218007680234
6461588186855499785530604861250445857025339670328717180882433403446
0857556008146812978614700974929237787956219120202340907442979139991
1279475130920940440010195560660846546259799958407756509886083862039
6095076621008294441508265048624731657825232907768766811031200436092
1866673314470640254440192262631253082865623039660883260679647544533
3738673422207423241656642477289983975843666156670374401267572851704
1486983159271132976906061480142142870402795104639594945746671542581
4912750910460488597255206553068192202260546277486306265659785351387
6390698362093259359634286160655078936688360467998452505581675251085
9233522821723665267826642742003140322865376856734997419866704027208
4822497814063293662415046530510650065454138689008984245883650805876
7077439574968259287920323634055062548837365587911632812365069996083
6895947095234688547385035794985802717587992266666635826191753714312
9972741362555188467606526693577264606151537852741351582951316696212
5646576138366860222729145283900852947513783807229406696338927470845 02
0518702027681063165000005493296267844288166877163912925368326885478
6585531468448469419951905052985371961417847810035018019134857111277
3411729854427503268331302334403753339509923874086247723291307393072
9158730543736238698583252248013823658846791443944951467590743150209
0037851983676664241887707535817387634692983459987953807988540625619
1061068522904815282057576103414761657181751534188926688090059335811
5513044454484764451376027524027823409145689454509188040627968303772 7
9599812267150402373199236756784507293911976575314453906859747735204
2770618195271164846089269570361542600502363675312832300190911876212
9988571793936467340992825480748406419974411088184813759817604698369
9457461301267036759490965695220023615194127042578893074396991383336
6092550509186296205132007852779088236958198476858667266883856726584
3167850533709407627473046775305654429908967950919232990187227342839
7964485353410219181839410103207579418782730135013462199591900634814
3799122097033224944771856824614754387262164279181658486999566438243 3
7710605464431899641567837633104774595260749266413263130825762486101 3
3643359231668257585727795264570441972023238493779322153869471392425 4
1176722124415793863107305610648008757802090233674817840504359106558 9
7489598735917813137206754028609980458298027000397769534376315340893 6
1547860985504756937966484531517728166088121567021272034619263283782 6
4477103327558057386986073143973176047589832909647316513342863356322 8
2763543417845388652738185018085695371129880851719916260473816040100 1
7937873494183583805496505194146238442415625872442326645683499232050 8
8214508852199637720471355732386339962334574008229079367190174899145 7
4669908768688123729812717671079098015341712726814345682666688036397 4
9264734478178732699192036815652135959843472304994505345275514049225 2
2067746790284697037771913010736590387428605484853105370712209921303 1
1109666280721399319326313231451809569617006293533052864880506334709 6
5768688721825889599910370895365889044446025006268325691360371604155 5
7039337111513585948366311865031913660924957447368926094658965466502 0
6674371981881763805778831202574266710596040368651949360968516818768 1
9673060952020837550837333869947860957864456399311187059459677864223 4
2356117174894629712746905292585318686837763065017647031845278556432 4
4442611842236381428954821473672743602432146534872088704586031375436 0
8319210312841057951719059206139355204399558574820103471449304436263 1
5165333663608894087844696218866540115433791610638450411590545901222 1
0306567959394997057553751321469869577569779513026548998830941121790 1
3994789031952487510299600465551557605082924039751995366121560370620 8
18036433212494859870392896507414313516442589802672647

648266325933978131252995301324363563757865803061566376106252920932
934444601933230364087689937764584637732235801840868435828290884567
912737592152838610386184178236427739047410167906710692502295308216
014012263089868503596862495842445351870911195844881673511091834496
774251912158894967293481768262707018563588247532673511374869330142
261041855852002921411207233234230319697367432679010385415714145334
190331342550650015195143050495526266703484297442038329873173959478
054387532583948118095350692669990563722897075769760914176804777422
814806693122580236353583307408658350848000098478395112818606633683
843458014138970687847759248424746420092249454803423647675132425012
318021977554534676228944353612005715152553551669802931899126314160
555771027328462174476892928728976784326807308471244949322552910471
825298090173783739450315282328122377865331234404690965346864714905
675485293460713449887432220967213948119132902336622203991355620362
095448515316402461136837604326264937897103519533173482543347422237
418465185175712391979234023087523665485778822055639686861860618088
210116073348809595335481943594190224449078318629949921155603698712
834877978739390592131861592246131143352913248533323950658785719731
429843733073020853729696528872407470333409924747663601453403077225
935924053190293458460927259097986851407046163036157661633226043117
863318216470649978194271438264616533013682571419374442624706612458
819005368220977777766360397439390972176393301969215057794300207357
227104328717226497525674188855432043259603695798713687567625872364
431066073039810335102415788124261031709459568829558352097712554723
780054850602155952991096285277571496247394001102190005760618957569
082035111612750791659006912463352622933132902003064649952961573549
329024372402139578558550200180990876064139097123429096295205555511
68217873739462799987202101088133233274247883454351493725079448359
561097594464382153496145840653704870533902423919118393473136076821
495107922910664462853348546372901325907909719571101328447266700623
144109534562880914968125452154777798807646355043277650859937646365
839595086899158799369258291573616202708149067575760880254436355057
577797613867407904637624871525692719255600898253412895425956208194
280213924649265979767647243367474887015356456120735851963473153541
666883312189815330290316679345845382379207360613849368846343406225
719563706153741530796342951518513666883264113035770643290066400889
017364280860232837044172040881990359787066431716080666597566535863
317780713645375488402725832460230045428374837047176204324699215014
834708700095510302984304237725984388980260598438055769737763286586
731996819088746776428413211799979183474077812480204844468193504770
965373083184702056621297035005764245507532884230264213899643504133
588592861663103582800519339383723986552580272385974815544109303463
473655023723119089789534537661749914567985518949127224433425030727
027378509290705508443743319796373333186540155763259206181485920309
073826654590553665744550018284363738414985130689702409220457212046
980178767375491534637651830286614803842152432650685940565989288065
276707581473712770643651148867917800122743214989737723276693746612
810228400609952790761552974969302593145743110293467321296005437369
384865368842435002065608064896338585090070215366439105894721490986
554082720755346302305774021344519666678158828345434751241730529226
756363403071043801532208087664706935940298450961361631778058385949
22979393327384658689617473380292634599472391717791386665412753531
826570321050807067116531431271089227919507634029114601800372963385
830397171488009948751826991749252920492720848713391720709400203516
663104840209892799581594226466058110083923918384962313779863416615
310022521997465399078938544845268053542147788735203710380905949032
616670387262674312077987368952363209398901149523041066464548466119
529579789349836928211769065220213766871092031968326286451257531840
897285518524679187318566194094728174226384001834609765606711917220
419243827631629221589010753860101666194594125209819081263268168033

```
3478780258528884831533005075576275445743318410569328661145 41624155
5163413432637128429541103822573810164276161676168149555686 42930305
2225469666912927204522926220867989277924749226694936583815 15488588
0108167036912749186441452428622867595235037210228677302951 58577322
0735237442626222912879582945779963461111984877489795095726 47498541
2917765126012113724255489572576582749366358387075691665618 02767277
1181404848277547953817469750851298315814744939019065059967 21756396
9641884003343398758916889650803959990960506176695738123097 77444883
8482206195584386010592673698303871693453994143219444358293 25816648
0145958945541046746420974436929746119145557457039250896525 02063860
0639332609605022719174809367757835960784892339193221121184 33160563
7156788570662284435750558098882695820460630963369445785789 18952111
0626075100225967096776693175331809230701117986387335454754 845342568
0199476290106253061900697981427527287629282259440315125114 15719893
7515302330391695552013377680420315359624443423737609560924 66835100
9029639361432987675565012080910702137406112946152152082826 93539507
3437460968875955898827253926500956844917958403978829807705 53942640
9465521625583068160762711364877519549982112723580714411126 22547846
6114032841920224040802656106101251236441006335956386331097 87054313
0547470697414162567880795294042657977161735291710285762053 67960474
4749383813241278687685712686807453043205107039416261278352 12661880
7515897883155696993792944736090166879321034108587424891340 07414353
3043422655770769915075261547470929753402890220020386697877 006309478
5382202821936121285117196923293573467101136104220057681074 50032958
9353928218627677134500767701644074637883723398837408069550 60501549
7545122109013861656116934350395700029308487845905725617612 70244143
6517752228456943709775955842335324232964605661164567387441 30633863
7733574004727995177603059943584361652007737593534045747689 07542907
0786877638633063780719956875333518682582580411940282820314 74461122
8214364044531466470273643370341274066803567907539614519258 62176993
7044904950612615334279081158082028653037561411154096438116 62307124
0305641379986183721570683132215334063349754693590638791875 56117538
3713985958603239772219687812432358864859770837175221414946 16698574
3779505138007424649392488815825139897831678081924265883444 42486359
3159798889542778078764999925697931602700677372398498125423 16149589
4469711743057519389077902287347460472828835557551053790713 08814106
6970651877268725361426610506166126193053332827968146208440 62057937
1635219381681915053547750098293231348068623285907040009605 55103672
9591276566703575872375682893896071632275122396423131115934 35624683
0326469348821104432823987384322721553450804688943466221622 70202408
9587845376997668828456002701553078611248894927100612402875 032974371
1603913803954956618035753804339663894512869812721522345204 73901039
9987674141344246249153125187100977226799676578735288582684 12306300
8457895310364093569189687701058731202214129434472510248528 19442680
5406545287501028814428987382380454065790648747266706081501 40803024
4474515140469307077462462683333960709952662383552559762969 010802986
5033459071260276811985349268482191652850356585616014817635 05003280
6950402699518926148422796271864468252076542496372729857812 41823901
1356678140279522869367952480766106203445029212392163431823 03500813
7033835810151324623036855921787351051009220371557122466751 78938965
4809060587517333988559887109322602825253557310306894520728 25419945
4382501800505646922945754230552569998709287539010230488230 17054393
6093930499834626250365449971333709407813372281547438378010 85414786
1631815603748468885051754813604377409176659990479806013087 43801028
7780682386451403020130027563359753218980719919445401157081 50746082
7200120739025691740387107151905267725932472475907272582887 64442775
2426184010149338328448130369265785368658857039574807944701 39706211
5898940911012736387649495734961251784366948084002741057449 21999843
3065809446189100668477444479616482454478132847458441928382 10277000
0425925179610197556394738227317003708286656857177490107830 01652082
```

6990257854346427334760906182371827279952705725934672552228558708 71
9767341700859235321471197729889249946185701028123254371961941429 49
1988787144230264098978091367843363838927030104122417413920656011
6568558831787970572048366498697790214305806713022873024032108516 60
5799721245070865042154896234907627307065537674025595955100222157 91
8781324810514052745324773209454000210119874212429192896591956933 31
2886215075719131138714180444574107118948326174837477692126428180 24
5709677236714909612815353486367265151300809694516260715103912733 00
2620557217009683547683686335918923786058288548391456897174906896 75
1339274477936525640099500674120108377155429313344014690486243466 55
6783097411041884255459762927762464924474998307269993962796697116 0
2166370206119887383675904123533490661952217202212243545995574382 22
3724831609839532258608062163339023736811589580477539439634169517 12
0464290825406916520639650162697439867154009553856780616617096928 05
5477936832878520438973193199241290389659758283150465900908123560 95
7423028273344955698144442444938629357271776174709024838985848035 06
5003646220678645282374132704808272727368886117860393278520827679 35
2561776380316193093815293500985312738094680789937528685680794302
6736158575972638872966886206779517424648150594921746093738684568 67
0012260710706674578241336811692347188442311968717648856204850005 72
8488175030109198119550332913396202111909637894797315570294828780 87
0639707735324011997716585466964331504589372155694310270860878446 94
8002508763121139432633807182735001852706229434659520858986565117 17
0342178391569854086292196890948697435399503271131009709674407084 60
2939494870962932744820056881786027822885761412287191689150658487 02
7988321445894263442522219014054211607637935128170608397748190945 93
6090845901405997137682149864720677272755829517926249633273735402 60
3140124677061841918116390800909115176833482967946578753834753871 13
8715806653768535073052032519664317068852142288498056682157927982 98
9498796699825760803682116595609466781380036402225101941319653146 261
7538665364664838364604876226823472054273354185650259991369120059 23
7542660961151273038417340118951599600034805883593419407583595395 15
9936912749844116532056306910427122788925100259527962248290241295 67
5885963282886388793096509812188605407207125585256090297263121592 47
9553592902298759601298674668557754044322112995351464994802036463 61
1047376144656559614961668911011111709188246936246183955394232926 72
8282067774804843878969442995877151019558718428063257699776416145 94
3457389815193504427656261432982423365931182126479312564095509349 97
6601388437093187016676091588658885189549423273106606244595218331 20
8770486949899555277191928726816791677171593279970274033998150652 255
1565682151577438678295543541637100615715418464087296882657939929 50
4823349477350107880360218001355974649378592222689442147876507081 62
2904512880579344855996485329447934440738837265533722194342507666 49
7248422702897160581526305077461583626304840763616754493011806654 77
5106539319236100243783303394744314319054771868499789417766928450 87
1851886410549440352128357606511938434931663909275204375935781169 09
8292573242313358545817176781847051629493448125066055472960685194 48
0955187825952016525348956356228789254384776664063509181966135793 12
3672861613404144576609940131744136813883817032141918317705862465 08
9813257456357411562444053813528445702601310216502184297693224058 54
1045891209938495806002753980599485698712796964951712706426867419 74
0455216204367842380599663072046910485536481404616298260613026631 8
4133419734976865214752479675757523781983250205049276214451880305 96
7187800630063993626184289391887250034773555014856911410678684321 53
0223451294333206172726526330981672385298175652165176644668819951 69
1862403097654126409315006271206527652015556298043373515345685991 55
1280973815964841827037975730586104993722584252093423165937564957 89
9563095762694305389822626377001561582140653267549400031801317981 73
4374186784759818553364364995777480868448363403786604956865123115 76
7442697643177900221681154226382182405790934585191144876529313934 05

```
3423357775332507439783957979802066993692146893687574808638417900068
3180733519481142884560089878302836299389994577386100468201914413205
878248450602410060194288510124618988895728037623864859697014140640
602875058660753434319531179031026305682007081488749156576168507578
743453403443461876003176040951124075893295761158812832739191426644
379729990044630605773376371812318923828801160309009805445063111307
481749886415434845689101881958169016781630751044744766608832790235
342731815077925911559496649285095012157804937705611654575849668561
815490827495910737268898329208400628080953553984647567791581802270
292585087891250408874272048249286815328685536159304207397732215289
519886181835005425816000229690285114662266604853827757808527693557
002631008101822563609538835953850411766822871707761061066629830281
227733239829135073933549409385487179144816729083003498138735230718
157818227960493707056061704154073189513174435137712953241892944718
403801845300206907501588384372781264768796632011006594806022332069
751396474279537987583197687271337828537514283055225311455566153085
075440038323751108414421704790235219390203043991697293751428524084
226764330383565110495254382106944410093494380189038160839633620830
159829908908073778470749086916013285939369195187088732839380975368
710991896575096155584072230386519684057026077082518184262787613577
932115099508120315941251205265340640516457322185472860569453808728
642027934164172941567317053391789024465964412836735698837308764607 6
879567002354091579522969585585076734181523977274010010105753704253
193647947641466580164737625300825794167290550819631840651068853154
928782521872724485944707144628628519816728603957071686941756756438
685240627600942373480185070440304383344447576097569720188015628724
185543971085801202074372603445423476208953436593905834169553855691
220953709059012086750345509462137004116690621378643946304221938373
735355819365517630804024851939046432297562567173177647417619 67186
977239277489364083897935044453355869480497388323791231124970500663
751582283708098379531568617790110396680440241080290274724556020911
604473065817986539221828274690774432060805558887848985428401870979
681851071508460015169819393984557452822543659125306709685848267108
032259742547680235513001203856833975865324746657060600419129569070
954976131018524068992788157867094418302626703100208406551392353019
451041503365097157936630303210113113376782842776313549185460876849
622910716962380986944599178274050654321449235578884802481779433256
905639695768454308518561452045896438646053963193551939184017523213
455502722477482403016679756840049224734549951105035176628371116181
695954554062388944610650549136329570764975533616196678013357628692
817627443953660285562158230153112099643905655386030812597902858846
286465950425973432147791281866683133690870110245900748147366054046
547710081618112969501411038347014746777833408383080193814942399482
822558848960508285511895309558404918877769255712745734838752096594
860038701491387254599053712154502834191539683645162555943536805837
472292618542687591863292079782242726425000632913888358990191709969
846407253456568545810764689107815291167793885081137855696657743660
106181607692447026588091511752197920028114744865093965142040888685
945478138214177085942421655396591988170408152511138106446512623566
285427707526387834240403433448222521461704631693498421263623194345
032311993261732311195672465249414197988000189092470030008917546473
292532982932194831112751637423418840925693173154715854379879003423
114397453374926755609590311145793638838288460303300716450143195928
222612218089180782562529041066103735230418999955618832098953825428
851312698935716803592199133892800772369988857895354402785793523422
614143517155473838375785483961836212788167473935250797126136355516
877378031685172090974565169313386449639703936607798396451116297763
078741115905629264789787331960416484367172805047275640399756210131
244170767618917028164698200434946451207789802080428508479869172339
141453491985042796646003475454775250628346157391228512619671307100
```

```
7694522325450449479553345505379387725953799399349538713426038446351
1643848346259849826035291796608667608692584648210528833849069788429
7201268343346919491772063846314876759826818736228980761712140888806
8689046906049880901324082839579817800130739869935134548877742617162
5252697860432191604031464688108312822526752102531600759967248913074
0609752809934445375605196512723102756710200616673639153374443686195
2268210656088178498213792220603779816068504333700303036497076644607
2662930118940076514866660163973680294377064236831455286161200333238
1435527730285051886162322428898287075430500570912473174132787832010
1127544418406884722350537968812903987576660695755391400881711683888
6163523257540664614226409784124637966455815548222358040372434458506
7211848638344730867061404549341776099834996687769386137904277780452
1052341973299787582011005076658967505281993264683530958432417199 2
6122595605603430833210000483767611497322153707771439661588125041 08
2758757045972466150813534992587974044713811653321404582173682346 75
5011101092442383335934555872377074860466306528902407178631925440065
8080424539294718227718835249701175333185559601267254046845261870 26
9811297387267258290167996464089639083062126134439977837028385634 21
8138650570350340445554195071664040618760672090739983710205613863 39
4965936987297560269781452955904287530303661368007436050135466297 47
3108073190493746079124996899994115480899356927110102774374260040 25
4133495319035970075356977536881730393218774051376162793466331986 00
4321307954375236221578585681679626221408284818334048256230454675 40
7460729640856847472698809821484921275251628100885324670267069840 18
1968377809843736191481196438640101588682935582262380686493149360 1
3240673894771053821042906005431439964027099965327258965895530073 84
4796895389207778148471664956387159303837812254391177268063142704 98
6174223905570782112522215265559982262148218262415724002579336047 12
0237257637875980327482005342667111651503705652456307209570528179 34
3620769238900843095888756203519709013501162287601885668677583472 20
7534388079868029706445439539601046552295377100653380995420628058 91
7191195268684413544226273394918591994509347663379600365477800565 27
5767264367038098531943550135317925141532073822734138407661698140 11
6287320968294542487308348834266944613337506005775084675818043865 04
6990816148105662562572885774930737458731299341096888328200972743 90
5901469197002285823573639648430538176291692919541743487728126707 34
8097049582481027893037222088603893064142210371491393740060442613 79
9376702211374646032436460942038553912194604284645493110679869300 62
4858686710341907314500597063480659673740298697360853767094726326 23
1565600985282011369625490928600581357082387169391492234900190509 58
5171616256730105854247240462371308742680119805943505468835136154 50
3954777181125603880749109296595991967387936161693947836479252693 08
0284820792437857454731943690567441451400540175134189421118023276 06
0738312000888214297459082745336618249079326322260454189032469808 94
0957638624479363498260878460263713291947166353410780991515978661 28
5859307431814390447750126877933685047064646862256956639125163805 80
0835301986675452492155451312096092439301970834816686252748703774 24
9770142034358585074208937288470600829744894119645427620564362662 42
9832953933610850107647439624399106234790911025264955052581666494 83
0571618932113182621776603849192534830980386017925631360923987929 2
2409287607246590284287153885596315619772578807471524755794249077 66
9981891427646345542837320609060440190597209522921405812331493405 75
1058703667894783121508364405021872866382786928131075999813196977 21
0784810701009527952151885864347764544226699284035283271509866295 74
8463697305999568228634404950643532235758046114598768122355216053 80
1303498948948439847549801336730868085580708590637629793615397345 21
4251843222105106329658160259084522208655010430647128150868083161 37
9482395003251610116527505045333013189549234784016373639785718426 81
7358382881847846948234059268503149830801950450278064855415436993 98
1722836663405312842064559854236416337195275502216758454652842071436
```

```
5062496783818077328026973922596075367544881991083577256663293334019
3035511877253261993893574401765054398629269809574875910147549 25262
8452114684463733873483641217559148369149672661223620369052 79404891
9071059116797574161831578486474533458000375991625041866655 77416293
8650677606438997462366121653306387852524135761526710069713 92224358
9238045896106289880614030968387638627746099623708835227941 15115086
2562442840239473856261858585618780476906201182077128527763 92743571
7292665691363408582940648898074414148571105147689695843113 34295890
0812699408302573185286920499598823316030094586440818252539 29898618
8714735206873307450787968912760493038785301098772176041441 62905419
1688285085414073181159897226867193188335239134596730233318 60825126
1630013002706713630627201871362675605510895473093555719566 65658747
6974938272286795385102812144134199328628229525558830960880 76411751
9847834308757060108341996825595038335246288576833352147978 12555473
1816686707446782429808051122104727782081163564900622026754 51965155
6346891895962078333852406784338377891261362417843676842171 40035445
8270422546506289048338223214800977568664278703358186930730 66998001
0387265391163616783385696128768361136226500035215574310069 24833647
9731086822206419906958117730265658826651865870836453024795 92137642
9392968612160261716009735394502595367393858880482557087738 33333344
1555589462685263165976219812370493666936242370463524642659 0077082
2997952596120775898275845590153713852179622279213396454300 41821777
3321067368352078913892206990022749903431843210361751848896 02157711
6448772228301753825121105711113463813377461492509199573199 82891414
4289069219603094071339782867527239653983492900700343384238 01949455
8570909781094649338233809658234200560003542280930178440448 39887064
7698946898376934808236097784954464168112744952875131782979 30201368
0356475742558053173516476701728915264326387538343988962484 64212617
2669297398807455228178946759317417981739934487043518725155 28932071
6238223751785641044902328501025541631925625889721192868168 90209349
1909136900594004192101945980920195838133522479068514282119 04825132
1694562070814948919036519369166202220306446220402980730198 88307500
0937807435958094581898326246967891596905158854604377129743 71982459
2830380353966969417216099070120034060174140004244007152172 94642316
7963537878975362185181074772800141825504427008500695583330 97502909
6109005943361811337653389660523337179832984378156183672900 78252510
5099968104655985015277929373878527158202988398823039601440 97417421
2476713650583477873439489633640507709811959583625312095019 66515049
9202824127638473449914740702017831323861130104006752870602 8501323
4264750395168790776676019265244539871518174738629364805715 86958225
3545189123332764335358935653764185134374876596332396372277 84556348
8012608980871670829662623315777962798988354982269346394884 99725753
0749862403797696895977507897299631578503773639054765879766 57087951
6255560566980339705950444194705280609568398383914507808267 85540194
8687387801582254302917607559733223969023001388162709685015 14885930
1665473851066116345917596214591952522455372631368783138146 17132424
5370499020783646641848802415237949364500557417853353831223 86456474
5531684625889614690040958295331768883236282119446025444534 86346191
9620404108481475883603419191473775142952271024565316101029 01757237
0892892940901948110707540564179860207228208930381898783925 50000769
8492338237004467051382874415466861432851637786187207212028 9628522
4875029984135191956212031556007054061482301273341806832349 89150273
4398457301483652226042282568607148515166807421376626742774 31964146
8481572503656041719142266041648113846789598085412514299930 74206208
9010876844884954455299279836548385805511495145468449826501 69015367
7969332457300986341787894400827465005055873350793545080510 19536724
6195463936189251572299578365189631631100480748221893132859 96193969
7169755312477552013430402978310496464792057749886033422870 10268838
8905976595841228096988890752301740139078532365748014087836 40144579
7888528953329248092686307985832986597479476930213620920677 05495388
```

17641060323886478320927063155375043403454673917600663655808936133
475691529236838820052450091010025645383149972487710444480592150703
198491921665039122784971271624887140727537213676381205579099167143
831396870341569065714766613865087667688989637024404700151549555001
383983911746003560040954834693066024932450129094294913551412858038
382903092427149171199636436780112266345428482846144505717070805731
981450410020194489559421198842161567541649392309125827383198521874
5954040638340600756500346739852129739122522255113300930596174736326
175972880708623966269921398365797457303779029307639335372481853690
727106012120033851190955241945383033446914882966138106627692026559
160515042235816414045373562911039783624112950290650818233961046271
549809539979288663817463140156834149244076965569051907302242886084
945187132527536273193280201757919527871435072395590258944687910862
654643899261283508702089522647269673393830122987992964335680540192
519234866605579778480934967465607414242195132085224596758430864774
360130618507695797996393251645425065513842872336442616229194947972
908070070234029458390398059358452586996572100843397432291075562861
364280488432297939471782403643466692630235658160421260292323500941
018392210804109882155091198806771687010169368544960201869165110655
234993622008882221897569191964092408901689530677800277205611438629
718845870277918639779038419152191121520178405912152636197320851821
102319855327659093330894877527283591885239905329979589391483863363
373286929171873654760072240214427811188877414240415046808132000860
807452625653022991760409036172071578185970050253754243228106193944
910437260582466425554801364903068068453282063846520525149830571468
092600469506643018730184020175299959482453712450438719624557484277
136808207315397858387740374375989017774415052508553800731486295927
872664383822755302125068371375986007490479753092273651103892157420
109243392760332042592871942624956792651248988100962790412554306461
077389745723648556276869797956788216666534132099715497940088990216
928794349610661632535447227274840795123657447228813252622293215986
301509931022268523882747136181152125942570636686651799219050167415
232328115047806254217948027368319073860052940783081069343833736933
1709038997903501071492954222633867746558294168421734101000869115756
597966310920215467507171864568410152206341943391809834914894381408
679808600939528894811058396652246794511441346471256131816471118830
453623403865213970832465028766471842452380226795343518600532111761
081800436002776559073310944286838045300254726949343607046390316633
676869320568365807853355901587021706127152074619984245720089630842
962745791075752505375344086868205300950277621914826387188426414115
481153046367271444714843852536766169074517533782053313798515263528
524342814737005125522582291928578062051128895355011431491181991364
915830125035833507671706813375913719635655265436257557880718731726
664539630536306075674971768424021312201625996296601404089006069292
257640507449291716661403817773815104115719517238318451260786986257
106642696714781849620566484037741536460745570057819409761652602179
995888630758083690557831375763180786116443206099427411729846336709
461509299231074328247127842992318210689966449190201456339969029685
823029166323787932769007152416522545722906388786633580218882729250
524869653473800017783712727301338656688020428446356405057298917201
286017949548271823209815224403647495615149271055904689701772516636
021166278546631717279664089447010682906348376375250608819619419784
400582261255762176289270255666768909515583514173341130440445442947
701746650167750631277882178962522387696847119140318153377546423357
315108035665291524325226734189117284376056435983518008859647853182
927737693523636339135977872885781547500897818004970295134515196162
0716729871633841647642187918848360224022925403943877671200086253340
9535591063555081658199985287993082953825468024148566475992494097403
663364661212657639792132554391795452146709985740531122351138519482
8893999294516494935647628383652945654746672339795514279146117785860

```
736002423372828925419204068249689956996272541141153594588522474390
359436886759948748901041040363280227228116752102052756547745706671
427782638796692433810384160718958498630078753107949048863665572443
823644671770699917515283092156317070539889688535090284904169412211
339572078147532585978615682928372752658980027130899453350212386661
097492308526729022244003916011424047404606249658777552554580616419
268745441992592874302125853464598710518137462676469381648815755588
991775679854623693197386704995839202180959568750057801796530453167
534327278785313922320158239424845187261472867065585209919014832582
136057160174504901423210197696313600002744890116606328595539998289
257284185515710679199394527129570003803063095128975253191082536256
955180915181977047881476418222346332281212027542900773072787455700
602125735837367484393083962411303455656214590077178490296933649697
723249452917415372272887271646458468659372640833544149809741632661
723401421459028093304279604554761185888625108944115803079063907319
753851596305389995869418071202106835013305676247554250213788919319
247140274665090609915337327333875972735227520294950513046440462166
397560878761068219612629237129656932186896246311211552031477120299
605067112678170156780407342919545679157747917472674374722177837104
897278600757603854831615553654334149996702954742841866607795279577
083601215825893141897866286579188274034521199129834413644937520696
134237952471823786712180176461335697453597250365888750756028581703
219196093421559745967408505573058707295469264342519452973669741593
005110714687861194078901927993977608396823821072770231683041381359
612386588073895490476486449400052254674457457545144172807457911281
188047555824086343145371605862958596872887208108512622058191242267
126080274264355543658098900997597602249090533221066289349478925892
128693966763437048270512743606505110923288546693096940139781977297
187230349364141391068234616572291839092113439059461122254525933005
539651869751071653570853374686188045247343166394024514150672342544
247726250680967986634097830339030423464209234824922345030305943020
846761045106363520597720256540874688166396681980000625127528747 2168
793591933315927969524198256173987009577509608159692062695441052722
715196993222081415773761550747406451906379831485661146408923090199
397917174809712174534709654056356043312633391089380962798419734286
169742921878456272449752045977441592337029154534576768043314718517
030222692563357995461687382998964307681634225959132759724149324185
248285635595230317821683774255979460425753922412291512904515546280
642762578070065662846390594001991603452253592528421194468332768707
821077114667846256191852490519559743989502558853086122977712331049
162624434298898704927233826738594913382144159340723157757012109331
299494661345536729365665735951663268503260809154233058545266091125
991835050379783441254436582425251311638990593360157908285268314494
172220189700006109723735145150354538117461177746681278408514687438
450588481728349796402154939800676804251325064680641651109201223068
443332340637457748373366931816153431747516579112554750162542346599
504750678982401836203707661105554617857380980022299957716434569767
846514807149211660837744444796773846294507702983587624927875278514
582821582346582508772506598770099608584120922330265940231926247078
194505242411497399205259606345720037726964604644181325735732019480
639147774395184258514817866701638652458331121537894851977832733090
403491957615948534812377766478557045359620718065488551579409060476
297683173536431095546253143599234951305523850381473769734393595698
966430470202561613445404603514867282907594112759490485488811579949
347948625738858717475449672401099027193541821670763346225936 08736
176661100190647945739794733155401032029075340221138411518425379902
260230441643985729702563786482547422194736576056477875537201552570
515919720759326165566996225188655243578681627843767994875337722423
006877976632530523916382018799906579830325873399291399437698960723
834954050477949465777379252279495905050589991601361974330799087 93564
```

```
8261547458918833759098769521078201976507517187343816957168406 13542
1193468163787026257264458381184855874882417019769569098269300 37404
4515762308921950425915751159449141103617495990692331996001562 380658
9874142058717221660204899259015242919797115027879003330687601 02944
8924148231491571849683989135882727229929084951697097438857651 55685
7512321438473813597584745770091438133693058075774295340429221 75751
9683523298030926777369020516319797252348090354318688366433898 50536
8066171266902102257623513847255443977151873726425212909994909 33037
2160561419900653739000137497165387474976484849995637706865209 70719
0852562017389649477520699330211727120139261959276612230414228 77299
4922483680690780023182597667338509112570128450451745496217363 88701
3289456342712696581520463213147311910179135689581055994583855 28085
4509804836637192813810481908215789171751611971600419985300337 59354
3613622534069589725325180545750805059574159586748372439971998 96842
2336614882758541418273557524430910886547791070492406553278089 05876
8936949926333761690670393297029531691406813089793447314203632 88189
4332007115740815275348210499132695399975455951939891036671534 05749
4809749023499891556815718676478665605253106991473430188354691 37707
4872577797293625911218901396645105027025516301417029380353602 73533
4983606025128542839697537855224826326114669925429693709951480 8696
3734143609201472790663386619388522760778104724393281217562127 90772
5362977867505078235330119515479818311271526217071825072852854 00452542
0911087045718992533438877443266727669483841979111025321966637 36503
6143903072417761955597281560919783022280696301269729171472081 95762
9716001576076268052227656938737947898185002470561040891549647 43064
4631792913841894274984881211912358671925406419583355453406371 55148
8734569020357387448695716934701896363046940036950327930705486 57942
1188320847088633747156506690645438335874033621258580123717350 70389
3366422870007690082775587013592267382760599597855751524501654 4911
1941363294644085310890618026895989841426384257568250107152964 73648
2887926136514939496166435363774817743220518814189777243597671 85736
9378718997021381983886312442635336851275706987116470265442833 74897
3677520258863221459306364025384348062681890089690493472565378 05072
5118965926545282212268696459030213152382372967175182374449157 94243
7819994074516307626717363373941113606556396574529821779045495 11164
8952918004388592528513959038256902512064453071650081726527092 04726
1772757665069430072389613582361350660988130750352354724610370 00225
0472823592456571717989867372498987787677252482501306545354347 85709
5929875383039677631029698465067904613510020495486600862321004 61589
7952246072244015661602738826334454148894696995787205105267434 79323
2781935106485652202466115848066316612011336993854587783656059 20812
5865664305764184064664279594499545777414922502283846466289967 107271
0308148082519376180088819758896963844491327058636434831292436 52802
9474910574079123222444630456469705840438428148027334297243169 09633
1441287753909363808296486601874828610345040033054772057927083 73729
0059398754535760405131775666826884739573100785073341232625284 68151
3000091753150129831634012657798037620703450304608920558374261 78926
3588379888241961408996741287912403248883710545991486916017712 82127
9801050533683701326403275232900475874011974175211196459566106 54147
5024374338074242658587234098709181586051164152932326892778035 35502
4708178122046167246701460437952415956473941801502963313488156 84881
9579080551120805261657437330380528953127159414669455424208760 75955
3604559586061662162302210200100483554358189686750330843327755 853326
3240325942590948550673620856810113310894559391341240487018250 69182
1287353052872162818366877233765642177396608759962011596111647 24460
1451299488530257724861988317876060296609419508960035683011391 50284
3694760076242185638538956663256433351858485668002975865700299 97553
0810863469576576592616158430224951714271920790211831230255492 48530802
1507075142623030472299305033880690462220673876344693673149491 15290
0378151647902259550071615650333812340592612017469924903112013 37726
```

```
31795730236487400205669221391287914181449296200932771373740119485 9
293295595577035406327067105671055388437767773319354223512210446588
530379795749513606515516914292333810764363990820937127981593350169
334371223100179667125831822722213266405283266191858994939053307557
051814524049078513685914436892839112822226924356979694060417141545 8
515999605949366556364966000810521037397596882328433583958606949222
915692284870208915424478105425811949279959254436975785548233047190
362419515355393825048928877223889443594605619943079723140763465876
259844638228786945998065348316156199044466680803246491996441694408
819260208994376544980958693589438888061759306636731691245550832688
930426719918825436308974033603552688701296974925461143871570086965
217490006516077946047553434883793641419756886529259323136300374200
530570135706740806577129158398438181553749812209913516884925463303
005805036352962874014917326075620569078384291051212201519299025268
624488411522019566364224128414969866589082429542694815427175268 73
268296559827578455191783081676731389016311642993931783366561164 77
772874580386517637844287522527974398283182131562205097365694577553
120437100903712541466381988610466722664956805824954217519094443013
583881245023861927628230264808571815001106357752274077962533419957
305042643492101516251349347193885358066010320758078127139516382024
200713605363987011349758350534583094062206437803235044473581331314 7
799535474721344722819656932811221049541192867693574113396316736 59
537384447666412710718698343678799444808421074704069878839366325291
966459200005143668346496199490796671645023296665824910010116879770
752335958954598362484740770290872089106914695644001109231274286111
819148720925928468876708716058072226691632394590178721694608523206
832502586651998957854255815056788836240471990548814467622692767734
808770957745553986220781285268032644062861080280605846766215035000
155691104332705654300496153264908407517618279945133899987249108 23
381582305902537145342756324208594583346695995855591584513691203514
605702019595742615053106608768045311675628858465545826163958713803
288244102561049077625180540885700626087705889386030335321058464632
156084772701522512390559491382442440115033813081127662342289809594
850787481653626253751042164332135701668464894465620664039771546526
313571443170118988266296083876181489497965016924348491709426331334
756270098020859962910110279748715332542415217083867853411312581160
000710184257663407089202472350530894656550062507070396723492803785
116540047257255114838210682322878985574859193763768063736676323391
437098692552582615693677367422132675291950447458700891290160100902
200787998646675679618756224241993050536402665462751298490287199531
348467272599252304047164370772202944022224858822125165568129374205
624954104991770794263159382090514043418653976132174574065468662211
997350187729923470879657024742108666972050522600253050194282546523
040217028178768438532361785319506500922212419715348986495719862354
772431819090627999519048289343393100049052335051558297058015289510
419906401571328242935479202998969080258508526150612610931425330891
483045475723811888431500194984416292896862139450143603436498526174
714329119982550891380302560514037735460776305662968961599850449485
692508312316321342120066474028631472175641120547645952760611446237
301187768107574774756750817584390559603656238948018072186713097766
661531077884702219189971936354265396639740597826834176099460668169
147561774340014533715629782864827797095443992969906363515252879634
065315679983851347300178458572901168407733903672748351319628387500
396273683558607544383338428822881089340543413602058226472904501131
641063435252460820533935966195805309289065406618160636794628018798
886908363031393896223893674054956726031007635881837861836301085931
186980795694768101499486724422352937955923432860887123155480 22378
563300699818218858630576799138269790664095306790490091908096019388
357928700496824756029891151491097500775200726485112091040260758068
393441012987267489779448369612604854627092244881897340098887261104
```

```
67567855948137047072165322950857084705251366603332753361043490113
45502695776538089866547947133683132011058082421613978311134262479
60886443905352423733932559569158666237669670982546696233950808127
14511385150893024463113225094752223599278734416452641127402697417
40786008329350535843583137440824864671143388176066068164810976977
45450123294851979262821031310105502533255229581968782435925412661
98209516811743862901794429465110165559912850624686398122162035654
99197558928943126265121840162784802308182519861377930722329398281
31034013394712461347571164357367399837774436184218289275647083662
49350295306401173080839506172228447277931045128944021709117243702
72319511878599178034676366298585385434506406252132646213183283491
49986498072643237846156266299355373717451045683237117515785611821
41923634639988933009521043899172525868749859700545928782365042858
22035755439087866583098601366105432578445210329198202526948391996
84495045373920359181157732050841242654901455898584658404313819631
72588337894020502191938079042634427542944007438995524217112905073
18784620886229761329801229585899382903749963203938715492394067702
10960480819293681848453085525042813695968361896743218627886023789
73310509213673057545043239690010739574154312686911945989732224197
35490131754243675709227892575038966273543153072155708017166820881
32321775271505055324167790259971012979142847349495970422354463640
36737603192515414625361587905790120717568629693019250285430475869
66500344700967144070846764299216831352629838549808024746839401949
71276740619939439201526391062026040313243615355282041511224995117
01840899942363591483394052479205259124362180358507802216454322022
92884024522085013569323936020498477319030454603129936943603232371
57073712285557678912656564703246591875992926529085638909683261437
62677990213379031767713422781152260180931873254441311895603863263
12447818932210797039780822517711401243165061569754311919403865769
76376731617076151421811030792416316028181801241557232699589764130
64474549698428677676249962888489889945854230712194264545767872451
86841362413677900731801259666441423461248754445578364106566615161
25661321841337825442891876682069821536878582323617439293328128959
95858477403913848721839443343351583411711683424919180759626330865
34469036204773972869034380548584964081594703998061947456911174861
44760091155737337111947527790393230007248393103763651040269755069
35349203133608669338220341911649466797669802619314611256432205668
01666260299260872232405947836400797076113244546741017607819907932
25513278306380800590402864272352762577600463932742356181113509417
08941399383259811269476775675435832898068359796565390346283046167
08850754995212172762951698177233422736442223909267753240078316321
90515607811917642398159856390559697104262102022034878477997056468
79138442041959239150135706091900411993286854681649452137535762985
00170074108364498033629474652488190165147961317205291116497106295
90108846251526055642889341874302252524867193744148308072864163228
54954153462698079396884877201106995967405003499827183363696326976
50927639622706749266692177999680558125400758429412520581336557713
17581140721158218148799919580519911735284047113692178595270330564
76111677434077943982040772803356728834312976577549239521256768834
24053095853583618639053782366501043370319217398402217473126378335
54643569573106409514729679844879201677470407268114674835115139225
17489121157036471801924719376941939665657644359075955564815121567
05676582491799267825358522474477886440630082967914788656547092311
68755768879255560106070078834733952343290303317666284216131486159
46827795037048171585965570465740738738631264181285387703050237789
28608847554192315990852630013095319165534029059298873464769283949
67640765972590833052413696654956207430922668952631723081806356648
19945397046340596416897944795157287667588662973538508536123096774
63973187505222924377214609532302047030209577853006210521841740613
76155668141430189965048523374426422589058476953266854218436547812
```

```
204592019423770166278367571877376337813129619744076407035840904187
039011089205021049078464182546630031885405204753794553603593640030
820088722703653707708527393402760259385119531265884930734062960905
394314556616651768679303848322939287327301691082456603462920963416
281001639885986449996654732755298769809216657616389070096878268306
247463951495744581771861790410319431156644300474453639907926697696
917088735444566075109133657120649605991395030172921982698367104477
562745699625387191837712796829565439866897371924021541666724994257
836037835416308935688213720132728406075267211705121340137534674198
224443866758979691228466319693329406092357024033831247290644120533
071219956410308554698290074653259903930978193412637880394961756231
664251999821884141207833761649425644459297039642321026340610693413
301711865811904559534652936139851553982302030826622895917807683453
234633195843990549249752505979666512685783447891031971185927646539
593727592933276582442573110281115168141179995494384455506398877662
444576127829627943376585936050362149872029573452820288490825788500
756183225864630818843591171472649812718037313913858972298540484001
808504649690893972145088219341452980167440518762657292904458078248
222722219041952745550254112542188310092898685857146081965279117416
276993380061023696678119932121593767692434237488695909932946958294
835737584771749352957692627931290539904111397678714051690904101444
976183298849362159270611979663496965072205991391933810054818851531 2
037402984045276633936964117696596189556420179142408140687826331634
746519043530297407421915703909072630536224574250751217848463631 82
420728945600995823334340709047534480505928774744713308343297388574
365691948715731637845093972021305792610883406417126400183548099347
834960414259469455484599880076788026603426219906254945175343249619
418758909047637705842685956077983231613725203913008854809263712 33
259635238426080062256844781902575027567757933593833924027742116916
685001980301891081185414430718621465530302530874395354598530349008
875353576446577107889373567403330613487754682035260517724849305508
481304389978020398978880332522626959286104576945778130759785915222
259654090218413420425478657480907355060675081890958161096307419362
389537565041789585133011130940743885760889230768132389734449083657
340390725086663908612335401295824274459320839327724834466544378431
410929633707044037893755259667901713669668107687478016546728892416
176197048280123555946188536297944217362440238143298724979488380996
573415026226857045753937618640793558817304189536056277669339421551
198496304824331600638370961579010717232737881176751846742967604690
260183087399984590943990897991506819771082181764482905012342204139
023443229373371569716930693800924144759611699145445183793902093575
224725791968819919903473479392573683217551096646645696142527743465
878085464103156988160339626543918599759131058221880352506890915915
798555339548306784396278857135072692433011331911593080177585009977
637901348754631194849690058999458259221919894493618160300612362919
552693645886903382663739625986732482623527581309467245795126415412
412333224109485209631491860603839354913143564757632340901893196839
427089353742576765205402409121635649144627424899370814921036632612
671055220278153190518120838578869906882868407676489401257918464457
335393942417816671722968123314509317551664070997893966375799162224
207214249604160798561861075546645192032231153174112441462325623048
205681535897413896597321145893400997992530885880960006646234601132
133940160175035982512267163355395923033608342753477231486044442438
576263675150079279550213540230485361368647550005570347499233841181
348175222302683487456072111770358072231509995929764353219569323709
661885842175593692971524628630010754082179814356289296381212314040
925114029980940402323490888550850564652027661459881978597766923736
268372754290299583912352079068455359065024528849648967946692923736
215623744201207894274831439336504915512617695583402317993005097138
414209255603380304526534147588999440762603213038627047720612443619
```

370777535000448723622343751397980383114151507549772043819552737108
044216678765545502989379840959464544556050978917321156337514714764
746841757639948573325933569601783467151221844480019046767555301279
039238968746424737300691852394143147360282410591109266449610979208
450327278154891164906013133850879613664306837903050955681593774462
162296689227532396333779844893818472628300472314627286511025782264
273782844838976097302514180719588049998035991739380712085611331271
550278558473260184921765260786685092148730642904189004924846021181
587501262449869393785908870029830889961310278167908341016249465705
921517492446726194694760392839047928909251029538311061935428244834
100675625824862851146843118999272381915878069021709654789240182171
504039382553675010424244890068537762906962467570285690579361570521
242533770146437439960599242938994412570196324119720706067572358522
529245988904681355176228361383975251535692452256654222427938203274
376179131549015506447412705344091962400099046834073646056869242740
612435588055201253814881296282056370142993530543726081574628624530
891239325295642963072313859333759556103774854510477647038349206442
125375003890163807672707750925950732728879830720381706143582065158
865531457885365532486508845696617160944113591633589449710048657345
9826181084001222109115502775177909877168039992679757605267811096405
175303431378626872608333703006004129497702986201367467722021117432
062368089040212955273107682056986669490000532591808534180937809057
408679075297040784858937168738020559039904437266948407465513277990
993607002272336648823952762574956993379013733260275285775298539883
377462130112566854643836972619469763089870348100211259000564253706
049103533483969736063181859272227608175480152370312117832221949361
931761462752448826138710632928673754687346993297813469399886040485
438872243404170657812141249588244795048383024971760248590491747800
371587613917621179670219572783841842292334545899594387250309344787
576957851919956658483945216629794850864949140465667951180072310380
350014875371155604890875426744329862109219797475717987032187482617
407986452913028957450456687692557691329968857316295951540791316458
47784545354852348863051225340037210214236729513378495086903930875
770202492310358157533569157173565991981246952657736011879064040410
160183908155066268457597263198130969412139350454770177115377086668
195487613199559430096634289645507695441169124574533345763646473046
195654690055997476046306239418126580426810279026409854359176746549
566231024703806506612795529836218137819047800721428636880355322453
427566867003076246326001320881814932479631902196202759529717181516
210146749516116626518267875254622397163987030418685370808346894278
871553394627472265269889727110420561227910792589361050848755842649
550068584149726602535681515900503377024513441602998925360302025070
857167319977848998064659093567037826150903772871827721947705987743
363690383058594498290240367365982345056444318917544786852328031554
409687225838835221861594548376776102759226683804747601937063737158
519247511945118441558390827244426571257300010805628577141456464119
093498524354779736365746496741040795906832889784160814453506619753
768856440740381545993869640105709211053262947386716629306661693275
387712273018755196360852214596070454906146238066785918305177392883
838759436655194835635990606288079111870096893801382302793393841247
652757288565129975851941835544080524877020940851041725699998265759
881218864513610777658519255610110515884452152637152090663359054571
33505076527019580193294745712939449565796982745255133176900306736
899919824797563896841471400017861903490639168347091811252573196013
662586308974548716428722565516168724379892124491917554313145272685
685815099366793687171590489674642162179371385070100039684863497092
261036525785587639293532018670546035850533721423948689127590036583
662234113309595869911594687963500368182519556885278062799589315103
236049820093404102336335938877386723912294854841690715111594928954
193425111938249237541398007886900982943379409965710874601307757917

```
99647718113656781346245752864854463853177586513755070682826120868 1
84199085583066600380975604976221767746490969008960907826927181949 9
36908379025434027918647159901394149964125138409724820565704776192 1
09482836527800960387024612133306913352273483198798358219288688580 3
99943793472120206499309220306324956985692049247590624984560328802 2
28058526934362141958292649784027051477068423095074148424582059149 0
96352236000874491488988772861503736534764344809161174882942954058 0
65968173693227165782178613700352333575899163679569065160904820647 5
83092508689815736126658002284658515727712474861080491657600837626 7
82293659061490895236324969894579987000069330722339610369348409309 5
39358013157148743918665079017836980029273959671550813987794806714 9
37863729381449879399644948525456526691476587757745953201246502517 6
14315775619657132924797634634731045291460613493443021483609619373 2
50628038021908172022170605165529874106676584742841070793445937546 4
74532458385733276038466379968218653556769567760979786506198516600 5
13695926479440624395930022588792074762878329335907389722971940952 2
83796882849115644673989670395628820789258207678945443570041584032 0
11578942995002606701042736739149922168869590213636205710518554061 4
26461434694848981076521540941126545107253284658296999353435880435 8
40454487184110857390807404809720847321736476370283041420550814053 9
25045677400409048841087545611412365143742421981399334561887085153 95
90995144697373163008568723411570224855386127201234368000406907447 3
72288193475096381561873576288387837231042390969985505970343518044 8
00513378721935824456951920907330608557079060000791695506004348268 0
47076606553067730245494534305228419728837715085100007735892136920 2
90090771389563736103131133279125450381751069451868776079049939058 8
62891268756295043846071730474348153781357102552176058714118996550 8
11525731712909165413272688707563662091674374053256674986043691215 2
02382026089777603806360339482598735942617161761757959064502240571 5
67550846293516467031970522339941442380155437369963852689158779533
35645260536211253132138319369856073237698123534000441725720969938 6
63981403752284272173682610191527597081021435109162141882516600712 5
94505582249227857289652399415087849338658190144177633241063448238 2
07016327424229477934323667958787841897132934127107784850381928514 2
75881891671185208223091445500824361506292138573223603836321625269 2
47325702540955864665974221862616126562907925832160777346382505690 6
19685834862840308356530595996200378830049670031628005875269347287 2
02940145033531812743502546803830959416357099863359195138516452349 1
10609510998808901508685377915447095034537526879561757063891238118 1
15369649524256429477533794667270498166702848708120302720864763438 7
77652740998393822140857872100004521670406815156680165161284692675 81
26581824756929509534593621074316660882186606880421597423549792321 6
17694963620701695803900735130362578539574156318749283329087217840 3
57528987738759114384914743828980566489974618936177268111683924749 9
21515647628280760089503477275175042238698020485884708129074126009 1
43555524305332073898864597640782249737481349939432248362978673369 1
34489783426883490064286925837735464145571869557381300798336462402 4
07602735191538685137702804843366996989602875086776115113193418442 3
46876657603401210665816131129251459688652806868684326984102103246 1
81324966300238155372683907753335305807287609418067784369872619921 0
32907573214755741913887170683461164668654189326828006600034323847 2
82952631885675875845126540137089465024415712294013423530267147063 9
50624243752039525308884195318749417618112066465057856871402955114 5
32704254802809274363647008450634387239336258254389101101545913900
87518866550769312139348800505529675492941016442133157185694184887 5
37274232566908964684837264131811685215304614255486740362399176409 7
77745677905867196919380885685241064040725209736996421045869210467 7
21801502350791071362221416329169605555116619593886840048735108101 9
41555790806512464246365237939289401061446249339271458849198501863 1
89988075052737660043213402274598935018009630860553124757036440774 1
```

279670321931105427183944715364973872337796295717729330947450872193
382889219886919877300736937155679832048846155581088699487900020853
818472031198532704180065992820856076621159196853018554270099633503
924863125339716415353097156719373323102926850104532287995535325396
536089846321182030442952237322217874709832793435013119240447624064
455426915054430476307929044201311243875232777462035775455974090457
903928655912686911442640517906550734042364121506797875260333175824
907937277047764433424722252974197583489727467951455716947809850567
640483074886392963258564174358094830480854401513208698292399868612
948495355971831604967637081992841928190206754147068057517241608876
729058245707737846421143307643742297572470996403300314156511019093
302365209788985472543366670264063735738210480847635269988235443330
611985006347039121597508504568115178858505571473981910747504449835
400820796267077168256409525788853129294840309798068678964308133715
538122143584030544007112811514246138029646537550252625343917638022
445819887756398466642540971993286459185449980995834323608333893944
238799600447476938347306194276840372731720712378292969416545531765
063217715384844760726997773124279943515856861108745555947352363104
940384208436436297521003618589408053168430222337342527851369523342
598924517948136360386588861321085748580258555260472235630352223330
786128726153263430472299096127096886739311941188073747065810221642
920000038723341579756407041317497759756614324155871349837038664 56
737581449065885360848744352693187592400518112902309079972257470986
059394188064753387379575454356426769286422122271987440442457664989
095037166191322251607976254109136913735983807126947913668942491221
450852478233004718695007146887686375982826457298510395747411309378
476356018668052515122277936609472741654150933970892214676267719236
443642216792233202587580146163268789006311913203126862353600404405
640376636899440233386944199357853438329785980340673726073279607324
450145407073896092504479374883134153026618960882900492054722588555
580832705411149155456027709374385853913215476257318025403619972546
022010708850419753042360813174730278435878475858332204840133991388
969013259929926559218347622562642854432808212385582997765397588 4691
443051507859106509234399976815429946433963706309651627224646132240
222769025828330186780624606336868398063570454273154551132684023273
360752423553423523744343608527775196978505606358149281387406866706
267396375768126611957066986524270099668084461704140079832557566356
794757653882707275479137257999156117306992075639736627527468692808
394180180472850957985060335692650445201062148383140837671675637346
589769335935176674219010654835981206603285578317983761613423329098
229239747462921975412301438244631434835550739735076333603619310161
074820618611381769343278773412802226410130943015997621985744427508
299421560620655279513264262817356901629496761394684504969346965133
006378189341215303330815521003182028582986633365296404806504241084
494069083738640283868296680872396244309517103936548903688173073128
171562598239554610264733430591105809609531743997236623464156354003
650064547145939948889393034017149658486692934225910521499882360823
100986776275100932853611608226414569339383456283320156354652902909
372659394559076784662433260885098214010334981694863072182988910577
659551718895820581601445454343454442757227645638872814720858022 70444
143921466565703535176859968469852820690412634876851969289199576010
212130856614820756578166096399699762030808186845854942227363590396
661159137728413422600909411219930388447664819068652549056918 40136
523533517211630843542148079499626940576620823719314005976962506914
278999109575366691952888975532703743787964537740554278278572230755
151082084769207696129620856572103286075940042668482985500961841552
348198352159603434931388305665167696221021642038655108740123808698
940653924630240347955800469182258976618919417028534511806158883015
657970877522203487272832658781465172507887295409771286627189547435
123836281103583944871791927750879953399882104913116265547423097763

```
41967210291459623497811834760187499200651736809796497435074373952
715923863972845539974893387361438468649100558497289824483307986191
944319231104152765141743202967776559873023315325985039415259672782
689034976734083715651347883152219091528992545246919324590708406771
314369987393390116162803330568838646602074202915627180314756946985
936867109751890909630465608094573757674945044085203216333037433675
178081908523889948054509060802301415377133044121346788397332736105
261453833043447060438438372223231170304114865436976394481605292235
612373098630392888445560581570783316247682346625305535446047485012
917552946120866358919356710420019639074574427357932462046645922242
502101413058396564901248006181025849435269568729157430982639781673
034118994585879844931427140313333629439696417810422379371583280193
781098503127061028317281835096473806056173723317254229379408531937
011096190703385277260942227943323152425250292149152510344619049450
932063719836411071351931039712275810735637581883943954113229194303
909883406072413426910991289741932206546998754899986421282821801308
803290222871929779857537568336492324108605264487820688213704148482
100897574930504070074327121325729043584554690960477101931128459287
027688582277512097653109891966212937877193824352224299488892666624
546036542627245321134238417225241928089577420796867214498318376445
973390473537389570648418880353343164744017779415554420755001484557
987077012383396707733368087776877648508057094099060244846456784833
771206384070606019052821586039573320647632361254028189167009447527
644450483761384600394736883335653205618719890606035058457864489260
338621250720210832856743334284383500161758801070698123488828160270
128892563387777843147203208378401671301112734010940190564496678457
790626008717089549781806866796979385555988197506439943547547450347
495583796977073945898386100090889420389046359394857291959287451200
248650815915061144300974663567677016426795899444525618432971938857
831993037353331584446236294113968724651977764238457833453215939368
368799422474226533375562041422823198837304436486619014246491235827
592775934457711292324980074926590490193003270911160205058434359720
050532610070513737303987928582431462242748458279853683981520415061
977985527668374625659628190638546972553787674299158133335287245675
738930171600124923945448727007104849966909273461137142933299841005
245718510885647881494058716816709589082226023439316024388257609807
385722782995557746688854149916823248106992267007115737640499511588
699616023432383381640417849746159887895652955372090234045748944537
631511711765420356285809766913809269628570812272258944556971245547
234701926412994988413920937445444579388446120415118628064476951680
776452567012887821405379821630624894480593933044454688606140699327
704039530178647741856087743755641693870918607866459889053288067213
407635439550304131771468944098770355141383325312755193985445239029
484924017554805865085864065742348412394277946253568891702439376405
306989167898153993719903969007455171780260466048374664812688081666
944262940440477692651094912325377299876723125150944800154306377783
318753240395664015027178943941097843957254506916055315589103122933
048897353458159752208022813815515619873665301699815198643336232419
882875204653002581802823891948061914523254359915766988949539500546
969693158674679021049010200277060104982991779926233128912562288708
941260547911189417060692360670116147281697025564225043894778645090
907701105631067748339065594710020665814922128414119103926514381527
047422274024026764455225692239030823825360079782918528623139266850
865890191272107732529557658509537869683113263922701350738362522104
187443906814888185627790762761961138963296561955804601519119351662
053795879030411170969352645855779633716996792343869526722344874383
002850228115062126392627916568409357244483445371020044095021589930
214993364127587251491937799407386714001341492087360485423625756344
526560618974444687223218562184433420682070597668029748218421275409
808128069884169235614988732439935579506124220536137083543270071983
```

240770396844895924647791008578235825625867975720243611287950614667
378305227208527175210889635956314424474608536330689181465296392419
137935824357498226727583874218278445692980104693956946583854910054
146621815328252254174828775414681473224920743626565916272947957409
490240689196237870161266952537570735899291865427575484203916459509
669684088777454757416631817111481752757533619661522045013343733888
498310949691298883924221353613105336667914865370994780317010563302
892938286737970204572392662764586950046144583076159126333055813099`
599336676280784456542123993202846837249444370609671867708023560211
696809766589385531375167285085268032423640525188704217732256212038
381864798041831984347189887627204511137414994839154161411342520728
693431507892352392001252753347552522022177703141181042279720775 92
204192835228195131592024371143670185788072005404295897091155564900
682869134514492780063769237424386773813732586148492813976007223365
979037331773366200587050818873262164998740422835029463743898925342
562437082790910745304659174425753132771963135481949756167106 70966
742883471068033843267373375977046729846760122798258110676761538930
510031548375035381959159345636790310306089401565178688570108642488
447516362844732523767537623039487865601983334467199674412707548533
993646969421268155739239998318759254458662734248328554773502358553
225671960910406411441430553406664459251747939725314409438432182742
378224517453244244846381765216831772144290029509709972519052388756
212248369980203682053257193107948776871921548398137173362531012824
747371680200852158860485933889748014502013853939806619762972107312
363775733060929707297652623225948635165799350418295787358098596121
470693331627374883792238051733778673209975229516772479924044766913
342197709869158847519512354794085255353161892271112387434194357251
436544636378447684952637642612675671270120247445655993474096335584
245929749909240681691133673759304884985267642950102999727136199520
190386707357294903404746855636222578729492194699492639600138478500
730291213479870511721279838872236057660995798594344885219428016758
570913855371229847337095607826400323989293435957708972890603849673
982387262823649486761708481147730099110563510819848462225123163161
604666796820549170630840617777608717012954489677424625738952069117
940575848612492430403473920992235182338985424196950479634655254376
814510250989088124110735136888629100051728129724541403044907815199
930037436764986656911633350853337573155536950749374480462353228523
387639811344125509682868071562239477762035932762579246903925550590
520576600376934389553024628542906628153313539913114245974576205874
362575647372274189534728556440697958167204790650605766903854620128
339834828578950251596914284886965999160778547319231523585280715542
129913116620618628524336926393954083693036241673576610775338576893
765415787289106429762881023015421194539615890055860074532852015825
570383138390190706007392110900091497678676538952854213005313915948
322752774202122023763785746322304553462828100919289512862720610491
867872771603665475187801764456469336505420615459553001039854289238
226192750231215183369890467451330711547878653266069693185139329609
762787383933692888194262597173509567135132229837902800512364146981
062916517033171038350783351785130962821754376320524260677326170263
060412848344846858178501563728073974099767744884594732220540746626
927211439744215369285476465877073674905705301562721581200055097947
542283094999096844237632221887444575591247486841569691547256779309
767291711394067169225139864877732534282696785691902696069987373270
253426415394933486445637776695461760555859948976608473346109064284
927965844645266300165818304556552562349235038718678212925880655522
992779142080143107874367438612580262380202261796939795666101155656
654353325840221927734896422240677326280342064630501618971109865313
117650294981746379823027511161837608793478454837294982097494556764
530249298519064128473280162657570650008969747107658608217524271505
827635563644761534410991747116424642680489763403372467399045838886 1

20222345592126441820487246223369145297129326777352523172011974811
90024097986841212286437340156924362930760781377201334138190858544
73230445755799395307020143513409259360300871246644859642488664114
22549043315029026255509471471058339367654874996440101205949707605
71252982396677187291381063787054733181189352013971399673085782857
79343083407945910632892550723589240918515721117746445268635087828
35220747043008081905584655195719939889403025920708980308282620450
43114117327512732975456625695551558295715764729311943529161877289
97385327336894646949888040781096041803343547740051896763229743026
78103848221924116577183749609200192497028069642376013579360494386
49736809240164602111077002131152599340312107985584618264904324974
79632943590517068402969866387512277054830308637531501023787148277
19692860524655887367522151350842568552840749299467664478642112205
21179838815521957259074920133293878254456650574821080722577976194
60301618808325560483818994678926240355499787401160407552080693119
26234324269792054621676300319863886229001611765887664396940350460
38689665492174439439254071242713016860112195095310156716101379381
18493434167253044068972590136197359900150174086128190493012496321
16714190099531230658425611673252749046225068226468691391781828196
86599986978381674485673807091430551531397616370154268037690261491
92954183794948413266505747286924782057034171251112072676282921716
30023182969830139307569103731165303509621692250904506116391498780
80703230315099738549602574112852447173203096422215132231558546524
86936945375058109487593093874399089801656565816339309397274506387
92146288320580413361822135280682557152787072466059485466625984585
29284541396856613236547823391615867487602486591318454939660644841
20531121345379241440008581196472261751410023730738245046832474359
03150589431603422279540477685417405548619544949457895823201451603
09657699948842655643787180085947281515224052754218356063366312543
57949392299376055536860737664824524278133521320735276433946926108
37407342736233631440788539074284314355414245739085661961990490089
93788198104213582274909741700219856210455545653907087442229796357
23902023292933497726649618207937335710588232491815734206567318626
20585652427852584515203274661250684707725071802991069029275023347
77002250408709270827251138935994838646178555844459439089596344215
36483498906309006045773675059216454139199970287482596484112611637
16453566924755369536373662681821864801574074515091271780069153453
84234348991580864470684527271193642094227997630205127770414163628
86226815810736965819120944562695725431762319684379566818306573922
66945072094160655533968005321741986279160234933434231364095469132
39361339389519355337588749152916091586938823613693309331326910717
97304925081108256567173173158503993920007418184416068416740990207
64325335006774879571166014299693775829297759688646349745045288476
49089451818638125321394600200172454386830958604957514673468288043
65941218099941193360313466988925850153582654841435534402437883595
45498847169226580792901045087639563413763467840528236193132287193
67970296260749317044984534332041833493894910501551655479858343414
59179192427810191167962973737218647104651702469506134783079420117
95824040297717962786830125387770496461377181264411109356306943217
17670408392857044570350687615956845611093888824600054318794476233
94641512835389458567379009099934319496150402071064967691639428919
03514045653307266634017810246795780333535406557336745253690465743
99582105206693457068092969827722825794522847532001358530910814284
77936047809419180299115945854733729882380188917386422715553733767
19061384618959939412816547938449072985465152178169862471625163177
35124710517860437977536895730823485616022644746225198780883264852
24303451174705364605977813598992400365716133943581722433957926120
82381512038916843915488980753383100707311655612118749743610463526
47765367910357794234563699639897700303052254559572178017100985003
29125648677940016956922383847556910759674086318996661176404950782

```
496553493227975824451116970700334106982927779196896263005572692662
826633845389608249700959778112594517031787408154467819742247936899
685602350120491171212661190368797977833379007013734774623937064010
527524806900939933719849862605607636878164471949872561905723677667
636865589053083226229298856022733853268741119640148903303471630861
923130178943545966540253679788125274547513784826633703403394227395
912592177253096629778198034941182694324148958257042902811634613664
721147105621624761687365757341833536785164515484704823247598679089
696828667647234428065500799820195080226504490085101085046944423497
347317022301697034526896061620412339412377815493654197097275455311
756544762487684172540005119706912220085177219456211989161912143977
127558156183174774142144618489889538244390791153891137028376927613
538548416444857682291265891990691971565014135085804898180126267046
224911027091171825380064126453394928113048722825975388100428039386
180079439060519403777684666736116110522646592892365714765809060866
253220742158894868329134838086038069908050337926205938989405047430
822471332402755747104026359805025581509356900996816173822101113194
984531686862426260023087726426068923777151945624321659128049286662
377309781519864981696562031877497763972401459044260017320558190972
217540611067049621475452895620628536299396692443783149172730768415
942786968251933158780723333084651335175096565788013899344589547848
777591718214170721847676353179377342226547603928656546943670204795
246732917691219572339889747651151838945326670640516875389576887109
189388471964083219410960419802419896296438959847695680086619988135
087875933994771289870707288805655600295807733334426354239058475146
177281957538481084906859478237249733070798561845588170384775743017
932086643384799853699796756234033808202409247159926474884064511646
046196194493130495275047789311355004979298424554812724868317431785
838392697398834528811335262308980698709569247839954853796404090609
689414075222405449192108706476375049253078936518878462258543334 73
103256914272007113730339545573515087994605307694238245770397590011
073653828254283692940714108370389939610261880939442418387215806316
755013747412933984670566939122121809855626221446507896112359693851
095894858433645300521693495594497611972668901720814880639626403475
629589679478386069327075341018589466407267888675595283841150149024
903100325782625038048018163855207986285050624747396449608864750835
312304082445150826827375606335417453636857566729240880890281603840
394324190985145989659244278794289089289307980252222895411581805323
357966337157956299527059156581925827422473478507224728086243723112
322237708835804232799726569761834571847572096113618861643064193549
008208078954035722597816527103664349542483517822148338040109928070
685765288337806597142144243781191476453837400916692309791536572602
605291961098182805414453135552820901681461962489053625117304098124
521050479827223076648243514276234463056762692615661100958838390670
719471560970440740088686862839553781752965462529035279945805138830
028160675280536282036318284523971927224436343288214813864990974774
815651984993340893517030923895353819803001571262268124854475138133
951872482231606530812743302674236287763188018603812319191222849818
523573130769708946922846257295520810477433267575878305088752505570
734618433197018952764793767755217215464448874440307303862458571588
845359711368016251063726858893282790852687070109785515825531864100
397480134817596167607806319485531226049376232014085536851989455227
536778327525997597810177280283037296768793252715580720875832594728 9
535939167466336300026908236195304264291567644984200764059427047291
192447323272128977960275097378217049574903632365369642001657072245
901158972511872803321536963586666601087818579552498491387337979253
920197378143325487581080985247028623980380792464260724113336467236
158957343810694201463968211897640351832337160142282227975047811048
642840704636334448476842714148748673561921378379029587100878467374
691464130845805595732938731541375057992937324000498239795938143475
```

```
4860309698396486742146093734563157470225490776126890218518224811425
9256449958415436600814345864449169288483458190340351711100040375345
3976376215174578359310308025328461817836918941097623746755946433653
0325509050629917566132413207813873842051879461078789841567404307373
1656413445080396764759851271399385235036352490580253905727434213735
6257152830374761939369981753857443843036960193299666341236632914855
5925423690283888474290390061678663146109346395597294487206010019745
5451305388135665372298183932228635076407669669262858403746493783805
4957187056082069966240243945337951197315393009773762145874294082845
5333327379815642146756939767237039763156627894004243647160276458025
9090488935329371790733495050810625504778060399143860431482851024225
2396182196601401268597185663149992819053312303910109927788867982955
2523389992701712777671858394477473405782247880622768447710549186665
0451737027013124770040840515969240271446780918613160830200688964305
9367028422987713788611972779254688055012886761704447592941445753345
2150867043177912444682075716575815218901205004350383995819082909005
3857498874688317038435012712304339234796630694790633697831317709615
9633678621235088583154271188255209388809278082331033860953957963555
4906839117536980934560287247338620416610640182286571528065291119955
3861091341365262167979920923120507776339910960146834776055393400945
5047731104916232311202572684943166217532654899485596268661923354555
3802591086224020612164487002263336881311697029987627330799367248155
7006092014601855547594359733478567439670267371610965179302493120025
2024771421051099658096096688163655609309854532316323040347373835555
3918640549656343130250434038357899888242685083771934079303017414405
7311246290552437752432844540321080862524489438183955680624138760785
5415244240015778438143554705924607528022937999234137418551854343505
6268883823387540657046753978499564317025788321184426015152537442185
0747026468342935794547246312898252013725246763533662820096645518745
4406504341863031030378061287200659887999789911482612867859559678015
9942109898158718401594207819805365135235195563865697031713958329435
5886240604628397477884379414702458753155192816239825599761721623465
0436523910357266624830896714893037206475930088261477149495161110865
5613713055290300617372028923899043340328978139367261574217562159775
7058798065705189931649033359789193490523840436583635921663115032405
2603811459472512882847510712796636041303512412518401865863967892745
6741243948676964026575997721294933068350331602536134050859894655975
3317485872755982005835468366909539204171900600653874835120214886725
8076687352356931552050529377125706165080121291438046262651477843525
8877779740751188470354145589293094314962991024525480382315366863655
3311263180429883884815724254383098730318885601581303789232861893775
3443429467959577708543361664241437065411907973989422213799816737655
5477152330957542512195368836036441923118325470482448362897010378615
3702664371385904200997521370722090684048910786363699398715855868455
5705891701937849985572022228505028364837719289994711411104219185105
6767007441575145644840597048699053989215382182794749356237525431065
6305611514868103257082417250221682752153642288274293561101612876875
9023267903591907012756606119694045157258345655068676601061768347455
6150998560271865713244215656933452613429568186193954223595284021855
3193190582064133226693631957102155426205317587561120016287924972305
8179563934831374584219069261949633698437562085863255551239681587385
8309344903949751997154436618371407989959996203254372579634794550365
3228475392409850457483079846747464602475375452212277977730245955238
7675250140161176026363447079922990618798132966590007080662063352955
2056574466189446207103838361312534852726451237860162140633061266585
1190306888628751563115031794471416036271169802931177462248068183465
4448825434566682050997753306510084282072304174351911057901409395575
5907502790566536197167685764844184123250421115049638173139319909215
4759731855630022296067338058286194250255279016607520437506026828895
8444274216819231989098649478881954155485906008017511408836879589575
```

```
761452008198705663103068287169201744305146594216682429611213719704
386179019226219105806843479203252196003859194413551091619910703598
958648585335054453651824025865704202381212093889178754313700342525
716999031955716873612973605800541175744749525279105145741810095746
904780998738821193632517481522211300053025284863966050045749577669
523042077317895759463551341818367792388493538164279487201861349173
227229996710957884788878581307490458235991994278184047085100827264
637710590326081791082266951829303241312454161786902092200722913726
206288416627427126454396156623532903793196710774172764356054792811
489085169582084382553721983195668332925935169359352007522296120803
079665123250512033045724288071173080936911854294238275625269783458
965974289806865215490548377353113219751540472343087678041251742790
038558764757753259073673983107666007960136956515960048443020669835
492076568803803885689484403135622736089120348901998735118120712114
077868632529958944691335513412987815305425603698374399414740432101
128518411101679957352700086573212978414376694382395960410903156435
504612094808331532522888131079384822104394220221934315290947511480
814110260813043774479959007102333014763577612224010506228264281139
994842191051753473305328433819717255231367836278965215786918708946
170562775035777848925173815596096367694942814287291743648159304646
210118613088466207218176605883809532477726705093998815957171387560
027025653995096050954866877593950888321238894379286512442842811395
643115676203188591743624000496259204174603398185800996779838328333
588889430588723930874946616067961161956457549891972541234176324750
355702036448911599413371403805166587645839834800460393937807833355
846708697598267526054576417463946444505176440083155855374304927587
627469019374913106178521462595623035725324132305510242893440436016
753858816469739710217111853862407651202460103680764095898891535694
422530033534995737600776355923282849038891148750432656353587065016
619591526216061639590105309139936721988696975948781796685141310023
771954511650361026472340456407299326233585158540298398667538051902
251650052518398336430176464469536575128858950409463295745357082070
309976601404105090718192595639776125697777852566903964902900100362
222803664975756763220955430998386280461087068649192028480768356444
355386050378524424716405586487644326817832957852496570333991207217
324442819092299668987692853021175533719552032681451612672776693968
515150210937638957825681474919826770213384739467481575729035920336
723103131821785156706487484479349609661987527651069497639651819412
372692059130379662287920662889124983553781374642262776527283989092
953694770809104020712370312830073672924115339164314237400633698398
111717453438423146922338002052633166795507731559181027237112143873
469345911916863644904294483790225719317523206755174006117411394278
313365246852835226678466374221049465266794065866380108549933184751
127075692280443020199082615985528019310050594513322911806458062876
895649189096100851468600871722361668644174507898367322174940329617
495739694280648610424684256974935367036808078168088115093220388298
758868613998483332557902142846543868221607824009414250304441167833
341541760057591688191667775396683896408302775493650516802419941486
653927507528119097152206409639838564851854441666134336727390150740
574194882643075864937769101199072076701183378714082125647340782611
612372123839093027517671158066134967642899078527110367915000764504
086099766004897077556378588153867917258390784385883986923188863788
250918696106412097133906234177719298111055743370843112128635295541
209272321219187961268189415283326269597094039427715577330025501604
184025952476231542292040397655115032023064858273339425533555835253
859310382670877098128068520901463824298996156515322330830976227942
308776877860371418040701762423342525414247423669062209543413747585
425483566395739907290133932698526768178156509210321445692599340550
048291305272314357515080212673978295876808057884397853775830754350
169363604120163421512082933539940314164638983192992216655345054359
```

```
1064484831961777809514815969092942977051630823796834270059015487 36
4412759704557081466904323156216202242188343731672943619813936037 50
6359307570329065001523762848793790825585339713469194116667242078 40
7994113275130098035849465549066109425885075259955896131456476682 21
0098087770647305169925433777177781509688617273210277242643906156 7
3765301079355624204631347840060928401884520137966919893791815381 28
5435730492493706384798188463852724747952971954506776118952238315 0
7167309898239092628824468871197362227360369479840798819619549566 17
8239901570953629040939402200823262909624955208493341162784059702 22
4913199262157370253669477775289397816032719043599322373481008802 05
7926746299439354509652263971789087557372279544717410720733820640 51
0726551363457809792941199477452228231398102998630397520496268676 28
5536137552182502682092879576023712104709841214918435486557631822 43
2758831117821098066476107132207806384040509881894763500701768327 60
4411657221094578817952879779442943813961878595802409953170562963 20
0587382094657430056800252616832658446011942994223356756545657091 25
2495346927644671571508808635318578904634378668853922804898694937 13
0726411205610362266191480073487945553923145785341983846583191805 89
1412046994210197238611799270340490844832867896225006020480268086 17
6929259665767560780950812814338309982474978741283694883753470950
4873632980767343494620182604575373581595616327384938153421364032 24
8346272532722493339244002648959231394537538140460466520443171961 3
2685088124963972336814861702312544248705518571477290303281728832 42
6638075754761831369756421301983615565088971260446836265102329752 65
3414281174803453998297539174271372918335452755693504112064488379 69
6091241123322069182680774778114806994946433624447674135681646272 94
0780530509256375093977776411755525753903812069716022806771111360 71
7969329060280169628195960084582758321997376313026034022027576523 21
8377406490616551209297910693628948191091008955885746767069737334 89
2891259316726200525800265057809889002541950485860507115494716844 24
3548203775345853051588076305195010673207527304391021847372665187 82
0753812554538341793323449679787659716381932811412434638774774501 90
1154624978057909357814256200548946373026345375418237051132791765 31
0580650834645680194870469700373186443650956237088166475640399830 22
3443185066685997310843506915891203292740978629674512047716795498 857
9572965071297370405731665022667286318544765372269038955639542799 26
7801351109894140462577684539784228591561790906362263065476992183 89
9429032323079211924824832238302139145386723008366405552351456307 77
9435938164506446880630099348971370561380362230665414321701211328 77
8020461197821781357706661624177952064015652847332550852498867824 16
6544037708152611526581138923958320654803239095085371779390313106 27
1206429172835814917149207698564957017912706903388638788746890123 05
6412535333544548452234261228086859252992568304063917710228403380 30
9737567723136642146747158474561879187551702476116143004155251817 14
8286937979778310164142002496787812418339769171573150779707030383 34
2010493261511473892175352408632748051842275063683321563688556568 19
9613242459304754204608110709469357682746638439350767704232901211 01
8167315313368211461184468226315673840597908492265257010028645496 09
3116803597802151484544233303115517433994731941278383442164452466 22
6312562251826385964410836497382949954995375855930397334556256321 11
7691373985710548967427658519110778674369081835795688945611267361 95
6900621557165384068889691918124615443509491450897275298514662754 8
3941491771193061806134895881283913776320157857643790677933447907 470
0402131384444689294709826721663342559235778285382749579997481879 7
1900380270655530041095106095786430784405273950574975976998616924 7
2896227404329500868526888866587599130083168062927587118010730827 78
2396553089647825554149775303106306721292184488329663434770452367 50
0603845549507836861588659168885688400777247382792435760431384641 72
9778195679407072700236353510736166469180410694550878472232715121 47
5707227901146626053563156114909501709110123370494678763052525202 67
```

```
9612223139197128313650257003374045469521306238554408254026969379934
4325367641734683930501401290243494997954596362362805346491536892243
2956525378792713690607455601804796901403682731771603967869545316755
7711324850585610982343571003519454559567259979135774438380416325322
2097908768432800756481322121551380158696572973446495382790133353833
3548277790429875765076532626802493349347260456840129482246782027222
3817430894544487353297118065836879361586634071891190356834236416669
1431188193225925052300834401674569298854395518453313000329340134533
5082950670310274557151949049340723275937342961871034452030834736933
8697984231684797386364906025228038096088880960471295942537506371122
4422779490409255349298465521535848632573209904573674109777820249111
7077272524427057456779394727324946826589156671246051703308796326666
7584646176288106512331731711852978689152411300804941291519309613600
2696757172973220961031226445657857686385015126229189870658224294344
6698464291945690957821493585205819903720977319691721498702495615399
7174389758548568845777399789964966008566178001552307455610854105411
7181957335641580873305471165718902008361863473523827073510596187922
7432960921171491287918446002638763857715124224039066958468635942266
2695711644623970322936114756880431339151523768104787051569573440700
9140316186679211882825064477206969415598431987614717745325215753855
0097601571035378484520278233495123763867810970887762762982414334
4969491851746433623997331585098267444840550319344702936555363004246
1482092831127280160544066443169115574859737134703075358126364509700
9999690838846594221073377031196040291236910507120257767782154264655
1436137871085325600590304559553424187375015178736968874172683493922
1706236672979597527516612096226588059293087957596138590067714712733
0645615269538655891263575691627603686026168267511808118115783163211
8628164863185070745714213640536096219503716581140522085127687105044
3802884473652556692046732922201753043729754053396749885879789004514
3228975583926649731965769277663776481043002658551153330634114472633
8891499127453237657585560228345942211193863046421137252236091119055
8394292920631822445599823003617728488832854560457088988296259344799
7456369155146483031100855157618396324331887316070621108062151477044
5899169975254022884226110765601597204724517471469232115809286058688
4085381406298697672998634252216653824764130860511640912205043121011
0663801386142150348799551003884589078026869783605755776563045923355
7961231740471533214791899305749021042475467982032301237194306644922
2087850804151879078984628416188863993078212095193275777845971204422
5473602315096877133543527647626669085848223762577880825249794517600
3981960431645575534473038295620137533948529626262774610804038156511
7179953578307327986196143129288911440328993607942687971208302164288
5033601556213506628338143436256429103520658924039275367616650919599
6441677684578192722078241475178427814351980021491146894437394337433
1943262138366889518516726743716803007470708248922875294987915662199
4149364319073859756227657044475336818854089962328837341887086952355
4473715727014867021962910410843198125559823760529205920105782844566
2032707415928975421975658185551411897935960015837786613113071599988
1360071961985600643039476615047103055517473709861862830824612336688
9937072010922350165590163387931117916217971530954758906781700589977
6602482925877148116156797015946041821127627108143695463549620874877
3399305966616970980897274354268321032315543699684985549888562296533
9058080480574374730170930071609317604048534336688393145261735937755
7713753040697041824137894951010163393685062438535914343126409781066
2248728269780589374577759525198609154091685924199118601333070169011
7768094758191640835200455293315541051041642492683968087544432031522
9266056427481701840814773846172941838437704870049822278721928306488
3443798282459600969633762398922528944239850033508927467956982240255
0639842360360625543741737320250900031050722629970929405605436429322
0284994296985386705534731906030083479555472952632664859614696778633
1555038758837829403005283098996098429664622408319497788419228645844
```

```
352652199299824118485703870907678891412147718940009618066949242179
065961534402305467602375397664902749390104165497611258208179751097
179750787482039022740707844277783787872259522423287825254481496870
576979191943021447444636393809564020079932352234186294064324224443659
167226194788042571951743497917550092814883360111224900504356753075
593263916205086219354591360624956273286339902173488045997232380530
171291604027614222320195662615055816861374877007621777475027585018
129947956462975457833460379257184298256412130593897161215435651346
539985301662387813495075313126213012052341044293609876327229177351
284221163053019578794625767910269020372366957244181543949795205282
220022678752841576635513898908495150834619755593588174485939738410
893556950325364073806134282324431379283902350728712486501609182634
980845846816365007843325899699157827373473237163275120724341425116
907673303272241617653432105999836918263574393617446078828972931711
120127670490996932647020568490466500898660600273531016768270247770
164126115591993077234415726942639423129642327461391532288349425199
447905094332122065651331255492443834777217413422540147746233258522
535781948839398903981863921739605832824296857009579157889673037428
686888392707632729057847517057592486708539666865433577966438315963 4
300741521827270640714589098268556753460081654571885870949772899544
072952053976540993951054808440936370759552766334050737600741662473
659731435326587680808233354905809139489524255974830864090896898520
302365003701990105969322080099025741769368152126909792272915242534
518017045412253605407229888901589109033907768952783445590172362572
202193043627867201006947997390266505514691936790479595835913014863
425213743913509539479471141691484127328413756387889166889056430345
855496645836850746645640724556535429890419154493466329338175328330
367063592585167202505995321340853379429251981151389114540174898504
222590754014192442535638555092560037356789143066589662761304147766
446283565605863173372497904503880172358695793352404951097027563721
487695005178015101189125856270586047978527428184977418031816532776
222818518122790266629594272831535018970811804857632563033489880648
162864427893222534199698472599033250509694393749715284735935500397
451302458556182537277845998168447405804907621459012089389346976348
061003960735017237323584598663132565861927607164849817531096791067
167263009894183997466911141043418874919610274018876878666786642905
879955065077060270729203016142279613210167081723205657760758564153
809346487494194516673362370962285310231803088982718055385965646857
723709052051975393778494059415608426116606422552105971645633611784
214629347000913548222286284564068035784278655038345394089402390207 6
460352750027899815768668651833040277296854117917528157253043566045 6
016077643403674095567383041965030695700302662552526923829654736017
554460462257733210152140683444243793605381058177549125641777562917
772374126305022780138245648762155268809285697278224847945870006071
781494641272624395582410548209017057706649399405968834372242634500
670404831312251927674343347409076740228010934554786883848494317218
050063032639809501717590860646556044379612610979268716529272472415
196376326705729771007713654611537834258883709699301565991371403386
966681241560606305108710084516502784882727945516389288827419604396
140084847260070766304641200594884089901444287023061776726399081359
526260516943199020519719650522269177869468659222691023262591261553
969781643389448151310315167207610512352328503215908186800408472650
285558099862936767632702100455695689840132430292854596981739151369
550030603570540193946473220531160498375276660205984637577785155846
314926056349843258614423827891827195931599224262355372865366
532876134117734811355718629796253074780591388357119407351940420537
748387662048875298466865469030690500041329131326447601951954348277
678246422805479576487110335025983258674120294750763222499344170297
948250151167225981094200601255270303817115328307073797673100599254
054514215210420761394627647093302740977875737676392280809437167325
```

973998085252835756358624411166796905227937663800059448701577626052
157434516218571721034622989085553905538162124329139853417646686292
641986500126633218445069667169158281267111916132672921211553853776
353601372713963587338347254161485018247526189390456432958132372703
750950178097341799650046062046012023672756875229585484536739441785
141630080949834313951258045323844171255847254102444715183666585796
799864675864849812550595248781549198156302118060604371403617991945
454814182186857323777013262890617636725290324939585717222722996900
650373610987215621918265922436647047377635117398266802198628105088
521748412082297364240247662461652614327478874529492808345545093096
883993198576456366197815555359201927448581083694291971794870774529
846761012034790486334911576889342434723975184650357335848853663911
465583286510382839768272273892776022048807599626959336722648079 71
034553733016098398698884144358865758603010854761938117486975518361
049950412288087336040040353509500409145909052021192303540316682757
296014831044644863432040342585052448427028808886223665580416457104
769707291439243386475215023506400622318602838103846482969298761579
442424300100029087808435261586075964881927431219777638537057825778
525781598558187802080600956174547357164959799050802163306381928093
137378222442442434798546234482661795042089819493664548233155517336 5
342935525936371871186710666616299524542313031073058507961463258 47
628721000348467198224056372787545503695676515032814243039936316577
944121132761213501821322271229418809043925788066972937305437004031
568154016801482186828301403805008635974767803958867851123800665385
850086690551244914720335795078974917231821608811593031500535021243
223549504624301770174183598718000134055204564428236235677176726508
784318879339709265639081699038259760511252148966681646349362526894
092955800835635295201131073504032039304589042228742055851494373368
044804822142297008600383674926104865003153682819050585085850386149
573047897412162522226379172174866387706031332566403974828195804938
356298934675019494725758270608492855840001777807698696114670423073
022545111912069288536599878927385205434944412993220128760209694521
053059850718308041342448447378276811327543760478620364673027400447
515961625378024683155065375564891912018398089911499002269380550345
552681871441149518136465549580231815652732268961535960195525516631
472088507767211027653317393874805166440038268409261600774429973012
302920369783006832954113305911492501891604167380974803479382409249
359765638408591693873352298111725985266057516325282955328667957240
115072967248853699906517024025306671986358740773555360320128229275
122109753992023754612459732324852672583771467895711339969466585068
577281835381113229675328542353318650716282077925397165914794583687
543716887450258089503617484830121840758809481474115834043984370225
808249449337987439102215989003752887318340901370844409159627408461
439133405111714417111841660607736502271752724569822325797525489346
767599543804404004506369675442407320820091437664030349113291267368
830378281137836659511611345830382967277106837832639870778430293255
458438361480492566222971320835346149770313256162342380891946880479
263554425889886452898006360567571985818009472856472741482902587295
419330845001148137105760228105245405074848643303982560355008321147
583695121628021567209959722669539704680086174379598323739192072456
475745800299652448543097585748857718809862356669658290930054455246
376405185812550767155376083554121355156283745959014113368447066869
196594087578697848757698297621990244474929941848499656887118737182
737063870288236648289027636156281879851709196436411892005933302900
229284985239638652883157621481714896305297441296530262353274449511
273409475113364328814590086920067393872558570570522960771399635351
239505849432921752354775351907345501148745901866611586064511133938
883941495450503792549036347171013615609463469098515817573155784143
020279812773473063372845507584025168819646949434773277508017333175
084110525664842020366717994352302996605606081819133353526156844431

```
90844521964391072792595048095277321486755932012038482674133699477788392695052684503618013384633879230269630159588113340591641994880045400086569682572377514974098980786922266918672389153258721842276001529770321475470400907730438208578223589240723141681652833998841000782453238051913781046091899238875807476552727877467094497331852071819761251583090922120521662333092477990651496511834733937030388139464899706821592499802159542944573033509272454317561814806950498270614854803944663815478729386008323770665086056284411449355147549405878725043248206266535169856377476100494626918389238935612508936755628325841691986569428870172878226119892666980975928850243513624737650344966401629459978187867634761870628253336697595370448407684620297867277465274707950259294249548768051983677572624250976515729880212278730982838185096325701396697584696079010905696032023160449170256439010419148368516736128359291381408362164139653300491379542504080076630974025545646457820006453471639715035183760876338974838951890091559946337265210255518987964847206277809928460415993194556191748454825579012386311930818550030876791597754274827312219840225331579196805285180579651419263384071877679104532044992484849192392526855058650855457436207360044455419141887800168906787827256581142878325849884297878100488796602536400211197534458687293530940716221107156678992890066119793074960397213443170817910532025634810575451764047676723026585850031438360775268209816612022410070567317560001660055425582122803467182206151138784286179942665099172269093844043207194853754937787917117954053914127020798556778449768721110433881766488835481472674139381666668297137535015983809188696061474971007439577620030626806159916534213973293222775409478984358810270875605233743784617747577504367743117960478759347192269297739080636567545685326358835238355316527568259408716846048089855752987473798992168269161176586678934874797687246561458735381603252134367419086353658738601572169896626481005897884523371061628234236981044528395085469403919317048870861439686345207831360467998456852382107683786139007228278427026881844384440294451209724331767466767054832519867373389686582449500955082974460048557254580198207277613232322881144831812344282699337666264440669155892571115026810516784480697446513429259562162448593996318059488253269689939081486389562422118629162404581264332925529444888578671685414060892304719542363248571769993350968988217246786257270700584899308346099904802049663711957320304142633982710686910257665855262823971326881933975995154631888824237413391535272231422662133239161814297624741931389943115429840126422292269360528981393261355168653922014450860419793477640236749978315417973197835295692480298597854986563254806401086849219147347214086667402462707086930113914043097417530836742622323136635465246517420634859366705971148981276012726765068630748946699289223320644337954056720923220351459050767727336044605563014928698546294072636210104544538514362227966204440825020226639934361896998129521306819103275490663868479077994844131392347840353199859221520293622427072899591599393970275297314031984332840042371593054318428616767023476482119386242602897006343620400115923978461296906702255970720710482460987585095677742746359574809476042790847716261833745793112700921831857351846285031204385521226921951931661447230185138656002177820317912782145789885530096170824616702702627748660581572456660281441194160466012047286424042115047271451666811751667317658988242929338472257397162435052899223297392575474700241084833029592427705294764558974690449358195568064397095722632083815663245098431350266153396079514035704803047174593998215234938755594830307949604443793685784343496877073083918961390185977687089754484243720189344916844111205128724998699093647676446650453869026654888197996751177777174058803469569600016915867723421610134349900996964270657774921785892549644706165874451969996572385946980564593359189682797319477556286315904183812901225216508476693275805444127498145308092482639782852978552007056484
```

829965115973571861091293928865004103403972184838914143524196033604
931001507499640034696902555960240570068252936797488779282682507016
563154427978416713977703415976804956306229506601666629984911444 11
576738504605273647290962463458024289272849530298710710461825068398
501263598571738892406983791596836327931159759990187088958039765122
906261122134224548890272435814490129875608745516311989585767623434
923810318815320807735909564643955028097420047654106469154929113852
616889730951717757572384683920045760956563590209349695986088366340
277624143836821837594777970822425987300986180541688046461945933111
026201132443812459466149920321690055606953979590283882395418639183
441901624950424670904463698701618537286435887185547782501547119307
840191398499296460450355687876233541937666833809949237864033626200
486904291473144683166932936446336610662876074979105993531241679636
612585471701240758261546056546423463844867344736598922090501735099
700893886346956241925864753508366193195778924688275936566973249685
243669501140062182072300590802028369813644953921111905416295212663
729906770369012925746165346122337935307300334609172752925745032251
443943138141765165616402104245414223514034819264459355796837362449
867885789900599914561876611242310706104172854194568537617583972711
302829701872557828758178521347280893197049955364542107497241815741
338848753294704478109775632150740540881852069292969981482404640190
372842236956877010696042243305786939093549156276351020767488939026
126773759570451744797064662812379831933482378701220660068399849267
573991465237970692855913780519174286360783241047726796961763236011
977101633949476749330654234803552862757002996359277112325959959103
156287032537230648132649576011410227785824507057714982498120833531
120355554285640793349800994371634851563604371705770197058096269887
074455259323352945604960412864111863563553413118849402436802037381
002536865614485897825381735659359009758067996349269357438712252832
906591792338106182538567112276788235131554834273893101607886870001
546993001567890521770684597170495690175777976457833308477747828045
291078202694481910161954427925753605515560005589803854310072371158
703237122700337369061133064975970051876875594864817357172870447540
411512718013222890370200761443258485894604717115618279484922301954
167366460650929627158050808191401791958252733674768036051262864490
064589680660149731573019090888606981119710223219615635268130543142
434369918299704292845916548154435297795852152310963397224850743563
177613806076532338614797264894175618271070820686674324375386 6891962
170606251673848035466213644172744979862722811219080840756234999867
894991000253107508333309884830742195833136158594965506434448771477
159169684644350804917025998301616673983930493195736674998247831659
924382209668398776682624534332606555052470619954276734317920142043
085791564538420214884185974342832150881601581760488141831911197717
266683947218915982515519489656532479100767966887822366200756668539
133245408807395736009639249354408795403758907416793408988006169831
603324726328218786585487550727027935293034536874041159355064640854
347375390029219847429135565405056224407190798859056522870447015940
103610223896642334143553695078580528055396563581593182729704362132
722465839526377830130998921627268488343042865066204158077275603241
377129150139799628665063170035025063964031167383154123253164874940
939698687474852665005698898208483902062443645341200763947930679928
781680014865600174202299545788882260956496088501617365989066400940
099652858379744182456393075892816458729735592682863697073027697933
875284667224367425949738248082826701258312539916192151225266260 09
899654099212771403870416952989260336332434084267908951521970848770
815804244569296376708197250126955756971906809391685839538448057640
845996891652833912862882747149936488380519143173404424296767308181
144203014194157528905525658830559409785944557248371151680795811788
374423367356776590141485200757912409171900708564469596484449168800
433573516621251641162335460046256194768114725616328751129567435679

```
200279149016524863828596150575677990996601596020154734074853639588
934537291238575018510239848602399163503050468826311417679080697977
237294579823001758899611726202618617231633309842953362953106994752
233643411411469191233488994228275945005434145103587579080530947457
966931914186010730454768757915387123504880976312881893733472938060
793261749505653883822647725905358346153412545220726898323684157744
882166404725831516674853687814268723871511980681225891343921434856
082201752111255575696225232968360946381187007013435637555326661249
627708302266791368481672126408061000345883190456468897560774054800
731728411607444714985004465756734136937411286747409854099106427508
920376003865632501625329377668849158971090296046186603361795017959
809292170457169234605063636872791689093371723308220736401362629 56
951945583111580856192347245969641804097947246844795435008161551169
611341591550690442915014037091606976503567906440513143839049528928
252246850120696222586218113524892689062617641187467307744763360941
280352682996944169246809908849434753690678886833455754630812 01974
965467677304350580063333178001553796257772780054213478505242015433
137151361555110412769895007605903425501326894062257449235514303566
379611361615331102974240930952368195329663054058133173421394839281
361143654785366173764334748789008591994600184938766217774307952070
978100699620379139490997724199877542952643108344180319523992808810
993921455120782933849028810514143537295521867031469159881077680709
864863080381465696417652156742067991403377375025752142066429732078
941696541343855660267481142994729217292554756515180272320537112871
497226481400333222157597178165394699221984041747835000053714877645
728638021389651896936773785576328108897373697479409167753858139258
641082515587360379794055576991068721268021911312283555114116509802
461079340388813573897641259613382152761748851508306908701399478458
783717677763352619506676881327927353803722768970898514011065403346
788180733921152892662172747967987523822905329905049267282082560244
760508182215964332016226464044021468755003665985110241869163095981
679610388562171124856211541666103723297209808625117144814226509715
287438796969452458256209742407132342135292698067313486151282503146
751428012923319935549936010978027985356180479453942399221829584116
741128678903803363555795981358018927195590199258775434174653292262
220088428080421793065205736007817666251120871621246261439867676694
132106095727452485127757676908503777681335286497623032390963336026
143451319202537694122826711510946320146036578799427615401902571290
484001251272533845175619104740089275872164982922626034735317451123
314462940539897489926909716952550856443119912952085914494496605235
800599047994989754629409007745232393562860485423947124605148238843
833375419793508905221422643444436585429151210774163562116364208930
938368056199768393849113726233057405840493591925443639059717695705
260090426395167932145864875766015069659163204476905807997869760197
908810610861068013308124564920428654833272945841960274553100694141
985792658101282278065255959393123922023205564570888809518343136873
700698285262991243891536971330705968355679941677365746812274039172
036491699215652084431004389409581177219132083647663251037635599687
918534311831410377902197587791151416757889570960667164337253348927
098907344061828975599516196075219672921049052497703066699806940749
389345688669772750248678841708352262765054591143510120578304492372
083491628921933021147534224660367352426850716661308848530396330115
023586234025917682542861183290516430882068081751893720623160546527
875597683914965684509516421045314471059854656159220765133316630638
012050803917013681268446883776835651207805640876129029591769 3827158
643631091076670688751630783425553859185306855876394739785713390390
835126699907521686760135136823756997561309273416635713813876419976
259973945987884175453615694166069262263038364810824130721893393873
323012326438241930652691525439277196119362044433961232611443721038
604326842030568647877786583474288581774670104589019694988420964658
```

35530444877907571606562637137233719070658490550239903718468089 4234
37881249000211234149695691374830926570753546937478894762338053 6485
71853863970114443960739045985423329867786622963429746246169384 7060
11594742860747614480918860342876539609913431104657375633656854 2619
91562137701483247349999564223295381826384932301383123710499646 3471
98253169655183694878589429421436020602156249029299949072187442 3438
59987132716609743667503170288614729634996371953291691100088971 3533
26375468579568869350225341145821051884977168892448590758957553 5582
24236682437819842067553734302938827016671366718005089225281027 5167
02768274084158804026051938429405403235141556195797209315802022 7938
29704291864721384022587621426254600043730073898339641765532272 6024
35575506534096320262193062609581231022132300637294351357753899 4164
07821179142157608101798909857492772754107107622628421976442460 0655
58194392502824638498255558391434511612954495974056990960000239 9762
29622310085668423238088011744300550914132973647605057296090393 8994
75762134894925581606160361933852160557536752364537852251264063 2639
54645057111665661958910281366762021999682792327279533256034133 3800
05974622631093450761891788955116233355924448281682999869825467 7427
88526895855388729693830330729957503463589350175794483833283964 7442
16635956958103465934852797230842563617373886760153914987699230 516
64149506050852244511062927906271250189774232133011801605034214 6289
34874850543092239206426355670841651132177570024771745450216677 922
84654039292337924215342900339702470241172076562214112072735586 3993
71163873218482079671309406067075883280770378574711346684839664 7913
60530059137144169445780206485072182917615867268032664250449760 3270
56677063447923272546998123382347073986548190824262215176580564 0871
26294146002294415734395626382811471118410305281513977686383092 59495
36986838792823414979676759783281757627225692809267627292493825 150
52543023565641758024237783018462414037316971600776953917704141 5726
94424509180270279197132531197709528509644211743562111344921855 9741
50032173318353936450531168240895255587235567769996006616343536 7120
44309317889936143032231397631995458189928684518867601450287861 6404
96398963223855972809992096386436148344549679763000000157121901 66789
58532873131714919840092419208372795146560001206773851091034592 7044
19694212680681195271392464288810497541825594021297910937591101 6923
90069853775777767163863821080587976882080135042397487708543068 7260
55320741783682366614383153002846629168735503396010823170433260 3565
60191042264510014838571424642075476134679907200272224470135442 5718
62494405722039145666804170441307367856434040823544338248908532 0534
15835495729363648846856074162910291024752811252425885908480892 4771
74279600788489616038392188874479269061939407312580173081717410 2420
49456365203668146195951612511227851856945786551174739011405803 0577
98962782163205495167981968798388249901431886103830891963877654 9200
02632063832111334596193616997999130124415872697171199960826656 4325
41793002141813945563393428435223289809296149706117498821858294 0185
80212776972757986097746985149080429234314460689528594718186717 9524
64622737800306740792901093192340956819658744225724238751872561 7283
04183056957760016929963440755160347284570002754323604635866252 0256
23287544211550636984137598573479429611646107602736638127966068 8101
54695477049361611748018937258544250468359775911900615645465970 7598
44208506495886156746930271829910897905773480803714438535404301 2867
06918006561278739887080285560932432778704526963180630749004539 5080
62425104747501959374680511113695935810476155047639798261863444 2833
73558289999399211137646590117969537477427327801925525919353499 07210
25547252986196317690077796077523106975843992659953268590022575 3361
14822368985993335437197743430568638664914001070459083333443427 4059
57444998340552198395713577827633270675724605205663133633446692 9248
00000193362308503208155318494914803344738963547907283498041355 14056
46680327960702117858034437288502989844846632330433258895554029 2000
96315817763145803460572223595598714058850970047593975979410346 12584

```
04885254787999391484113796872899159120436277118581525608569 4043971
01114578849331313623408053537672714730560164745535460167 0668797799
97809966251034371928386327443336374892677656037020263475 8321395295
36566628505760345623761323609596659453821082320862926649 3946266748
60711177767854124224685632171096643202580065271002637085 5243187134
72325966685148202637413640271685816441332694061722131778 4054191850
11087952784470428608920403808370618655661058854743055927 4613038270
29439065037145643329779839782646882121344896779825574597 4631247046
37966782828140002708453888240106292126606906016373138249 8916783834
65739024421594110150473887201143623280464909599237215711 0119200653
64729154366878849017377610811577379797921407093339155405 1974872092
90433840677969386770493684262696314752386848130702024948 0534248761
46748585752792874649144246841836172046227320120338343539 97595943623
54790703024285765856030548301084703689165861557111128474 4293718959
95887664418201842559962542245846181974078526999279592665 2989604506
92702426177057532416582773708791837070629059768047106470 9241130430
27781281154949181914017664585554879675842754306975889359 9287709702
60428335344325201846389094766529127487715944431994273094 9873494374
85014535708717124372747495559041689911752139130991867353 1210873862
51720138204965135051179652796073960418300754723626538287 4902697962
68341713271904633168083668346666429616778740430984766426 561699851
24434104480284354374694557330621879502560165673198371051 0714676042
26994899540198944333394065848893162831352975788593703298 4944788516
71629192030547492126436522611761548960178627141814421826 3593589880
50167453894711683221662533262623280603791848004226940663 1041692014
59740745237877590605899268580089412231371875969951789754 6232806222
94996200245213782661680893801910855730601438473080346105 1147123589
46920895306464468834900076224542939189039422337172448402 911240010
28730827969325190481132703220439618881622125677170149165 6952680557
63388397525755108971000542235918145633430569087565099650 7429883818
07047753772164953535139341969477042030876579576041601078 0580751142
45388938818320372509485181011154262854588377635205953778 6013576732
52449582965219019552733166019364499601806841589982660053 2391233931
50640301449559463856086186188302456426378434214073226066 6123958487
57347272723835865591159990653901545993201580458761533738 8331230331
17483476982287629220215706921715376725852991640186842979 5977101331
44811572580729277661592374370645522509810730823313260481 8769051888
38166864742034354474137733295030762783969746444270150610 7431465258
60546566437560319314153098075965649400895759290102378749 6747970385
97130461729936470991188233759402832729722454733366519276 3449521706
30538006593535443441466430527057162558018359913921470032 8679653064
79089281783564669368726811460261978039149652467632312682 4044581694
57095102815722707715742381799956200552280885301394456442 3819559528
58600207190009505185198667641619593650515149328281031890 25812467053
06678311974639818925010201512648866269572853261935936757 1810966747
31351715899126601200445792815577939259058994805010690112 3778633973
91842315529894837314831823209631198494452380818917489768 9029267685
57576989284533263786978614832264338577831583707583265610 51498160300
39921035905106052107849226300239506307565786624200122759 1956688887
04839498590376323960300024641750340586173254788614172806 7473937800
41009730241452500000148825629927355800684748332232795360 2489766732
52409938827154146722418810278148020854863057998145526686 5726091924
53672044471245098119552818500926654283794468936558249407 6819857992
23048273396142046758466049591685206357704208810460801575 0788496638
79356883050184821644331273061535182550748886087143000044 04047142842
24922541108072157395792940714399870395600201172527054813 4221678113
67730366764141864425569030104172096832381831233729627681 0048695703
97034259861616202657103377802783618348318061806692793778 0334717519925
14272024794730875959449057974545320325325221499157507119 9553011188
37152534358480658330148305161825628804280005671223880482 2028233090
```

127626500279853167540269025072004787882118172285919974456085946305
954807213745531080523290932139754091349761753002727998896366798057
691989330776557066630976029942401850392561189377814293196964119075
558698091648921755712750004353550695878119726564940225410191863405
851652932659192339812646226258047528486164124221924258498527536250
201864440983991681532325801239027433170004157302834843846489472691
164778093396529963817614896793594731059466693516860843007259014861
776468014212158711599577453549091408934374639492427341238452586169
059243727630765343797017582852338533984276424858622498803997755077
465387523993838688574903189876768352316937493543112940827017016118
664361558920570813126468947903430123315768165120583433721980194222
343068236014735404305913885951812127441784887927290202202832942933
508796196272159940599142698441262653888339565715206351239954588823
845374070270430321773123992726291107070562208930405644388736575809
648665559421659246812662229242810998341245870779924446650112591816
680289960025948063094464880845715290665082139778206152755644521372
438126247579611445720423654076416371760243691697119182922751099049
908363378796274782153898937790571079898325410163262926714431762387
442013603427640593726722810000061942571877280660705654956265656583654
402481554252319335359831075925091386646533688304458652944240244181
241461278800958128324385902554322188509669372449102521554438229254
794112108432664424729855037151628029362469946768872154265767623014
915625447181759430280795604082463896114739124401995502817826862852
926534666115670676869994317156438956762152277126250798747416172364
114451658482153940254558203675347347761569445291846971939814978601
304981462300481984184969281363029300389667036245130062245454510090
849447717525974265546800581046440293392876918195905929668268836775
613922536061424073152744201158262561382036883502643110222625652034
193245386450579977314673967242549645753595316201385589414278516118
018500067664923288995388971642283871352531142146883351447940916498
384560830928801390065130321093115082348657675051769173656590118216
901138126425869129308117494302111978232230644581543091515493798241
809292694901225815802671490849967163190350691695785947843362093672
789062484236440503173235294376090510212835896933624091308778992017
078112228934423450487238314935015014524196345155091723523343279 44
942568830675105513018164772931663895290317990217648822033129079342
082746151227362719085887761456759815123197615291344367402647867671
606295496822328234634319683270438901018409467013392894199078350542
822143772703553333409247281277167123337944748656906339598045207729
119091035556158024691138084799665731749666205671560485977427423197
149699984506843153872064256587603181169084133817706639475988337829
611859275845260724782102353562354757145246033326061605245138422772
157272776497398395653342374636921236527537308826213777132108174099
840157531453692544638531187926796048847949103511058205783008107500
605454574948054701613244512497165304218041753212122646845356439931
483768238875780938758520326808753213525690289144614698326358534742
860312473614162764389088985940874026779572371288069023553404790316
040847249352311644625720418887652722474748254095302792058087458122
631068778714487586534758543478050822509733509366067407626711137275
821697304755667430252150117366863454460777096471656188930365611 39
837040205025121993244967156218763899660577339675321382431318705969
326129507043238547472811900460521943705529459103325321919534198466
424785806926167971271234225621240569016619696299766422423040227186
212046543547976597198136188485275562577337331424343569391503044382
664427974084777020667959896506686382910473100606504798589472662781
033837077493911888740462418343959051672434166891873803183385611096
456754831949119653054760582069520597537377872779969559980811312969
549065704515239906463748490332351313172006686313791825065869107 1649
130744417746957422325226494978817531934833864141538616859890645581
262511935652457295350667377283664402165452746663944318459742337144

310536832507224850637843782481018646205715763218250621368485322551
665213716821683266021054894550498775888759417225611700174416694084
213912014017625671140599869234069598323361509718275598833878135330
123925785764115895118619412808913923103122143521328944659066844104
166289090881594983482886162614213379247506761179389018867329576496
103014689500945823234932588518533026412820132652106736935963083400
604162887837423100719365651370099284312699860074066865214429629597
695295624424581205396539076490660250823998624968133989057371875776
532224920684642912208171738475364705311892348308643067617627505864
863953253923664790067220915168513660615304715286262002451516866558
694092628875005123265558548758099288417203226526924089637646953663
539882603350231873745523245505344812966326500081997175120389381025
934087512353482511358338763859557623321441821851946554029361536577
358708064904124818326570687620973418587109718909839032538389644146
190020799091536794472978680687397990966855811369765271053068527617
807044533843297602911153025354941106567345657966965663422279581251
681494499176716416814058599860574296121068715065134929540235334146
988887458549167518269054881199897801068136904875464456187814511656
564392096144659300938884604500672426923088847655528562816914493513
309732635612437410321785972197583237714575773202845438061804655779
772283334521003211767023229217643365986554035820577972905823424558
117670516872882952454879236418448952876522512885047808762567045808
001281244037687401694921335807292739448150979914757719561879589276
253048616366852791308189106438917028775981565341253670234200369487
652133027268195356618641215204849101020771575204387478450001152267
791808064624112034549940997916806987512513710719790693913592699883
346477152031962116928538442698222293996963907795705644191509106496
425346480059793092854378598681383356270194766738065661688635225690
313986837737038764612214982990406838915649079373946475695943468852
731045052971288818773442617122536818738539898350572123172919902288
928229413092205107223993128477049604452273180148594804904313993265
390115677850524744785620564670254170606207401055740039571353076277
814116542262897935234798011390614854226600729563377495025990243220
502390594363581141718605388209292196167399061829697521937169417610
928928803997273146698253552570641999807891940830403734763234190413
682967936732668121803241680268676373703999399274921414316731 50862
061139390679405715869113708304363278281895490019642377947298188389
506403295869348226564169971745439720443409163068355369715895449065
096925850049666278804813270762209324969756222629955568438223135878
097365592550532741161001752591390021386204081923590453368102009180
620874304711687159389689646273441608595762010678692321578625654758
059240572306424613484610566667205918399108016628476466044936925616
322928425989891320621814897816721742924375587684685729778480900985
838734265415048236105366626422709864967420695819437374831271588731
474387300293061453071789324915156933167202565106056148613832615863
222084631338432378165573213577861496491507726584120908189878353461
562278789186874485773979148571965347198375765928569616271631632592
823124213756274446115934490219389991657895495450504316360397112447
895656472094463956457064164381889822543784667301926735950015440675
358697563574143738565779913307537079735095015056053637672574540113
365235510359922347836349680421395946432269926413928424721803221161
784018670785407067569333556715874180208590559894265160296895773519
115753453038970965193391880049558320953279778571048458499424549716
326963213143779131675310640407743543574585787080472921189659606398
400986533594977345635345431684261692206115315497473610077283787237
639991182126073891002302948475915950420425878831793364592938928437
503485781401505118078077842727121938885562737285012545078087476725
112198202270340554794539493502668324760298358641554075288254636742
444887906530385276174557838883680606933610639749258875932900037217
519623836714363154124207720144790060934505154732641243170402944126

```
3412733541826441428632606393550226755527331554292774630621572564 39
8218489862380633284988299485292778941658972402289638335800787604 33
8558629437796418615364155683466708554414813755140596800080551944 29
1260491951181979565585028796018999041308002410801409425095255000 77
7183707849026742805169256190865165704160809631231429688543530200 00
2209803320782772890383598702922437907639396254093780824091279224 46
5091057114222435121039570521714064087711454184803470125393597815 94
6969772968514232508459206081938358682694571412777885607520742180 32
4515856697216254096255343047911260954235672197426194847430409234 35
9211205688704632127511335396516813829093080954442474318534419962 07
9490474147161569571036906988735746659546332344994176770372819847 8
2753974898008411955743320554382670142771729622531932905949705354 7
7451965665759640391208715661085175576412307028579662436658892411 94
1759227248163264822360328087064032352561945527440388289843727677 95
7309024384171481046778603296388121527344758149953540728150529699 07
8902628315918668262881564559766055274345191567075364944194991572 83
3805087479435636124889838076892864634417380297659765477430299611 35
6275995041521405369099559568670737994737004663597801820392083042 61
3301793062617303256386685895180747454780055244499001052071464457 26
9583556394371568740171344032685024796460890116076899376900987790 21
2792376311885493910115127109921773179104372793478735020971903338 14
4350629281120006357430968199686180040291680814854872932935081981 52
5248786598496678601621996047168864328670592568677653628534734982 70
9445552933847889785894236144131838283736124286098497460335897480
7588706299882282467458075824187153076268360200886831950110604752 54
1813889675421870066476096520301422142299866008661810458580463193 90
5332217595390529481915799626236542941992222371913107586849978856 17
4918007038774554093509085697606819027223518493338489894743277003 97
8744319036394812672315531905361174281732885966688286910205801895 95
5118159330774570303621371551603465625653873545666999279710919711 32
8596550923784785936176549513512446093396597621683378076720614032 04
0497303466814233408521204471641670398884535365485467638673801779 14
8380601104671229707000095731374256320586156731100944645382336185 622
9918745049597905574212900113390807312589385291132075609060425914 44
5702432260166033138610460270274127500601376244641378828499985885 29
3286198737083699813942793745177603656220123399374853021588368312 18
2654365964640037625781304813084797077331092832933427815953385358 920
9161353906795381209572636346571154148823729718258817886350671274 22
5256818961823600065572422444163665135420472318552146828124947107 38
5741767379027251369668102257803310069251481020183093550133753037 94
8844960989428676847329400661727414668387254128733864643936018339 32
3701755834521905775836975854301550320665471443202370465680256659 18
3449467388168014192133152130508152900135673739072748930563988898 22
0981952076777836121868647069789614950202996275555517619939425064 266
0373753189548627614458294952590651399819155408898465488081352603 56
1309890409213931766809370381007593487529006352110717683259820772 73
3546720675353067277016542354120230187627452394606860928951388781 7
9406961210165551488203229782841549872083669461943799133315166034 41
6073589453776416941500038167114749242735615772335285697910084720 88
5397770937025132583793389332394255262614065308974229970087812957 15
3277158870021681374488374936463490288091182127811943499369755723 75
5466274450361891679862306753753298915591374417012326145031375828 88
0818406760780902768557381138429407677458899932832108459815149507 2
1490858473470101558728386864643208104362736504299831465464593961 285
9304437405282010850077808272438336278932961846626003534212092250 91
5038140569437617163108628408148609597260082676958329956597570383 01
1185080672489987042649328852223497578221761425318011055592464002 85
0080083147225611919408233975551572031473444939652057571959956210 28
5353411093478549553851557201674541064429015955273957558077796885 08
9147472613943242667739130421354573852762767934961442562674473956 23
```

```
9416703867214161144627100486088044130127811196995895419977748254514
04818698808157540225593044163210932630579137856366049164196218222
19335661904249909988019014026514773832836162355521661938981622335 9
15298331940331278197117310618122864552169304495815967203061744146 3
255947555346812011780998705655266607636640577927675428907321597937
44511084714254820280355357755333084772294467440878313822263534030 8
72185861667393749806910746809711095615906304001695409930790065784 1
67369299132574465092187785120800809873212773152277379434380470239 7
74011025529888119003271439010760833514557142550582599544174131001 2
00046664272236365048834213429036810127266096493663725292019208092
70886043187779779987392207839803207556582168743920522356978963624
02579549186988315443255286438154907626242131970196488909209364637 7
56408661012130246993795975639994392350514274583643842849403012520 2
346097431997471834540732307802426203228431397292790017975212743076
391095402321914501557580744772293677488247863792415739611274922245
18318793293999730978827785083714240159604325720266859506549179636 8
60605887994170806652277502001489953277310441496193088084071449874 0
21022410049380307889967413617524052936900175544607854015838187555 9
142190632385263807921854796226518060331577754843461332975803604846
51743948873984767552890735404167610493430886319459199359029237355 7
14308991044595110019215160594671905920154063474253922890007282317 1
92387080959383859553208257290115033269206350171654400468043310010 4
380085910690431937731630540526858992451189498894608259977036536993
74688892992870207235331525599411878953342753106902330731489450683
3251523946127742558051249385007558868686628376985481942649471370 84
0737941006631403884576257948588435398275175755929770737255711006 53
4169387013025497300308987242781881827056291048577384620109313171 49
80267752734942328398698797510420434489397521753938533186055172732 7
97188256760905505139603375557190541307334249721556392722618656933 9
11787860393499675544863057052331575020778072481587493275567794317 6
35301980921998615866170721892887918587714500885260665248930383295 4
06665921033065900402413579109128647861353490273018718902490482271 4
31143281463226586052309354419817320526231135532182886724211591076 4
140461478696723337424880238927295168641061415219699435384563535203
67941398464774830193047146745930469515704863475826789765444706293 2
97483263304307264483901878947414406271058121280479600260429778502 5
22374654625144427343156887049473264709546654437026734290866120532 8
02487138986882808798349439102173699009231024438149587479818768509 1
01113944794171759601496037678406281015898541765019369278195837504 3
07596414237022838578244936867866421040078193485912750347193598314 4
69643480102669187450877376072754156376441193379666163882328827862 0
41739512677356840769065103103729320626192342093582043501071090302 1
50259665714164925295890786175351734955783637901725309257862732391 6
27513267386851794764367654155530787780448158131109679158857800821 2
08629448273239760103324675205832212280510480176891803273981073371 8
27345943229284803674560009095259544402450326127291919329013876437 44
57968225403385339841839197850473684400117869801602488072881577905 2
78266437424299277077784539242245700449957136176247219707542241472 2
62299086221806355252557799512729650400720987974996785381858742051 7
34739097177279119932171296750278216902437495538129222474345758530 6
50267840149230297848852757970510524393282032352780963344896936056 9
61697544358837608620203346465940586455299664726895712057221739739 5
36055406101279363256838952352160018011998050530822656395605375960 4
35766360411398625038360451034629627841606289194646227553535433938 8
23197848073163598330631908740530160225053213317672241598113069032 3
04225754772537711495439807184064726777525224532322593441643798090 6
62247504435798436764287652833002953018920294697113194283276953564 1
16055353643705094450980842612339049179399223813242964588881654783 8
54528501269197476033394822095531181513680122144002522993536643955 5
77323311922122338717347242439489129801871697560265761528104003817 8685
```

613407455575531691709914941870737340540099850007216027812661979454 0
335011572616778933214763212890404755828534748727923977233956730411
356357880261508558976416000468972330470732348387793883137857830681
035775913141515196536124024286942153672497961170741044279452867235
677389374261342475186729365548111247250287907951791639089802084146
510390022309746365346287184886692923760380302538430272483553157144
478568778643232741052357857999855115235793688002299531324306551791
874634201821119848855120120450396475250317686306480863497270511360
466551718477269768627721058597390274066673257544998241730786235225
028460425798906433852197587526164415798740901061586106943346276874
917594857070399448547987822484745482129826057181565075513709789856
010983165589031704182083680789911748474624851371357848585325739 97
024793532299678399460711146218796032740450552015417426472532451027
071711495819930536110636045624686399213296237378973938266990826 9476
194237323149174213482430840533916959713904951532504763973836445413
014731492741052269028134519644467962445239294180088451480094316322
018426059929353216202327989996714726620585727755318882394052383922
190479136250902712655180709863447712118149153672942276368351729924
626979350807158248979050557765793307851684827623233637715526243883
436131495496098089999489209101116683455704951204159394640638835451 4
371552789455595190684867767566977004647007408580065887533760242223
750960849595339580242294171504086592851722899712131970668079619 56
137891422196493562955640404393133536485444808529388830399857073054
123772081193407885631362512159266937418266337465242622361401909586 6
742587207582392132762139033654767029093479958251931165974957522 011
697113117246971553132016011895454246431250907874939347654154846498
033813271878072490770747606713769357484742996848439275878159542167
622870903678567912162652800381105749694353271542303686347648256680
124099273056833792622173744449974854734605772526464587978007468015
820480094367009450893765771883362641458132655508853918314174085293
124660403928008906381788117922414330553445942271246572512529982549
143630365860250846721885298069396135445858924351673743268434840529
622185708061384966814191280737156808754003150516310467056384735936
825479439798878000980741895741494569738244089574486602136136597799
798046784104307076953951318018296306412963070359472023274225066364
848581340458269625019431454098407853251619353784523507747554820805
178044813324520855095525715083355330660884410739123511373875522474
868632821440142899823270172392279373540051689702111987819232815267
981467627184969789710540232635303959376494722737678565767352742910
098163467216743242107760502944022583763823150721036432339832436480
848096745167298660885986779170418675612197201448129439335436149888
499026371094411438039570410121124628678475101220096713263979039029
910024715971301265894759147675893775163729807985991700762362485162
600033568871370072560643916392469164228511488687920662720801093956
590597906842816732363402860806793303241385899186169238169227861039
192769103941409580435815594388903960703089070016417605462671339141
685009323802942498048489966296755693125290095959086473147339784937
630217577305217040453298774913233517019051725739569682897813322427
892163805629187727697698075121487758147765491287503349944615574526
903929485508508277875911742733300207922055201562839378040221073785
935490757760288195957825391983903833125644640979748524754628722097
366338787124945332239760421365912995324453983435689366563570191348
136425400229662982034675542289129337631174427895190587108193991794
620606197532853321803663574596558870094569774272092853484439954462
636830402909648876782878824956793256647635359535912998343917001821
914289028334430573560913308915344883786887093427697493531332489960
214895147318625551035546994503382267997147627631537795056021431179
379546103882884112885460159412320269907519363338622738383875307323
204093751761596008944026902836972789203453833981658057737959711518
880923152296875752196266398014111766966168262173311570908691421899

```
156199967839352562369953481546551054202483683010513761874109734291
153448460642855301591527222451492494236529636136866100406343167005
888314494771565036900110309962639538054831153339282269848436905759
274136389124751991615802787850786883760955916300876229871415085124
256110246006953522616029100483065703439747626118535073206221931184
9087227793098639330603228239534522205832739891711223786631389246098
0212755462994586174729299736759117542249792791577817298614489398
710234581982102296376448522597298600468382363889292350484196130154
183520436052075257103041912083744591355562143011000908750577047781
992188011731892042090754106982758872805927003002943515162008229520
720165605440858120622092458576086983450411702364508152642477311812
73447034191508814863370995876137809885563585528492513929817276610
951960686047698159472534300768450779939597952669401400771581779942
808160137282330408853748988066335181933174346443406499668879260398
990540303140521856812781712483123670325580771578721263775122551550
478588081508766682646811227212961700196278768848372044308328034179
946982956536377153852716553866992694960437886130961610241834798549
747783063991262638166623253583048965520948244667841658757533644462
299171226880315296282752278891151226846596051303585307545460708695
057766804605833959679690307845886956038705284630221195005610917183
907644032772678969484614617390307752774229968192455395489741471499
590555215175364469098710622678332167426144271709583786524215585281
0339708868253684441042527207838112704056990732753083859318105481906
601709194930366444548704459301726222083097246084003036588249558243
060518201800532123747433354886433216340175235692904644018679270767
162012802338372887722491351445809981380549607227680865345222490765
852736126343905983111761248582921877146621497169955794980051001132
768410032786743790375823273610926511306478911345369925378470567058
073695952411850343522668328334389365082049721537904179515011444723
446812568207430436662614650446763082107990791686191747463197547097
651690122667413580456247344557931399751781600362252091683142606527
335183689226367453671508957043115997306410177688802204647508195308
854527188480740662679756664722308590762961250807071552694556812355
994163914714236673639837667497000296467480434759556288583153310371
852186819562901571658049623167394365872464920403775452117348537229
60684057386259940431028958372388153855678444950366448397025527393
3774917177259034769757222864295001283795097399232938542984795298669
670906402911301556360881669629593408340622859274743404491725894054
9901342685240420963204079667711681988706060363768197063743231230792
043444486859174043187594034600022718009381402297751662708222587493
905870267947624679330505923216797011069524556203245215428815289129
993735383985376346948874014942002735431553527830522334568538458791
244033022910425037547613747824765074150938277737794136622669230101
683460475338453757982798232329024237148986126245157106492015674332
996417861584964403429088042812656885863697042096275973877581738121
21682166733996255101689760812921678720372769877314799148078890503
769730571510278125520161256145329162262133802006076477163033028429
844042226750541470563099785987018447612135464575594915696312403513
411408662881044856545201085191448195655480400256188070781154040189
705998020390954507656643404337629525523781854022556117339673005332
888603277682092263371531923074708682249136398665859148149120960863
766148692577158450915243061857320235143207168796142315904613178661
856028172039475671885051218961291693459225559001000968448902884516
059247010709177420726286115474146724909047847779680857919590178399
489677146322482666943724782816670931396636780028261153784181378658
122132855399146737086347800143285885690768263357202626958122383132
890793157375604486844998806658499992618932631898503198663668571052
708536959657959838837728860736319543843037159947980591492863594267
047678111360815849133560813847520869710242375753491349049104363858
046356006703958218693735212027321076311170333971297802166719559633
```

```
08211655613724366813950807721073928171755724844400770889979161206
013177956479547071188427943077947986740187106173982476955667919233
889793661390845572270708838693130781135240598407802127920527231592
178594429900085850133687434155746752120410726765863821131413973592
009494266623262467411593219072033300675788248907718660372111309452
476024539323268737797863537598054524104208713787960015679399237217
697303795798793443614322235519617743949820848966146004219899660098
212841849651883862694971183535583039648655651757019056488208949145
917925609412417546342856165670841714843683199353372809639452893386
417914673437906707663859847812844563994060646651068713686424266804
836484147038872211193267391713461374335491692021045040324438090875
346783313395935676377798232779553231960757013276103064178336951792
49970945686333278701540764728338027817918322704002009447014610178
462665167727977877556739031567845898300450002629999503000378443770
998799433815354495574391485267680423683978869307493133606073028413
367466937289976607486394051777484069220697012005793689979070414443
876603114892753119301358594196113341071571986919372203448399739378
06502812283824245052578701366191242179795020804250365861172736037
195389499545641345644552631305820335839757630464324604152545817209
469828023563243640171494862165976649523856065113140180966394791262
791017694508030322292235412476340608890591841490842125005804633186
8578277781885446973621764851053532012786220142366827154888020203992
46778864772497497971166097419865168618262833326829870680176179543
768885907517739469271679751507137795481699065238552970108039033847
179531258518192820392130326781480679649868519455775024392314084762
793127057430970834952493884283007651777916794280852571783188315220
375074229490084627508889823940979805609127505028439932408093696043
527859463655774807378891315463809249577676501427378553441707747191
569944899470573622961837161877691839658654762944810483957093342977
24659639457639165424807676653366740527506767076659882202998122480
696053328987259984934639809279392892227879817303342094688935968605
214582162940638188749452289159384342985032004605254247281781372642
588044804189377303593550854898813997352602422083998184615396711836
98076705547050350898333854321388958753664338783049639599117833548
550713827508623093676613475494990466141084960316113683991814451855
878572028263464534079342501589512391944964596678051060721944152527
853149030075668965551375906404676579059621453862826243849351315463
725671174889169746237102040478886658331565967043501572452835602413
625523320676890043965075874878578341728686425040109576108833419626
984576238051381202370889045956435541210380772219011531375100164439
003699678954763352179287119135690889050315984289230803933836592933
831333474558647030382508196850426785936375131288632307811009920206
591179994018453287185246304178754673722908539545911243821970821322
954108592430535593238865360643790840217534320609321838559992379233
195072503989144941883893699838345866982366731360180522407728805039
954787932183762724104885448518679474774808736990064478288811345030
139119213036937804368199768423694957223564020011914799238266419708
377952748684032234604813498637136652224539793015127772722345967546
898849704725984541829379322541347771193497123286040516939120252253
128526488795159969849642603615148236106899260619308411027405492925
650249201805828968279590144481920848139949388621567222259694948561
664791331853790981919105806963496420496802349178082174069314271440
611206453391354424362523234265527030619748861651054683268263063155
971233965696048888281346045328698787859539421299446378326801409914
056548763574854741869795981282384690861421456838572308295651286207
138868234255073386468387028755624074629346762147903776926640133346
753617868052292871377068137427765119279147538184137431893632561898
425950168663424901645408230518330599477378413875226355405208853643
585519855349270988666979370891243796941538983928816792684373252601
3109213666837568299523505763079375837645417518432097376493447683489
```

928834438271072222747488202215411029354422792185524604432405347830
329956079948145295711453558725655787312906275597160475849047778742
947862232232492910574622675568754295686560084648207328421249489831
887010190418727652437208864119488519409469288518030184851534687927
529429188190542922085754016195720533289613718728115044101911346409
602033111868187711831953577610797923718771679000876727517123802080
979135704277414973609226831143560576621319571048000756741949695574
394795496953790093999990840434225025140348537578098387427168070654
614945176697470090020098305915465299792065275027919592985747830319
040877254387524923935282808349919308879813420693638998594646526189
567893324181995283370882245590435301154692468701917681378791561268
526948857565646971356731585604387037069044331095778692217550753196 2
971479794796674724360883070274507365510196309946995135987243588486
763858699749294496659374732550886786625722992535012112960175686906
773238956011506621521168869391488016690437008812832571840723450728
505110930992159732877233841999789412150316807027042751972130612059
666969214207179347711761118120828715821238716794798146086324727794
791998074334226081295863165296324835568105491588679674857318859527
860504427044659389707390968052840298506298308662398014045017833016
587876504152554884923754578203702168627148996488656933224422103440
770122158889247870415136650030963859423052861899181151940494323524 3
765268152427614154508267463702840221624301212322522382907077848759
660106848866057327298165445747158115856437370251160632139271800818
516170340249555588977499867295504177497241104870859730725527964507
205253331715819411507403109603120536648132310758903566813539071334
055510494126350565127507438759164095367609031191475509536721558549
231143092330693543283090011803633227948025326601680799221141860042
276548509781454293527915134130982535685754247574868766126842459903
988894910736302625680669525277541943820789686305046418477215929093
631275770821233820093357699610540014442157022932283902032454512115
083141909482683788690098571975331065977845571196983854025896409129
758202323809674314065541071779762148206600603723287750923448097104
115030072436037838315756873763358175465349993607586165593468633857
754132221279839768122575562637503108381959939553992848595837358067
224044471635180153824191368264165610011177414554753238437967178325
679163847622964008429889193021627671276169662582605872977338222963
967855583995463911630179325172432684976006355692478576250349432502
882986016078242144938402780575896792830534145813993634117905348937
961784478334670570220681128095620571911508864161672526064635299236
996281114504452105291007441706893930573387498138372706097805588752
465620602170301702331128183291223047474128019665083168421162399620
888599190859419986588252621076034699365272557818916471027444584394
945611596922822769660827157239615990900620660623367728828962451661
544512516297938267981789754424141235382840745094598883282737742334
592924576282896188499864964698454915282555949784271924808429875453
657261432391983451116836306213965725393202022836223898771099721699
712159507412111669578236767662019597749871577824651534104845357293
942018615218235772067602830409587894758066674950989526876804090644
614720921965154314625137756549347433415446487875536488477923469939
443703217219561835448481003951593475845981996812634354748275624005
537884358208667815155151852870905553501934762292715167032543346333
153437276465068363439409623560287277800479359959817041187155135822 0
568362682243505066039246535844519816958528474414028273234005765193
29214432813667031127185641912322014878093510998392570763109003951
901050644177864480010127284001750821031467428245957185762324473088
060371467744408449169044647957667788277945124429340633091424673862
899688903618420332658961907826512991498762944149674781791896141540
802286060430465399521398026256043519092812152852622195355944670772
502912714509126224222581165627072193043153104214229746395395383325
398880689789302342978314785370830991230428668753207966279036606310

```
80481130885805000060374421687029684551491570566411325851632723288678
86504487112622350629241424443598961006878287684754905021623906139101
16534311392611708902366792999100657988185318232694187034912142573019
15213472108420794516922461649497596138122907130248939166596018194331
70183343477635181324627820188637652192552631528700999830826288181479
56598143014211684229992524074255372356155005102519074843197534535996
57791916243570613302807534005051951962957266898769515418337921291805
61662833709937825570943353122175346460934655349339329099443131495166
55208604965455048396696657084766820737410978140704641121354216022562
98477569817188244180570212058409963668886534001158133044154576556135
91474361559333142575449394634144721429114694339985611805031914523796
15377054951007770122557269165251123345583133971266343990434716987944
45258822842028849064190440895006302009187476526265778503583945987585
94814498584368924791189130977606518103603950257675177218216736360771
70616053613463093796052683077101901181007090099937839591249498909817
73222747054293272771762942059862768371091123701071303831441478271680
54285576859920697465290032380928424798920528004129435755190671681352
84303740746999432664156207169477008469256124482760266773303249678118
98464839070032615807983730214284611714209349500876230888716683135049
78095965446278561090678248711090352771350307723824869030244076801836
50972131304757068996304964849048382783843182664189222402831421907167
33545963841377769508635913555217305070038566630248067338745438413959
00275569907921069694697146539681632317814769256151511623125477993141
66840116523063005961422702543190828838890283590369235738825436885039
11516809654104258554876516505531113010609733685353152180197316234976
33223368315981063974431922856181647872375172579968536068939707223630
76502080286719931740771736679490759153196154217404234662054268774019
62398633091212780387010053214607956454088031284214409721225291116556
84075622078267223957109054053345710248973509806078810791856308983291
50127919406813732874417578242677882958440014266456092154004253482767
61355296739776530191073971287403917558969008000524603380089819964053
33720005103226230720087892858747687291586102134285590277972022602964
18507207887088369747285716747033403822105788833455922797202759124740
52979440289565200333391944872775621203316564479856445248527378784890
92999884031856047913071780148998101617263975132255835421555376895401
25567420710779897207991170207255398164233717656691590204720751513629
60591981733163477584625469188907028777317405274824328063060237928775
60785274591690472328011583213150370488063739736213794622632267444378
64091718401398887477873026102261098249286957537818486860240969183406
10267874623283837234252332119169379085945502139876683331842205227922
75981129053728576639702186841842078752178593466300601611796835197405
92512268041379041374363485123476768322972751709670325801162494134749
30998552808755123274511520113031452323595772951855963442794885924396
75500410693163939753653049485001498987409842029086872810487121924916
20907268440842353435458518179383447566772087129339339875697969302744
06862503041505194319607053296371035043515258815429914750542962218113
47483972523641574093627720188239223476963381251449399055072376408625
60908148762130934968411604261008431518408157128282654087282007704455
80569185530643287867186893892890889264323348866719411938418304431493
37759745780030562326438580265365510829505117996832422944957163945532
23737234594210875741872582753671657545587379048896788297299014015913
69353953193970107937511413828623701975926081821749286595541080824140
98275281920380978543686042838687922510573507066687970460376119548465
79363249955025687674720210855641471983663732573114536684865976681160
11865633402794803546475391707848824818119256873351413710804754644526
70577133734147124748260190193870058301962934503124405858886441773507
89995322204532021378767991766314335042659860372050497575902120865908
30826958790790387721273733353530679138309182966293454419755757889351
02426553821342902
```

```
49600982338175956897955988458039046932282582584112451640262937792 9
42877242666458590721290571590683565381319301594333584570718737842 4
73132193037081729564126909424983246846490200159201414945091243226 2
92894052689238253820826090152509281928370128233461921425317297476 8
50896634056774045819082666625504141901108009820768286735344302745 1
21790259814895763076572866808351608407135624363308317669890513777 90
19592378414766030998442823009019050409925585011638208616436899607 2
48917958713372465972293664411775126127906909017231514717708237917 8
63398244162700449278207371734646513612061849655726921437400223480 1
21711529358476760241352449727492628261875885377342982123508862209 4
22820321569963889175103818148292945548040859603471796816604939865 3
98639804593319737332442053581552244193757131595746228205084050340 7
37759748116162169490693308629821170532616320692263156178990056941 1
83240661268558269279376334021162201337000684369271395062965542745 9
38936704095620537030854086050191643239082441723154771262033992609 7
24368005981592804064900463214910884243006699491096565032088792480 1
15309017259355541883365182621442428240098358127839852943085471692 9
63219991052441828637066199069551101650112030728771908738942692691 5
32839464581873699544232387527416615717286062497875695195001262836 8
27666487650922468321211821735369294726360537666261378356735288769 7
97484103377107313547043471946488908808333847577987494846430487838 4
83032869079184168419247875864178845281337051267820141257491899491 1
03569585850489040369956667081480291729953976081721695168860875832 9
36823744251357359765607926745781537407013557746518031795159578991 5
61726338427550017768942502135067102937906782220801913805852457306 8
22076748369664072957440320897277868474953581533651295538208179575 7
56404592531435644184843788056525781335394014743781572109069756930 2
66797463661304971288438985386817946674549192129335410059238904426 3
20220740459339558431342919309028638818700969931795666829073067279 6
69689055046908570384005570201114404436783843061793320681429207672 4
33610723690145292577835916303015133134529953935598395135977460193 9
94319569791063197338290086042245838769679937099406618479673157794 3
25986173983219181085113286158167827282563858237239395543172164902 5
37575527807714582450456832098065901492082255395283492045560173961 3
26243020901474380011795469344630777932988041719232236414476479705 8
60596027786206238556954798321268974535782507600572698439899875580 8
30303028217824951078969519945210477985538930638544927905007520565 8
54004604218934508393088547794995450948701194414500012339892705241 3
23132426159818526424544305945082859047681725850020919375430813108 8
25428179332290352605739131360188692605315058464947078850337867700 2
47866659354581164887805238659515978284448436602326662232356666142 8
10200663351563455876785538180399826224834791672456704927137033793 9
53462002066447096356070292623402280797588531252772691528772965218 4
66220940750822332635035704091248875908721575689264910735434385592 4
07378290423811015586629449153536460083475126972970947550849675168 0
74827169987280899591142039768363704627903296365875774830104854030 9
09743323117068108813249229322506847642092368876141645163395226427 9
92505201846349965473633283638029284806019943612805478158768054083 6
50405857293214371283810572064306282421751180452589420898299160092 7
04650393372512205523179386114995098763164650700553980560133757081 3
30980886059772167427433961505494290421134860573366908865147798364 1
02536429697963262348862694390111321014799718271629882345328848201 9
75790234393634916228444043135213442202663687628486067470146617548 8
93223326008708420530440218219718614111949117685381370192085288121 6
33997690449471180014340556044667590126741760285057855077792319585 8
98355702572511368855361873441886137642815690110942909775510649189
32984599502254334019534976294959280758499020464104303618048055284 6
59896672581235409795023069021773691193055266957305330452102805753 9
91571252462734549227510985275436080964544266833909479039246517990 9
72580513352892803801917060560046442029346444839637157617959976403 89
```

```
9938151506891165676058410210410947221373509236575455640642617532 34
1896847005202701987213645330763387651371434143928185305810786402 43
3070209634115945421556797929652735616433336374897342820451837993 09
5568989076073246213071415575122579120835534429951784657897371983 19
9427057622949578975301614862512133797834280931811201698020661973 16
7982676927675445764357970472962151314940242291804629461065215815 04
5689395955858595805205008932428284315475175647273386225339102736 3282
3568092274663904832240237064997082020159960175854331419938345816 78
0737212398821642595928527084135494237279588928947792280878375871 54
1969917514062580315891054432459183988504100118425938963585626673 74
8718564301084215252289961036716616073178058185367425530168853619 21
9385738612998854574724150786131448932578524146196610406362185130 29
4624061767242538509940576134549323190420023436988994074839722711 84
0832056422734413062173973033423359286849398136388968730087581979 63
3876596699223676777025379059169819358446063728131958901816071540 5
2796763711251478982569295144996371501642407702808960248141219247 97
7155245324519657877926633051824580407349345722685649455619128862 51
7180463966464433347914919711441991545005858528677243036309148666 73
7214228467821227306028870428687661402645497646902816170452433819 03
3370464582947221089638305941179308642442872565960973902999443797 66
6090397086103531061194423943633143145184060747063454695272630800 30
0218226888185191902503852095418859785459938056357692139629586629 23
9611834559432455753013802270838748893064224204709252451619226965 95
4125381421191625412207210612730400427441907130792953668195666708 52
7349682018663149018336692051344199646852285839745309233425860392 409
3445533985320085465173631054300060209028702406065499415329451467 52
0467396780192689986923985154643353556188443980258567821368441825 50
5208782939966808647081534989822437028192353815778285559757483407 17
6804170661050421991660599978409477024758857990473024379558142237 2
8376798766387533636181472743347906381163590371153934888511092458 80
3125169532072945082252575019024083719401585739851104690757516727 03
3496211025646638611562937485064404768343826553812609115027549993 74
9980145126612143789152815285449020671124485227723783928796847126 07
6686157260697869378870414354247181740530353521368250856388248476 34
3396837659921413943110093740499300866233598406368446493710465279 73
3369761456447351596031440099434672059237578912534211630072797060 10
6004147776018710201394368376269632387593156065443212744483902463 27
8631372178783747239133637270101333641904312930584004063716780316 65
0486038546697648311374160718065901692327896781766223163536880695 34
3106707730400606005368984468596528387972969634129608971112623681 98
2942755970828038238893510777258016735948080026187792254656782060 29
9683805569018773062426316224035144285558300024513285804005725134 59
3555095163420882740022922112243721625345430479788876468039662142 75
5788035674783843185884280264511124766209417351293512889477685875 12
9944870409070639807229312543919086972500196942685697477077076187 79
0196510644515784640203628090632539591172221883388439437553233992 34
0682674337063606568105708864550176338702905356051770989389301297 43
9978382876790419359416142959800184040812433392243344259480401431 01
0263265374954108357675078434217269103719327803958051396955592278 16
4606862362519750595474337556362196588838289208863596429999037241 71
4143785129415138608475730630157949507828404034947361326799997101 92
9644213486119810036430715266558987767733756750587723790692647844 78
9295629662086396572559749227973863209930300651386380792027468097 43
6207354416427687571107678759566779581823385239018040938030070362 29
6834029059231185681571092306734979012321544515828356945190807311 57
2819838036952289091896168964002204421171618737513445350053979780 83
0857482748020942964460248096980318798240815314068498815762646261 02
6012766922821076571902020909984460545782198875529775779280543762 63
6420745822610596873217970008612386187349311585232717375959803679 36
4412162208525315550958659116221109153869540353638850072159065870 048
```

```
3443385075375643484951372710231325063588511234725878533437501036 91
5220926192799443014634464559536840890935894275503969846716935728 00
0762053703272232037610240628897689638769493856454408645854177910 22
1511612913446290299429561997674532295206433319724944890834830372 42
5756870748155778563982302698100069320221135925170481012619459062 28
1210982713524071993852029313707042690757947825927558347793496762 42
3331098337683346065860943950891049981294008927416813332928273736 24
8432625379088072767934019692376366117184416336523164364836552055 38
0538567580432070753423383459276799341963786986861592957796637181 05
7498958865912684609726552998802875947392641940456382185968240049 20
5445139160941833832182799008109747414860566886835555359635685967 12
7854680339451365400023732505299753529206048256356662266492194785 84
3786678685947736258093204681140291554757383568086462595069535719 51
0557537436239098341193243409661100425393593511793027247797781041 493
2285522403691981536721210847170793405158329187459743884430418400 41
2474786093868892486865893683184650459144698754687929736115283977 16
9662027858749563443896303095631520383274236829049957052877859497 32
4422935866717889272506402281342121963692876110163297532855268173 32
0113987643750801034129933932353570245037467306293848790665552209 05
3395209014490278693414090221029046525956607478579775057974027591 5
3543289887853284322873505974310832603234400466898385822835393199 9
9274009120292143043532161127309457122485065464888412077605695838 76
0514701609889141251676418824090983714864969082675813656525609275 889
7613756553052015790983217267120108807593018165003209977373633663 86
9817054185991256734241593373069586893615172500461494118450471283 42
8634334829489539006599824918110688954331859168559631073452258587 99
0830724033765040198088403275254074380502157668275886522265035090 16
5806833579119724863224257157123230641538341906168905498408502687 24
9321067849985324226631940718811442586592402724772979746237058238 276
4324010860667380635482380070348488923932224269690910481049909621 97
5072319303745370916150712230654413768899239736165142355703151238 88
6877675003040368564044476925175577225054987096490342901161226916 35
6204102077090740251505805705850750940180319059290756582588895802 47
5641860313418199369552675411042240571785335328429004855325255811 06
8171310291472821948946415285460130044724331284808216747052698980 51
9003338207691376834811102352033005533491150281368507793070421730 96
4801304296622728650540686553246343369599645705722386013637067268 967
2409740942127155752038880983370056024798648736448750525035733230 04
5069425737046052296765597403140987310278363186203909135168613087 02
3033364521479000079442976755907121065213965425459396408722909348 16
6631373285149474275016949304953164870290446420487890319760141715 13
5255468626467599357576700400554398180511486741929298959977312467 78
6376284340063947630501075383724610406431505287644986616609708627 91
9177170369492307107209551340040668044449230669590805656764534448 73
3664583128834826409835618270104114357800205039506734560494819245 07
2277340778690527997615326041305410359949103926342214219694600324 99
8121983484297113205177456728015143877718493393512351749012298962 38
4107177749000656952103345074441683245178059905957638704908449757 96
9984360277632481262268390926686002633108492236206503470579934293 36
8312385428636335902395129288561663754090803110854683013938661623 78
2992964192935927173015234014429890336668258735117790743816046565 69
8709833695963762971656347855170902934102342905871730920041468185 57
8542271975032401400632870142641068043109343223365231504481206349 53
2863740439177289941853693104272836785705113059417612011932050577 55
3906287786137179294077165669030942763699071550440087075072179588 137
8001140767459118572644379366307315220274160303272751810433407935 53
9878904243936592949404930712290775989290257787938980835703210488 06
5991609346744158444452577623353920879014085548240628465906068361 37
7363270680606591801537740568126155567744063954542611221389011093 64
9900041626742317930206092178107439071097111693457167589185327687 21
```

```
50199820960804572789047663912692666114938298554411578241713085978 1
85267993408899061307196376426688113372566054282298195173408339400 8
0398300640888799722395541376906540067631043695072830565837443726 77
42952115782653041284898278984331818880310641358086122842802237442 8
75754979483802166015748245949796835510481506580765041355315853302 0
69724565876804882274950314248446403768936476176380094844976363490 0
59147634781041164445332897459324835700205546339506037652126187393 1
64705264978497165139650094210756230977685995456016271416562468214 0
93960314057423769497446915690558815071970678849012562857918168802 5
89728979997429883318321097032792080613609605230327348431918600915 0
25003263483286747439933262863058421940507961882106693469966504280 5
40859463035447434312210352105850453004222842017771455724372381944 4
30944302368183946759377654322453266725795936435755976051555374582 4
50068026723691348756933901620803658202781573351743217730963388888 9
32299527288474286933419947760820481117209100395925926968277572082 0
74630234574173241823719111591973903101401584312639451869481976807 7
50283785986701559760299441904629373418397162823401804775905175737 0
65452859185814830532708045768872014412397874853699090218293594834 3
75820006226674735231978356641416139232723084955424570640656725441 54
99778842574921125064961576347304888804843634622171742613746766380 4
08722621081720926590762217578570280877835530039252939065378208298 6
36291562829596447876181117988886566815886223404504023146905533669 2
72263560441954335629105736740131235619873309551791754383220418369 1
03969131078669672880762661811315065431999872286134974556561773411 1
01342191464944700054019895484294132088189301315835013730727670484 3
31959673951739494117283554400330519336783396826046996938421568175 8
26049168035239046804222214506421592335797669882416125457330447007 9
00547905349207167581792119386110599259942512464280301607796067229 4
50748240362693270186256243077270824467647665710714437858940196386
28815543437151211596206899288097449560906911856935553644061504100 9
91544411505888430341723011102338940961426662017285978243644373289 9
72115305280836498151474628893276875242796497165589407236789959203 6
36259665545731397077893999477778538853201860293921287380459605032 3
61695880797427939016160603722419168171748830035039956776721144485
74306173079577478328192086272091910433503179306477310873507413299
20665597552750580577674212046446453356210996731051553636684521243 3
95615383296076720956916120255218122915167767079536468155199327524 6
65176675900600107752259289030352484215625076879125818630754736841 9
19895262622769012961640264985742260499019301165520155222050497115 5
61401982166951734618669074197820275504434033462447816601899573417 4
37128109829380747820969370581044846288308666534965833584684286610 6
53179362196705030861368599333250726985681855136604728264260889302 5
93221624277199887245480322399407230041640982934583379818021196238 6
13750358804252795358295915699813003923816432501771411408286212508 7
36462302393968396594689254330722344937247926891079009967134861475 8
25243216319340379902567088391714312027326072519333341783962284759 3
47044420986317044629854383593645320754535858353839434388162391794 26
27757686108258696205582889139198787274620925931039181623966401123 67
13140239896453463235588970818932460414267043942628122503962011359 6
78834227709035243869431641267443246781131457139227894195268738212 7
74508852538273873263952454350192101241786819659125029975909051405 3
36611717480505261729617284760432541190278891923297427544507006093 1
38035438816657884284640843110368721453128760003659379159867496296 4
82627618559352932788161367312074691499588577913330591154510640387 4
63884053520363909965802865691289991352862666860844024312189748483 6
92771632739503005354086258385476181998030504511718921938062826820 719
57685552656781506027231459780471238318685439575171114815413852005 27
07649649817705741795220277914820266600917232973574449099112838091 2
16309215785483182089989872182541661435738608162310606243240022842 7
25576140959286403225009185752335402520948447086814459961507703825 9
```

```
3009935562199105244733687530150623116125607217584206127373205533151
0953199644608297700965373245304851126271403254735452984577259552339
5829545700193742141678022863575469326109013584365408599387401556556
5976282960219080693433619830136678018840716628004328384584335620338
7008847551217759218622645091710829545635057331146730296006185711420
9650139563301552390459905181439674912953818727524941681611776066770
9486401660015660406968559068292580037240798285983432231714424258
2991579918036744574274729471557009182790650094447987650171101165940
2727238680078841891482524313243316126951537864352762806895456055629
0541056874623782637924897840877403739604650086238380517399876311600
9535277981714501749923244135636951315321600197444558137412397055949
0794916763807467127526368024609849390340982838957076171140934811462
3932619974758791318853971404788407744097078744343737868950446113604
7948546285199324302306990517513389138774881351925296957367183277033
1042963147922223534045805795881102335635892629214339641849218554006
2737462483823405771897690499240888210036884207337707175682870844588
7535815733133452205567819815643113183902490143688793652330122677657
6158073876898975773879842899556760051032281420571800585110575011560
7944180194521115735566230081141766235971205729486714156591036040677
4698674087945937496934199461430656743529318582649577613379482177521
6081187376337923499184593813578088393653069344203029663157018022346
2229017264046784589433769261461917686702252744039715254687334115466
0904605696228649794354849153595430505486850693679547997710589788221
0478604557694901899312323161772088251244983126499804523672326772525
7363042484078679921249438015733718634550773598990007589656898660512
5421279666841691732436742796977525983638233022552931107553317344613
2276931609814967408441249176346821786109590253260768376982425466703
2371835951357497968042660021776565044221001606806062459867959077133
6649374471872242692389212390566703988778441009575759131454640764611
1706372538299847615100538682146437684240535410385914347355964077188
9324256861882073109452169417137375233305795278316878333934771007633
7105013109000692481286478408114101303661115338753767059544449563190
9626890357095064202941184076171219560384551991684840415065858026922
4262166288674343647833620304552570073080972415914299792907158632700
6414607379081566374357140387298604445878619901628319098691031074290
0667454177668177300396433596934519103055297847414579559783410655511
6463167998684087849864134805527135804572359115617964351527588201322
8320565671816222255960272751949287332096198066314342036761455816722
0846639485441279888676195050145243742336428689534088466914646940111
9587082825556534210094178099310175355670359958839645221649571385177
7101216280723879314150401816268801326770606864889138890203261834099
4723881037012873227158534009488561187092386891693276126747732039733
7262037010132884500965657502571598801068490566781533451727041959111
7205004003083429437776857284633542627191696502856554680068610063399
2542607085178725431782098267011963422059690780651723015519751777099
2947725907550151459157930777466143269851515108715998240526652774200
5034278750324078270354369393275779295624724345029895631154064593999
4278632833070303723752208939366948639166651145150761934548696997377
9176086131165496177880784355605365266635040498508557772880496929773
7633971306639787965977336877965219021466919379872287774355867290888
8151957875879562744665109618207841150432796729607898814910626912888
5599604880564629815515822606937798602997756669025797378919789064544
5051338843659947206724234951379075130401245096848309020127291802477
8097560097080126991027554915970470940352605978207554176910219722368
5733504940861608728660233745517961561525565975472403520293374222888
6102017487395765686556422307907797530014733177088854154257918072633
4757278755647193572134910532869934815640690440127499964123537040544
0193259429822593622742428342336059532898765510587890860102665668800
8786472057277901158778437609863253589663045805702939256274628087300
2640576369772028662330769065222693982472611839928983919236345073400
```

```
428323405664850667239826041216361199445545373058164601587044606172
117458095818867994453963504519488099445982546694155841496443694978
178213384029033406255044282922555835815547207890004080198302785793
429488468307154676837938395710633631014064388129425177922351 89595
216495145822954378251280319521446997850455346904018431111667210490
396989325545803437266407385993384955987164239972320079876818494905
553064852951674380838106700783325232030453663022236905911567446070
605096647381534524028986170711627146968607848937313454406473470857
432403467976989833741201956463994729247154854146288263490304105045
073447742876815278510101526942715965290023091176681098151511260790
807987747592220358783508975907601347751207415182129172050506646 87
556014514361373438702489519990395755997028047108746372560334693564
930963929121070821092396664870317862465563979376475542916746764890
553930221116038602984693345241410062662570480227982440771420409523
331925994360649570282168987822215008745880686780900031304897715449
846181505546131321476771842001457533510215252337002666991336809506
864881470835048462157831539584692092974922398482833111723402239215
653994176112447211220497009559657276525282820473825847256801610719
252257586405512916071295819257949393888247907824904434128783258147
803119840933779645060899848063138986749479895269735021881941983125
771333291049257775692019596238171191322999774184515397943621221136
573450384664480561528361032836169878467192616336813252047594127645
023873841505014765415951574165346693260246779479366057032385874229
993701013745211577247512461207782552665086955013790793682385347261
978003813900249962221788737785628726429651083005686042576559176418
711091183254380010103878667504696935709610776124584559557746380137
690825166528409510520910640541795190745272858788886996608893715108
813104365687919210543284327671969884035863253807911193064233683929
193569977267117113573991079020792362742676738520122344870170433939
505455330812260964424349855237189703135857969089585294548969730992
898736455703609460917580265565706185575335481201878461295398227918
993688735224048452690216357290403735615147711617783681210661192005
717123428519565806884630068856141623022278290815840747350798866237
609505337256484395724594286855317981242738561972706344774579091688
649235198559085845113506481175387573849656484436214891452689733371
930626390336514717368943279825609864137172842162675876623165555463
882120967354360922400612898182070798821258850530165045069017949211
433366524134435395004417685024615179539325445221692268521508942484
622298029833053771388322687192485935711866191861642732873434156118
287191723102416450211323214371526185076672239730729981974784051395
244300472912512871454095849462092110199918800800230135726715501250
005869589498017691924652197529778742182779118754053367722808816092
683321330698230980267412339201633001990405312725498944138277301750
711338493997640717822277997989443196500895183067684121454231810716
677156562591357820322008887475799639790271495679386648080857688808
672456242966528969779615652054219766645754587543395325510788245218
362819533665731420158737763451745868871971076006199082561237275492
890488238290058749153387883973688187633204811568874566464707328968
135483031769919532400175821769513812840418308603824078255108239251
760881742568618743947357575633749189621097161729052074195747308217
562400751285934147410247483597617157187744181676146747591202166859
267987908678727655085341471611169748007448610797544149494181003320
819493913702773618298197437139713326854837799738816757722594765 32
498448852017691986667148190469989906344478604065400564230792537926
569422448657834769854032418598187721236576040711942844609162086102
807609421384913821511829193207015274972437350218505576598514197798
891447879538091131571342691228253797173205677439939066645088819946
454266334536373026891966291894827866651870141089809750074503084762
931010446257125068285728423536500874141963991653868406296422335686
556693679222134282552447527875608731588970320601463678233842860682
```

```
77363548711800037667926225291960989121013672279867492286963061915
28420764579340105253267732203073505490975150141706390782987503673
34714766154315699154687823355659660693100109876146802883233534050
14099808757298358397011279704881638416507546730623531550163925290
35717956685392161875059877171539275871091985814580210163859338527
16005129094811968412533767966342711481616624287407051644982975161
17369322034930136848696569529627759741594595474134525130134646245
95633909778903795467256956595489428543259703150808997047817245218
39502209418680858281085746862677723805346049546732175713732498807
18090307112450512639912073156792302606268092149274498414174414543
96178002469754102337911254407161717098084114701775283929449237028
72456274248591697184864022159158362024717892532833242609297452363
97241012459442879145536452526583977339820597878446793414205854862
21217085443970562145499864014425017172837569780338816516359552750
54038406256130692824777277352171560423827727581643617790937287493
40969326680780470093950816594083301829527334937557734432034406540
53812260750831713467315110191397553452533987612323472329459213529
31432746173192776955043719122045222395896789896586342824023073702
44212820930498157928520539625308532596191003663435324121923727891
02433772970996119629674283409552064272778161603909085597144710521
17926698396371695460326294333381842043947035170956981277658964341
11071345768406219157324655404513237099475813347921023818975204059
03110622911319150822221566802887858789802402281034758612567051677
65976869922098421074449742997482558867205354849708630837532552360
67069482618297787707985635918165768076689698429128638759312220542
83121894594499049018777678781359593296405729186356838704899519797
12310575544660868194679529506337495367127297628665290271655057708
87271948877577771551253287467313316692526627915363764964540918231
57436844813868614117981478085537164290009170093750274028002939426
54467737230553521625281445521651434495189135400513027532214646134
63332310642234820478991123299954115622368215617100383469819315306
93744779948397736176715435654964954607719053957055985033257880947
14972811115522489633481100489035585704652801152119673368097274186
16174676292369823295737530684322014108846976684129498977535442183
53820294627141844650241509372485652010116580044407654070489413216
63298797756679267567954490817046569930293620582917718626831492213
59097290231273819359734118327390332678480601880904067171453169774
24861182105998676621858339967551145470958612221943986708394253888
48007307196559919949765793550711369159360904225054927480544062406
35422470761351508868128974367613810566245729320596267531392428943
32561810376229958889329351078999900055751659317795971225511652743
07018716101155179447830100464678884086121956095808028610436542880
99678663112763880327150910574380501000028548769907797350952723472
73770237794561306542649914774347149553377490536467254156981330008
15442512384384213758004347952900372149207412457577589297613038064
62713667001538432038668000583531789737120300666367016737152128520
77654724616855714204565175197232117832433534096889759038575079288
85415918122925624805145690407933401279541600743918375315893899882
02122423415834393617719001014784499153839596336797654875843718807
51610247254319062660892000579306422520789710235806219463599546407
53149158596821482648819221274534389035775023632895630211707943327
06893581189727016016567082836585159703906980376946183517842515134
88455461858036095104882770703139049546680658666805382589982738230
03261047733294929611977594630539737366240100402734974691107426047
71879230223457075279200989084480949658142761080506452067053880444
30401939825063543428853228771241398506294798588530703801088354296
90954799800051029545604188196326338766117401989080671863013798066
73926664477513460089979800655916985170575547326303245378476942769
43317289405718141441339158300029012796264875203559837789902495695
07237462161581380776672656624501042773619405975415916916339294606
```

```
8931537216297626375086959040162614151579616335418702538854147017083
1878886019687201657593390493413111056990989271140994661746345863007
7762815943342042926455847360107295881786419101625452765955391350506
8335548671454646413424975511249411327495544412524531435627368722786
8282826669782133773794072882611575530875335412577216966468245532500
8883455686836324886855399116699207559703678918082685824025046364866
0532736125868916771361056054736449698812411763414714403165250981933
2170268359302683607576354055754854776894734330566684730368650987022
5232533830142830811337199464432459414582002041033647135727782702455
1562015019452871507794056264861505473119590214966582573484948908499
0152029143208236535399202469990278185701002223403667136638456835211
1213345970318859335787804484007541222194200404583945309937392038344
5784627862740544816664951977111142035350041810205551607012948405900
8114571344838757096804966866203246459547947173878211440277545174144
2732345087354546060153605325144614982797900207887394432022889064055
4386178799841415943028430648141917778630247254041730111061128040133
9402676123202268289120182394543504120476536842940930236074634377599
5954306207167785199417272079866850916266937729042383606488236133533
6298417073183142452667634769165565940576493843877672459123427735055
6521966509351443884953820409403776963737203655096643161208664065488
6563107743647348855083152967191515887570154903300450353377296030988
1096702088290348633600900649016688902254560112362658012005411787844
2962287003985718669586251299188893298912822261905799945736253970766
4910578359085822946984692353874168237429020614421773922893882383844
4750910393083814264358001206529127122733772797402596792941971478477
0986863881017550460014910948168010475796268881262883036846488692333
2607987731076141865965712624020673369798536497170292984015152092525
6420316539124695424858300047675389525645545811946747824704947118566
9992031348166167462701733306748356502133704317387322979266500019199
2698347272656895975019533001645690557832296648883843254575830229244
9346637181826666781246559514112090595471542961042105914827769494855
4110326788030298249366563459459723104590598962191576369692368832499
2357076170559010147981661662924321296405083291692092476227599810099
3660407086775540647338197286998212686946931858514657355105358144599
9883853795254184194551313296775473857394516024915985627743925674022
2747709239866077317008172392000086395705435954636019993118374280133
5002159153906197084174091077424806375589616668810893183977771094577
7069948857120001455165454199466001267605110036683504595329022429555
4514279238826709725962434722590128816093142754927153935545420931099
1685770092801949888763262866824848792323662287599071527693992933685
3771911677126330907719038023037230776556782302207785653717134308777
9516291204251592892657914624632654009015818186450313277495815898899
6712842564970882435564661182746536847078507080166942248039707259611
9433946906283804511994079500282592908481066990473730768989988918477
1709577441774044411650415697113960816460013808023308996176809810888
3845653755840410126002043280527670624489934626523106860966993076711
9196734924880495376008060660573026480752562380737941829078712316788
3721043743119673488706382503794176712389060600795885911362868988655
9575806242103157718465235356133963002345875484090823926532835903411
1539635236051649875616898206854308955777781910941974985897913141388
6270013276902646940009743463828583401836195308531796707734536816144
0610278436163018841016128933553816957281791697084540703669198861022
2471436033117813022082607385874003587427057190622292854789022289555
9340499210426475197290706972918267128583409984056967394858685915111
2189322711345598842033286080396264187502459398797198884332098874499
3543908826637279291925634337764091215263430580261932898303499979900
2807542529142549486002788210295794805135759164477481144214739799366
5063552370108140341653412617824180381283060469172501430569604446666
8425479104395550078626581532698824530668369125893062407616165669166
2197463505496654206527927986146297734010972884068734339445299460166
```

895622408681163534165643081041610252761849229327096090304446857646
431414220190328728488843892016470201619268470920309657655967960876
570126361307801210544903990702317839722864727106799294452472126298
634434487873126315260070121339463514325482360456005216276071884942
211612963864784525918361471951514811566842073640941941083340448261
521210736831047317054164462931335363794416840094941503958430999282
264187143071523217864719654254136270340733898948520330007097345797
739359624286060368455836890724931367817118884735945141462694081985
427831166679197929070480617685887492123181756611550358080526705738
555583635415428558386607319600037159182788130815206238906780390691
920838677716677543671224973127599653346030225692016411848401940984
714529347951276932743528313421602664618491104571453692276652090637
933230313292348332927601946001731697104927127041820944788663127172
665648355861137170599969619544269038818022367452031969340992552453
346017315537158603329393450911966137096134952314411461898862230817
521230597330217516477980143673669779904464291251569466008707759104
572931857106712540923960077666939522418767799256103441239021458950
505618283058497020706078420154806373212851732727529724815001662116
837296552388010012457148410716445827998345944100984828415328626727
589433549461500196876115367892762029974081912426776376227628387387
298799299774907451257865261559319634580005411046354850535992505223
142231159860657782881261287098270870180618436499314069231365710403 4
792287492342410868139607304633340424512944548716526183149971976407
743924221508248874921544281612968330295798910403266953671050063750
476749847106262246527934980950710535685017013024388266553591542362
015824133414350650247549532411438285857297660715088135703973894167
054454535360199896606366663197869100581063685256392764218505881993 01
314358165250199939326054757205319781555135618734200669485025185194
292358554736747885108211396042601207700697392122587550949769764 40
836793020687082025906788052406675671398803382773132522190161620877
507058781037513425288250793094562292807158298971156198266744922375
770680926804957119580293637580253717267079219373451228745830382148
951016202281745024486751885487384199237612739085030674237842826380
815759657891924264098717274045973940926027440447784261539817840530
041468362631601844525604424909238527837846347285546502366639505281
219763255182215184336214247487908312606536255835803225317335955814
222311728322406325693122078648216631641329756327970713760029576218
437378183718341307982953047572632183794691496789228129437687162693
079030598465898945724062228518105869738018247295457109089821871769
750514785467546052594692869761724353043855883190888020077891780606
486355148656482989246712092824962857657515782918942062656557930465
126871750371905767479542415479227976241597980117327709134538211323
392997111653729654316331400901229253547341414725770122315201311707
122504621997745316356788768464659577361680734557772369302145194871
491820337298950975012812295514304713867808573107588757477606540460
497181746441021903419654569912313101214506034689400051526701207390
362376659174624722777444025395530712369630706973005279584000996422
783465267549297379633377230318150277729892538411462527097053961114
784119033993887485375121545168531346616329653082456290845187097409
714728164944714841814464857488053136399932901353355514233603875269
565456457279815097084432655290415262393431309520647085595331288086
137477077175791856838401030473304337189407809082302871464768023705
011830268077403210666601712770753319195365202640302389002630777268
220891196402099101476752144976111029718754020290491744941298591721
479014235277211225465985814600697205437122231168773005687179940830
081585363351860309962800311649864590319394795439823403147339512942
270361857916637874105278068972142775764125264084229245037580632279
190973785221582775080503003679094195681717366839715001842381694916
592547688836877310991425230925439917475481554221832067888907321478
358870791939878308788820558677593246698760602283734344592239306516

```
00528567427842243886535027549554335153487503399685971775104271414 8
99315495380616286077585502768732963484491564326985274979204426057 5
29641751633473844048671385828900890015281063540744190174170654134 3
57038420436264230428948684299365685766532479085406375173836608684 4
24282645175483795284515474879375047550252651125541590766356258396 4
25445041723503753274515765001299181347336120222848017260266250611 2
75300333931908185608781054176872738928131710661100204926050640672 9
57164411629054644757488872428053892009528519451870907646171088093 9
56245622549621734639630382317131804595013647108802528300611883276 0
91398265091231740152340438799719409611713563014974025361782023187 5
32009497485267094325893450274342325372184458823578117780076471568 0
46276734800581769097373536667163392820567565494693508242590756603 1
50186907560267271366543965475734602822183416378782156932459098778
40415085430184497350225921353260932959265275754323376724662587641 5
21342093250088032637566826011170098781092996196605310974811210582 7
99723523848696027811484291332166869447639353305896577233699467432 4
87820254800517931559790082358603965364928163940865126660292495079 8
33242121773832459462778839301315786556809348890519770629300222914 6
28193077574952147467814113555884537431508686094768087810483432034 5
17986329497260538940700562228492921124129409383101367964623551311 9
46052321072359289564805241387676693718861233589748164961489484875 772
24546535440070045695502969505475117576918944516393204349824982133 221
88995511777758073201666416902370108608479968608758660122479584338 8
23921807052616553548023582675786496986632700759410884567612547818 0
17133757289410569168530036782871378086167843891012310174358086354 6
82229595688848863929034882824818647972081671881637066713508426796 7
07429905608504028923976067725244232384842638206583043536811792839 7
73506120797109307504118814575290435421957986881988548612699294541 2
84230285298196761468516061604428355889434301313761719146157347189 1
80575421269051101560267720240611957910113279744921593808882132278 9
06431124210639277667330590554771520307951531796673243919207103309 6
31809946893611956873732677525860546710463612431552895086540461811 2
42146512072564341589612348588949698443989041028324794519635680346 1
45993086320715697731097243022712015153638558918425333884709598977 18
27084527485666816774236948148137250321199474284422187970514980496
21634126483222671742837725539734676392041533454661093043264898284 8
09853662770160747477680182444925198703890786152901710869831073748 2
26766009182695268640171602951151174647011954437152020344368162741 0
34048518625003665147643028118091477896505574581888735273964702865 0
99366792863863177581153770503997316652368337920517827363879239774 3
06320805156887125212159203117171667459103738490185995968325272231 2
44262972090608509769847490338077684398812883138169682831009321880 2
05707770033993330580760088531725124659692059474086856182271513919 4
53352200138504684559387844152223596380554366678113916366582590831 8
45538162936194056395557694762003759417396821660112477660481902016 7
85866362722922637520904439026946060808336475256969883692278064725 9
08787195371104788397974542085964258201128932374812919327969679277 4
31615550420388020926005275386927758453955953813526820216657889011 4
24850827138209278672185846000938563724893348128305603648954593861 6
15337236761826310659862432018440628483644029860735546959242989955 3
36985138924889696017720495331472755164292464951203571613124058047
81349924659428744560104621279483242994742379794094059382477597438
99517627266184512912105812804284034026027949281201150802191974382 3
72817515343695076376388966480776289135460537087642971909415025428 2
70609025872981133237851224300632954187069823620915029402770692073 9
16945409033482615537315633715523285519987728096511992069059279686 7
81616762973863912602472524464743872005223659671605688516667862564 2
01578879006874202675664432751350317304496156974369451188306879599 2
22659738279595758397719715390621904003271193814526810752752872975 4
74010183239651845139995984254181471934492429026786162627620979109 9
```

37097710181469122481086585718766607503499483496463890403948053 5372
24529434221067354171689922632285669332017831798585314100611532 3256
15598150000086313178630176846929537679174019998717986635191094 5818
91310236193199110269414175848733717501133066334654212234267377 0493
65081607628100285314748200513179053738000656637692261565153110 0030
51984954781877636660202205797287394533815270737202773231408322 1270
27858885548095868890001654006862189221478175511366574678469475 5323
19058055168353313809975907602241037543634789622817758417273294 3649
88756815649567764403677135509505630874761352028804617772689660 5464
07266300985556868333576630528148098203589283370016413255793779 4580
01194896760346705301559426062262608910338818899442382167734803 2869
11024025887856655213028576086071648058546030423347868529459693 9023
34047530204355047172583137070141288620079679906673825313011613 9413
00332029545190592194025402010656592410907280839269979953926346 7526
01817866967752947878032595574574544646826937640924622715508588 8210
37813224608684876634726405540123493332289372652348817769879194 1327
43615928667520054928431183555825220329099241431883413556746090 3781
49155130651760514265052899031971715029061050312071161924301027 8933
89523768613925371154146509430698778530269535283219064565324845 6278
11388302931401014674190692449611075699631112761753662271547544 3365
63621555136797599040523384667350302852211573508484472264010512 5628
46730502837481308322119269422870031372314872185007027116647224 6117
35208856230164056989884913164266144162253638332706156618511868 5454
79383683854584630692824996389579626909424571270378838741278399 8176
36977476617471524880828595253082449206967289109801684249262820 3776
43367469837921234215222072584275530394793315036621698644555405 3087
77999501378082701132727205837674623003969552982818456392064540 1910
03883653285750702970882540397227662208483184450833424987858141 5106
11900648684591033282223829102640237749847239103974691309393772 7316
60180279564680075642846209302850392378695291490646604037895327 4889
06988823640334412362043839792265951738256335158880827517598078 3142
33661440932797511510600129608284878174694204828342293942611280 3939
24861555928204470442199884602060555774822672900929653760776953 4190
96206768364793661243284034767722047195507803740828999079547551 4116
91932070407958239311970880693495201804275587628579988312728489 3356
48938047330531945772791440079056014054171885630491536061884946 0164
54196170530956233675075710692304007923710774110726335316676211 425
36187304503960419539313862023813701919076250226534566945221589 2573
61299132239938482060815186299029655952092334805473533086903672 8951
26401928631306783741445424309686774451073217316911948074521237 618
86998388710780713413573370392465748035446360874222322719791920 1336
39517530225120151410833587290205591602101517581143299022705939 5439
63513393581350949286583945875898778430696231268581406861083067 9536
05480817145214954669746144910274418014826984740974730288027367 0341
71283275331161341153190496295729647991488052202643821808440327 8188
72294962700238356069498656220817462704875752887568328334555835 6904
65594776780177314553136379188972282267485940854697215924459624 4213
93529441198728484662462042087009865876759965185600927753931715 4466
00836544475414149170888317504162565124851393534414761502249844 6773
05259255864139537295851830467135225120269885891969033111495850 9710
15635261547925657901148327433005506743589708186749130290156094 8677
23954972731004436650127051439535500712539154681376476877915749 9068
23268607238024663048411407695599084849357459308969072473872562 0959
91572956128500777361928363437362406145235211403258851516209656 3586
88578471231687544666466669595463325080413788234306913469761425 2759
14411352941652373899297493360172092606442692658728554888054886 2327
20365528110982400584565420204800853348841160018698425292926295 7792
20045961373585685818114893580203372447428097288477746415043095 6871
65794313959624075190413346769081665442804159874034600713120084 7219
70380940158550450981729650812182757687620594295807719749826144 6472

7865217413256159119226125015329897812078044357592433748186992120 49
3818527790966946585417768671311405162790841746755684032848009340 0
8868375613057907694681064452035473187235820902802508899513044995 48
1201604425640357689045634500059121800875920147756025215212236115 40
0183961793727215636519736865809466710613611572285848584081339331 68
1392837091356536136639556065520893951883212802917477194059469417 95
3206628083505105849080295382316888687619119819029466109825391403 04
0708125747318969548560940446833552203565389819389759723702709249 22
1425868886512953286422538494445026120429159161763574732094329676 63
3854764896987701546322485664771173923206385409638467241698450497 16
7162014148902313368306964208056244216918638241157332436689936178 77
6671291349218463174866411563601485576868028392421097638755670716 50
6485040657694137102276688647254930078085638558618308128024449429 55
2669737497344912659119637806071886953740506237948473931211633571 40
4622147461573487724240685181274226424189425478671946189319054824 16
4721369786322024151980571191597737825914541629823208535413525790 94
6402663627241789429332335032001757390700010484547103793856881345 61
6246203469870476693361509001128263080000748970787217083211823607 143
2518523197719654942419412922765849518626785993931468401225561894 13
1724683611347352953961774378124806879005815247158439614011590185 70
2444311218589640393439396620831849971144917787357960811579705646 45
5091776406443627304592120613240864397506708837921367046955607512 51
6880092966319579698103529367632395296543522943829181808576521183 08
6980578608709635086012054711780161422675232095853737310114148466 26
8986147550603371497522610061158620819931307366583730670858683772 70
0170698767778458118973006834449230641262369730066566623350561466 044
2547697535810488896420280534886902960808892388509965811964751222 24
8273590338825517094051449136238174557439447035155639969424791166 03
9973119041851635338430565352171084088432191604769746492620301351 68
9702752051677499944287129335959870811904749227428646687486521821 23
9785801578047638685613353518891179059484101330124743630100863432 52
8328310581728590205804629370553358095941533608862826533543693831 17
5714075759830965743938978654325421230400525387175077271535433707 80
4659649344673882868475933046420823828675717408575782194169440021 33
2782984069264603846347071684794139665097030711223544670023648876 25
0953245875656027026918184015630717378599319016522139510375520872 94
6665466935633631631566630201268969177508496735876620211296641959 57
7449737800775290272604134384650270536320092598955151663819133591 04
1823800175856047533412314035891158319532738051260466621592095523 22
8803665649979426455835426779821896053887075778185877084336955755 43
1764363741294228274742625460381517250082677934662961237176506587 66
9331352703961659347910442397795402040646561742416140602355832462 55
7288115121788216585734373731639925714723495015071752078656701211 12
4778774056656975505560339375965065790874455437279044343556894447 42
7172282868634920500475162292666822097533969052196780790408634544 61
1824861809907252825736186548360502978222229215484571262735717364 23
2022494729764462270852154525525788494083266141250045897831999527 88
9056664605956929097198009353034240016943218280974546622443895007 25
5951633244235657009356016395713025219010843554362764678443212709 49
2231976516225937001111208141948924012361122281901444558130721585 74
0578174214970913256624827354015692137165421613812609465058968185 08
1296258987580742838023527380752506541512464531701271870974217350 64
9479602279336605236964660514556708805439768067404713371156864706 53
7501041286655914067097205599718437436962598627610902565455382413 96
2500712342421136946986344629710501972283609811763902671289291811 49
8518602037103163165062017384950405550777003049404002477277531844 86
5159152315189014341726437797016305468330988169204977571953187972
7885416520568397677166418306453051194535282350168190932434741932 45
6640645803244816946766330482776639303803302952336798847245906606 88
2485742548876205563837511284364152318631025678296848011164294244 3

```
80294407160534072114076287726089089407733988690128926955743475120 5
27647969118823568870912408383699214360147055618034994379459536317 8
35825816624995670824500047513955696986207096276390362164741909450 3
90211148603385310983327782876420731855456624696356406173752468516 4
62697252104767287468934575535789103293467617999566412415746867836 9
94908341331372418735937068788016726457171692368477432418235926763 3
29392534083993430499263469089242186843614649531992625005516971103 2
19865643484624456298314738497984966583268529399820270579324940595 3
37477514103329793404429633869620317899291212281963793404543844980 3
48078713314830498759439816393697887421794282226601384975579010653 5
47723645979813039199938514648439187255989505188357775127196867900 8
81880803496935470242694657810962014127284080844133217078762243603 3
86271274554301331448709936273745530206336643809800795137417208355 1
82656060243281579407833701864896801517978371986780613229742665333 1
78260322852998573803629704924865426925168448234025474121709394993 8
68633507739608694020602913669392636024580541355032647126235364108 2
77075872396843104180942524452866622775908830920760329031991487205 3
95058531431540838503212299652666061703502430083505771758513895434 1
52856519082164465207837512733083452462161429850775336238964856952 8
87557228205157704584647787583545795073505428228960947570750856696 1
46917524851345714979120786907766945051475350305630359824748866740 8
86819868125066654409148397267939349942230552796173143453573119519 6
94416890801199424578824112554691038442795616302426238274344050818 5 6
04448039902430891097305068206354785761443667415129058162114632359 0
68396078915723913625889076258500400459520130998741335669583530719 6
57086810860251287834108570098541640079457114094092921189065859029 9
40807659299710794482373768417134010426021660010660161270479357051 7
05180083939415800260103929745006675274634607976251743917430778507 5
80170905331951809593487359660031796868382300198324846645041636551 6
94618016432751655248244544744585488675891826001139481470583248581 4
10809351284651392785931954590885710894010326998388953460706801200 0
32232235278345418065951042703184403446485154244807705214777701988 4
20355702664337492625697118422801185052109500485921764796296781179 0
12601540134499148518311295534624937991036280875917277931089269100 4
60082447814082597187312116186972545362517306868326868991544520278 8
15265615917923480792449811474608489629747876925023397486945352519 4
76346298541290906698106176000184586495964255206387247739046306866
03426501271648519785810422610021954222128596556293829958064576071 7
06150188811917979713959243179911433593516047679598808072810386834 8
83302603348678280325892543639193254802800373201690459803372272916 5
32679209259468077388685841043077449487779298758335392141270015813 6
96127307897253789128546580456999413190159934260724396300220749482 0
09480927147259669925258734010275921675838335369848282591443747207 1
46029119387727616158051263307386134714334621254187637692409983929 8
40853916247575552847165098232774368948050913260326188917728286995 5
64617853935609000695630189006862376674430949587510377343329035326 1
76820033064908270938346249370947461902853829296945814986904507606 0
37265252284693607831808628697074122973843114047078844564777026542 2
45502049904443441758119118666616857382540665063556294015080574827 4
92370909254002219339971655304338006283607646764591572862160746743 5
63516200130731591388610613866027935051281959599028425675317596802 1
76808697518755127733378242555872191735038055157484949313182592969 1
66710063811057217831235287047184819454076931325495092499144586451 6
35811986024363753736077609268876785073094988592402334303628164412 6
54165426630733988450258221964887098757692723369296877452188812810 7
60687221455191024115605246983953152428883458373346467676592008643 8
03730610866582549338842989890430248942234630986657901629439574805 4 0
43093220860302058369175738218006523725201070208969559361591539987 8
69999972422512639695906404514957143003482893153750688794520751413 5
08583940460790560732403177070486195905885581343464335562142516961 6 1
```

```
21929086969330076274725616055779992343889328219294145908962902575703974163691649040153735971573825857795304718955326584638384907581582325561227488005994554202861889049638585140861096724670804234303904661719305525100620134597087614214473708638901419058524686724424415515495120548638816387699717509406741858070770195018753800765845675874736420705814199727532238573605331249948123807099393254461351969456072590734944964228355044408249008045518210411988023771131548414078153346183364126908672285020991111018071127957817271531291641642826153777160930633132446626702686430766751293199790877007278614031020948834716515933489638705541737406768791097474437325368449343800135691407632685549491571561509773814277282893299225172805707038095833514731675325109525318584031483165079887160527219417448037788079170429398585578632288264737321394348935135544046005860707752618860331315450068373695512032552676857224145827972671475225387024928273077979070133326116670200670021278111594654213900097198441006670762215778770986494760535272039647366240222130863206189718294409563354420566529368648706005699112349755168390903038570345009303384745748155414413879836276261301886897583433827371188414389772915122774761067947211958644276949605557476334130548306540034675650734642900452747849324173674016607524127449024026562527440024743213540897256357004429927977005391301457968176433392003563831283026380637877916118759939416257887886280513354212674958626535893562095959325651336188106282173175654724078858355144633105410650046066630218108871188453249014424392405265625915030559071463653830111453585355750572742611023922947102596663869032271283713988257812657486263615837542829447712286872011195984161256002837815382441901220520940937595899057420434503380152148132276526054250784909664501058613100690740333039542772630188136748753392786588314226249419854912303615039789639409898781168638730116337246248086752500385854842841389665136015046175730816750319264648789045947736904121618085377638005905459017875425773939256457521115250930736922429534179667735381011406365373381924034534843522142308030516699706286236613437255459088909419726114203772326378508396527731614052927995210133229502752480397178667292661403279521083262318618615345673982398839453406433230893468806737506887102316938625381173406715332184897199429241644748204228929892416544706229317349564830031283518130971833677240577077288481309842069348534171719963673716527002868122008738751906082511697286363929950605258161411542534628616869755600701323615613407107111512608528405122283352929987375176474130432578416791540877029609789091647827949589889627026422842677360671500956483336322805134718808975883322589682369080785910889605272562671735749541246112194571263853207292812403279047611745510722628214160948018607575012498463747660782626721321545977926092852690135656776077258241190242437843327564537510959011409716816237531001526438883256232539045445155521959234594511403274869437716893727187170118787284774917553449637125505670252178457622594121783358873263335009105540995263743444402068885835601438251753284333086784957409211809090602306541287011631863526740669031961656448836145690905881322117415737154917067520106519128735097830186167849481701951750755959017422547175336906043142341511884973711309944396204489701312494139630523728431449519883547737580168750533435958882066825829412015974083870427432997435289181096059583643782048836420854463226493070418121340367296478721381193713404954398848563610969580184494583359375235610934798660174227222544259849314224126385946063007498505929671088410269520622755557927973966519115290497249540364961354092680784933221492681851995801505449375348120912316769525132495959694072727436009705685197513234325131609592712718573076895856152079903342587606799082468290682313641032574921622273302850970130134924750653555612481616818703556416375135049991064270675946522960587119171145907205322350973799309319445826336889073119540178024783352195923896138217421790337357362242628720334015315347
```

```
7455625851198646941759717418110848200108848626210274150891382573 77
0561894485188205079810976842718512468303141359963897678723579675 38
6021401980606141272205676130042896619659763109784193592179866408 33
1571246801178377337244960772012363018026673144865379429649891590 73
3167107008846672093687060362590037668504778139670390422056470760 42
4570213454366455115498570926244974416804854804184404580094526528 57
5938764868178147949968269910134695966087500952787855569863485106 512
0134201897635791188976988900784297334961759494823552599655230206 53
7358476125698023448009344383094355806669294956790321481102603711 72
5157722417261104632428263883435728000160092027747960798129726468 67
0413629320415876505903952329401283934573914827743386439283001781 02
1845370658767040041223143503784338474569225745439436154989584311 47
2150793288311917175287353461494842516685808302126271370966823024 32
5417594712773653453953688536679125151737896165944021102133838299 07
8454193586133666956729641731924377988671324506343601300742951762 57
2164728964818728370824834221436357799647307808862398273083175622 46
6015625966907800846051865590745562874802873840887721965978583940 93
5501895116360925827788354947897706685594827840678059035176067895 22
6455364170630078349531339743479718312960350837810726194194075712 27
1231165865478760827770678693946611273397157212599569501715217445 73
1261160476354580628175543610503088003188622647630852619408961115 43
7312420162716734916544222316205859959479752848651141162797837794 29
3422203388370257671017373226714747381980587401219914237649806431 68
6593272058155449919623536400056853448616974545284284803786726582 45
0003358194744907006766933398075419740225967680366899263792014286 1
3650596892517954494915408713463748113197763685858783922973573564 47
8783958770911540362653183992892468446754424805595036462331698905 87
2334268369210028255685939200239952445228982401679292120176780508 78
8070028124996910954790835194922438429982698719690296818845244466 51
7212191290714716676193543084701997636932356385024195765957814490 64
5998205799318213368616858568059491576496338605052044521983604450 60
8933173189737772224670818684373054315922671926414665434339884595 4
7852820299248921928384784137396733190150699442473271113324826881 04
1789926904008091421774458440266681161379838742208351789669344513 81
0490131979918244053414572655199381686717410832895397332087306518 98
8369267457237686857937611044739059300069789577503822647027424606 31
3978490555711198090839239036695211636256410115334624200122539301 609
9338992069349239049477397316879620725111417576818258388702960835 62
6786636160024027186804229238395064521822622793597804834976279884 91
2800030985820867984874323533797545580975545015052712318115828407 91
7459739251648610266336358857794962557594151986949955646144142378 00
5222423739833768581047863198400540622503780623163098461178667794 04
2059525307560667531128136777294033965829369604507753761415637950 49
3250196015470228071767007031037503533131478581500963603077013478 85
2787251290459448312734824685942714029303174938200108521714819633 75
1112719741464911105696272873614913246205142492982524951750235968 23
3779715934914912242563490725821251443067157182155962964374621719 97
7125093568274890088929740479628384146876844327769549709046104306 94
9353841893895280053853041793189771774334181899129386554803594047 75
9567302183578365070941847250602402633529604063693827461903176923 26
8075262012762676341741181821066648178710827361497379182159302542 055
2443478495519659426602138727855128805751662415272427163863423676 47
2204283311018051259838496770307924794348026528140038523554380875 05
2255502227177283375878967787900539759991166107991605857481295238 54
5037545808421010609227498998664139701805574179810876718571680820 48
7540812713493531136420065012815279756982594479368020320964922598 05
2141472498496267329521295703246036211825834471258340885752043572 42
9060180767034798410824794342006247266436866340763257115415860093 96
5600302207038420670926568025243437770468334902893111964905696728 91
7855524697460938099365643632382877369702228163872259438709035891 68
```

5023538849001686195898304712055767495247487897834040328110607696712
1572398128427985993883330462890520310759322754180152480363515926
51229226847775006435703119715808958834272550668238747890731162083
2137360677284940353956827781090572019539529014953015678116920342
8787307240511387080777073007607281494199515842115705497432355826173
9045900600312354763447226782082938789496723827321895985778766831
6571407349135846791333089152412763221058301094491427176365222627
213779502527996747060465789146047206804918359330205999834996693096
821969845847490656633341043149512562459760800377658916537508069334
687611742884646827960548872431063275504354797566456581740641254904
5551717998403432685959798362350049803157265921389529984897732508655
9364348073702569939399518038888962196559068556627618835610521417844
2972643758700374184358194911354849939312017228317285342630939853636
68469618101071709560323902758982285043407125808809145919732862600
55365123820516211478560730612253267510460588493050395240931971366
670330970331026205564073689291189342011645141826348133292487969813
310388174047908802280378945600619441218882681379891476861906602967
979062552836887871910637641204818594628312878213514671839113797057
2769923665865714256615072960888959968701722709466323735264568290388
24400569519639903039149255697762824655857050304649717537398002963
6969786768719558327728101820260467303109272478408234242431112625333
14788771318676070552033239939308544925850930701541580289103643272522
1720311633713281079729442314667632668394353221289942992109193690703
92699268514389155672459472967608560937169004286821697706056661095
53899552399784917522391669888423828557694454951211357666857907428
98317502349709667067767812458612306177797263401458102627404829542223
897384861339329688119080106661525075203786936014539417772031218165
9462581706962575196340721945446398672637105568634835271356973719053
77968438950775242948018831781649783304731601771381224088222898463993
947266096966653075692214997372644770529270147223743744863351338994
9283675261118597779563143394507610096553718096032579550722724918483
768041681065329774827440174160603629003594376556872810687377444610
094422950833121089826472862641190403983574426886481812663380009175
971227925691695599619542968230100628102523497608766442655474939140
507776092794447919302748674652973989859594299918257432518314044570
3875516005978858829096443852346752974953845525139562401461830181113
2128231875026789578985490127335530973970711546006446570381440935533
627385445368618652369698702639276459517392904820283747413092219552
80729301852409213939855476263465173760779737902405183989373298548233
0918974521260378930384871619396710069555954778467915201266896708855
8881649574242857506694290072122010659374479619787900463943261318423
462798565837919057886174317949570453288872007750802600370076780831
751921994834225365246642396446312332633201210322184951932361931341
874955252793870954066921723667743568471662400317236573866022576253
3127134541990221424572939433955333101453152815628728434230324586053
0475649600950059956956272192630599362228703768476260305556562128010
75128500883586467120168430581524932426483276798781489358756263535633
3774862053994934052722791201293654659000237873844169381550842319393
7018774894220757083789350007900059574469308694920813238361990516383
4991312172875407540570387431082055769888036409563972803791422154683
7128206784960469118314289711464466555590066064913139165859055669733
461733364367939316110630435351694752466322405335373214659001570835
2331090267334891547091584686908657811996066053600878367488772472283
8731199507890525183781620159508379110762622451113198920994773409022
1169300964393764386520799109507226852306234944874406138980523680593
0345435643545781831361373762331150257307404542706304508962653160293
2190026394111912915487784425353628660268603998482382316769472402673
0795195365945134219270327067805876136154753768156679197033230825903
6541395121715443881540827087653235911409312751377463830786908231703
7108543431864712573382029306910526270299871075241605071497683900200

```
04043200816304560062423928566837947800403338004285789910815289023 3
62951940783013962771782660650045525529286219062253100416712991835 2
35979740101194687759517662250725217391165093806263780819153951426 4
65093135665846238002602879469890507060403270719790564867764601734 3
95434333309871794051466636688243661473548546276691917946923756664 7
04092586656519113327666713964951548271430176910107787414948427159 7
35816335309659471026945510303159387379195082280410713906290485565
11492460939481353273857635994169178769556160410499466899108952024 3
06747857347936642113137337069915254490143747712629378801227656856 1
11270149142511504091080038110352077015707185342472496043311241817 9
97686246263062536583301523194504588455854891125723083189215016962 2
44686924988414546070240879892054150149168012408427172725508116945 2
44912087689530962277955418468363444420181743761949382391266235878 1
86131414773140768459106138450914349364935464104721916457488099391 9
26464546835421395911997712780669876805310576046928779124607994835 1
22402300154412937452191097437024674189358300293549577387027508269
55944661193660678377469814640998366381623999183264697042596796445 0
68724607135180514542749344579109640628579415813579577743615117459 7
64087929232586182623686707891083974435688112622316185270576864722 2
66746184551748063490553368954918479029769365722207781374887495714 6
13315270560116854963929942882584719142973298429830792181360984677 5
38264268999632967560868130576518195649669096539774122505247670627 2
81897300506465232142427365789651649502686583422385777935651690693 6
05454897199098859431033840215097859163714216675646486560100112789 5
51939506892653340853857195715146285760418298842343553402550327066 1
51132886685726032895125013429543205411756952885144617959012388356 0
30590498684217879830868525706778071742609999738975463953444945246 3
76120572117723847312800068831982567333250219081494207967248022647 2
61913419978040805010174233509274484571049067000260067587616889812 51
18364976282380493184768278645988032945865141444294888506961624029 1
05157771045697184983695426953105551253384039900492012012357303283 6
11665911664530914607330650568551062335792656334837500049718057416 2
14806099380417679552710317392152551646374806100843177772312713673 31
79122963737667000218823588716275503355473106028789391115068076081 4
65293136797957161984510618334101018229730390314876026460531043087 6
33756065866587706907400278743306936662112314042013154067544137156 2
28189903605555076875760418528327170680662093721350810538538975889 8
19719017687992070502409714098415024720551713705949437676788800883 5
34892901152878843742521899540915094650049453062349747372380057048 1
11685620209627743993469500455106028538585749487476909468933686183 5
07421238489152167748303697354059107610692395180852928726357384072 2
11824809077766680968076924300676956679081880626498773916116918997 7
00320750575748822613212752402494674816879525327756013296045682425 6
63856987731335931344285917286248529288154037226170618550054092179 2
12415523379244193933742389915554368447548032437751775749728676865 2
53590547768121975582970127976577910070824376202070520824177104704 4
01603110760992683829005277725623038099558227229123565769727924023
67064972299064581603819382223602135288978754184543025004608076327 8
23972217388934258733961066156775331531556555053609003436188006099 6
34742321863726909980724940972868474116283429383575136689001560326 5
73236819249141033489424335174278190217618657692111657797070470740 5
20906177094183261314281032299642331987600487643656908924778756323 6
33315824206685179943310356124593121897707011564870846930213465559 7
45062393443898524761467409963016846032657090780610776695953730745 8
76140148155844658759796668438302250377669455463938480636421381982 6
83862118693686809700178652736851478345138173397647183563516072346 2
11035512850862578590725687869854165878119030510612247004406678748 1
98385630413348851460154077719338694861049131158860477228683569113 8
49046975882922885202574920199001846983997226317437303631079331947 8
29734697925046028361000140491173316704580058244004865801969208799 8
```

```
5731819308087950976198239392713876744186689748206309961657042953
2435432222303892642399535246972809727481226821126000559278926556
6660475738236398925531792231923668477888367339703098233446320833
8151512611895728421892310584040661215210969488077105339394284611
9650702014310437294192539951422554657630538155470010519412155619
4019667201337605838264292752476528477061435332314972972763940570
7568282532317260335037011667757723349729627054274912620336008772
2388570724193429725099962111560076037295335470402298128528634184
2479885553512019959078225226476676548025265749420437452597337938
2494056565565082101901642198373784878447157499094763226274161878
7404400796786890605405475050139723746137713761924776416940057466
5926613480203139459182731099309401245890060726458725762221479840
6421972311573471493115735999346618285380770319944117392438258568
7879232111615839268716930296559152387275565239189330630060714245
5505838117807335221263153278873175609564325649706506191523285776
9438078224844384815560767342979636627955233273373107248421331802
0046118276949430584499488436942311744715095009846039184563944479
0584455132012986756072063042557854781459816752528732094049614836
7636448189537432617178295380258080798497887146477109497797977220
1761575389575068888146130387854585501814302516595349340565181454
5000301907751514397583973757010061899060464417344246329050130193
6389142656538716455815475202750456728742434769480239029490074267
7402494497492237955435616150883053115708950531694483656866373484
4209758716447672503551503492847742854150050517975603141678522799
3734947745634225794320550219564525879218743022867702104207823210
0790070697836874240666133471005573175585190142750142553101217332
4183801336847397558489013872021201068354295315226077055742710345
0022650448911902067877682875998888603594170036435610468962836507
4959407194673310453826978322196986483158536731436244683031032944
3437046668944077262609960296738399556953738362634064617425752225
0152309176058608263062949351438041734929425429206464844224735625
5299900206621120679738396618263578247023029674264164723058364573
0953576429729902480265736944441872973702419818034194954295246736
1179470348135091492084284614062320420305317232938686926902266318
0793921475310978797250404110926479353879884502961654521193285153
8689247929048984815185476857616241844940683624436761223788192066
1103617381206270296551929103948707106432161204183342481255970454
0502396887705200843828577041450511481452360458219694396330360205
5502234387105724358884942588059457949534641814938156003449734332
2097361359010538386402572275652990414256873429166021018718754147
4322483627180987274633659662889586049349215446647508431360485103
9611026100823823256681720426551108971495591918682001406308625535
5113956663589728566735081343438628952190101688308273011668884810
9047145840753010031456465095258668788784004090262736843782712488
7108406074669410858404831811691793293912392763974628166645801579
6846319172577945206662206480462588414277361791568094148260572300
3827854595199688720383040709450483475521639809632638688204349089
4450317992205533532965271238526563431027307881626878318492790364
0061315035504207479236355819523722819528340638020827068888175328
6193571020408859889948870525229284034709589565079019367478712791
1377847129465863106357334056951213765408176795466431160686233057
1336777247204798299971903675824890920899446566780635343802261152
9283179609451019506688407329178690179988356220498472705225172811
9243319503200551477225312128974667674879760881309691267053005236
5928950725206157492569502176138175658586226431177400737278218526
9716550908245665234653649227333555448808901803747582489514750874
5551409960882188394016703904625309997246986564623673661258198147
8702780369758107260053759830061333678596766254600466275323118319
8541759444888698568797983082742589766549142042580958692300821265
1047551777641077979336949336512242510071042770396946989740821951
```

```
61956114330817543379843499655433517966956630848375592522071329 6718
97581496426393480818553342112072551408631992909726890328988418 6821
83760353729310036845899592523352071890491758736964773122447352 1172
91212662516530909621662205536108577349807506640647831342923719 4445
45081758062693986856990172451031338840458166065299231989225416 9164
30549930615065700382221618295491632335984990699749228608946432 0719
49621080947579421808787681275266674578240638831054648958972116 3189
56562734955371434492734761925780182095114097381737080693424160 1539
26529675217443137028714328306302580222647555932499658659063660 0280
22078661098944116272002726505573673576375101632913974851094003 7023
49778336062498583606909799681118718879917778646819217853275904 0474
22884705523018568052556300017462327061661963617268445876372463 7164
40704131845998894418676587063696289239464111455154099262911199 0549
26737069469288125961671391784943647192330850441392482043747980 0450
10669250527844701849638715083560549997340558043430495482251845 1462
02395132287695145859120311680438145222335978811463641112190831 8650
73462836525411672968057359047901097528153329717436012481012824 5003
76002692351386497398586810996766531244027007798259756638300009 3739
73357033167087285863311152154548162373698400817268339933518486 9952
66625617152573952957986016991181176788412119674890590091175267 338997
57562238577074872478571656723682708747627322789411436078618898 0413
89505754243074055241394003634537420179950394211068359688857440 0488
36192003955433238089962738209613735140123099461211638070536744 1090
42044209589114291619575443153063786309906566472504177313214103 290
09797152181711760619490789543948941807986714766945402349987185 3192
77943492533740378112204378300251150631549249626871693073618762 8783
83863934802163894825256123294750436296563333161982262700311698 3652
39946526321481051284710345908533925992437366215080295116599109 1024
22527697629594816581497763325140987475216035815230149407989959 8446
93392070690338330404915886575777495569420890261795432049026005 5862
71580149203428454079102432447344216486685399700685801518279738 29994
54396675846353468190547137184032474104891580533516165020541365 3343
05712634176655192059037906271281827277066715652287852987028692 2037
49549531284326535876773127320392172540133340382610987003369351 1879
50029054320448573522660762409529711651388155800481260556474124 5950
64920213233466921842069722177385139248406394647867141339971838 9277
36967791271657830887630274027734941812820604705397303328628759 3720
05514104875099777835646063135700236000382267537227948188781287 6580
44070095424255861356493273899525329480180718625788778079916068 4179
23842626243672180665555858662521508081650935096188484010273887 0425
65982297166457135079841318800539037852749866322373586893410313 0770
72444917636586992700983069573185318936501539810684006265872046 1784
07106825883517470358299707735572243404956177753273399428564722 1119
86849215585809985896254532524522524519896489290258050816281471 4288
76980992511336356793759231933622813610963050246523111584583398 2740
28296523987020599577075875994200252688358977950744297818909122 3766
26069628242946849463312866354597602888841767207130450075964925 4689
71643431222722987409670010837724426811279339593295466370581600 5694
64648784579923627710595506470621583300732766761550351677506502 2207
26501458398175110172558700509866701150816951374605551584188686 1513
07827403221161884280623429762738902825559170846628026651753565 4509
96429557456344816720180721240568332915834448047601885172986488 6914
17061342249505858609565909454244410580229189002205119134147460 455
83175305093366465114644058381613530121645847008095308197238627 3019
84094215017696612205678343928514733167415327266868507603356145 5221
87305917974880514071520104050246451641717526949895728862314361 4759
35505690123192372196890219471302849886149885216526535933940476 72351
16132399477387131988591017157270092674080634352700229307570700 2573
60459200107450466247367123540830257190566153349557765352426230 1913
20934945256389456449554923869664036251734213414093999214215292 2055
```

```
59372379505832829326154585083373672725774025126250100199692650342 0
72809300545177298315063490801425750565797012788399332867589786048 8
56102194474211819316634494606581004692003431097782126034221127428 9
27913402996352383991394983747998772094527433766913885810542782137 6
20621564919270133670070452553823856537737633002751118906778898406 2
22488349272712080158217442113552839238161521733554277780531592993 1
55504536098035830353450768398093725554220292914624496504146086923 7
36257519671674402508781600518729706077472908246624970938805606020
85012802384504731014923500143190443502075919170611436637686597299 2
37102517017728496439355445305006093398097527209239464681067802279 2
89456510068317342216459519498182815448376879105172647305967194051 6
03253407788030021670900835244351702743355692437665702890756547136 9
26498111719803127890909848416872850506184134299207477869552715145 4
17694918585194454950076440816235981493818423844324714953136296664 5
69031589350632985815831715689057367380045037236041575748795612786 2
44758519618261409322846210888076214577279303626679247829262080043 8
41743739574212771747163088962260738993421753228660725676184294761 9
70190954246261208921015823601461757322510108406399163354304778820 5
00242509355276886692702078667729971459397130626962134526078397192 8
72447794358234866500496870934801241492253232665542386626822122764 2
91249352000119146231169760782574351634553208832243619631356826668 9
94948640580064657750196948593970568155001062484937762685209764848 6
86761220357110048957596689640340763925192328368876340031610625619 3
37046529797368461552766412070152044343581375103638969621209650066 0
27851120876840732409922145385778886329970088094482799143651804394 3
96470680559004084197075577921110929525696259073617758978865181540 6
84446924733503491105800311484014952381429599363163117702670812756 6
62417518927304475703501953709404414093745094826691328128965586674 7
90555011104576811140710876679931530796091619993670512512322602296 5
89803028077993635019304643084618156306815663243640928731632065618 6
54459515371793392356174163014508584133221694044908274786800613884 6
53697507200877086351259298657199345779579532763982521128153515317 8
97914822093173442439447691440405833116819015016099329906502587819
24169806428146345795525872963927754466594064031923314152186006099 7
30051047574950718458968519499393718962162281364680191569273125595 0
67145163307432446339540634141260658841663371210905719709396526075 4
88767977609444889080105331650262322932845855369922263969284955162 3
64869111149757897165348122070661666454722656650394103665236097401
72587452575176136250042924747763268580969511038498413776020768368 2
54989243032445148585210427846215319033193856392571606723717646767 2
58612333770681252913963612873371684346841893627175788497542714210 1
43618259460201470422840389187939362462429016114117006624308794767 4
57083880226250250314538940486199246305837504922924065954610128863 0
68887188305953823407163283924160362075275573695394775303345400142 1
08085311933764431901485288477089309750489534056410315186366665611 6
68967779193799182589989455377707229411499292955779025573355274370 1
35535204661509319878807358211105907399323602861124768066136339579 2
40145548850308353545606565222470632916533116932292749785927518363 3
13769427216803876939288693924137191990645988173249450324902565818 7
25848991031505292811198207277549177054017852298272868867654294026 0
50149567842826293566903869797269754564877180809167066067114735377 0
04277873130895805633754235784583462352008172391935518952576094363 8
90334997059663491584838563249021563616529849851040762405205270386 5
70811390182862509044010659243243799228675963025896177491262028654 7
07616081954765953043998889394737926487428900252410127850706984150 2
00225602955900886403888262374202902750910080767320267161087421369 4
03999669399676948437348261388892279941775338906016634202990859668 4
64642856664323781328565715989621281075471633608453736927317688114 1
52703188837503578930251340255894647276417327897138749283942889992 0
86971438751771652680278960561025404273029061648414616639792528596 8
```

```
878993479965571579094773515671039498473808254114107008154004189359
024408319670007112556692617102824782589729787670392168514360816073
750761320700448559489962056656771717452152154203916826508733954131
707795007848568954121208481360641080321219470136523538673911890336
724182123707753789228155168617377181712963358868664886871151121091
591992844471264619638114716029223304107932796230393973761262613880
733784883218093112166251281144585421424157529311117507193185260849
680150517666032242827371868696127908931535465153246052458740453762
585761317859079655513988269204007880353981911483021595643220528525
408885047715065592820941273421906237590514708429781424899448257406
433155749819631583270724974381181457359478042998198286740184346680
261904945163137372856217386780933476306707393757304132961614736558
838913106243889576303094964772322184119063679201644864910917683873
414108777700091751591084263165189558299876220898226750775600143589
641941219880614564906835357721321107742483007672050561719543737830

```

190589216519305181257910603639754287072252477877891298876713322295
607840486797405900969031466152931314283833262823581689509935793876
327688918752076227826411922971560295821028691147486119423072358992
185942905762057478154132846168740241179679798054051469190874439086
453682521129504039955641938193012596561307860369994099284472926932
234011807884495789394673813330625896393405577292162849884437622666
086493504470061552384885720114090978276140256677148155411444454780
818747022662832234918234914644221623054700563762669005608255052206
621812778515563862154326861849256043930977021734746061248519535558
107444291837682694792386184451869030192217203928288574055778324236
653819357574865714711261136474742842484983871951155844878444278333
633300010771187955075199361207814928201787142808895983613512374646
493402007456239745007979441940527042976076791342981456263501584027
022381311065177641788421637844133514019150549916061201969266475180
647572355475996883011725462238676614966204874826059277386979056535
939970591275036932333420412195462572750606113950809339340476042815
679526398659936145466166999501232291785987019081897492413501891391
436845965945416048210123459309086028618931733361083631646858780275
225172836839644034831185941762060663689236853366954013180584518422
509746396217658331330534710062439859991817983440475094957934103067
010052873859484335375566469246518107147078218287254462133378158931
903142873180023011180775882668836019450311325405528444763210447368
050068411286676395391170426327676555211180131488251145180106597744
661485774166750371604610783090483434463836616589346744417397315700
332292443747689830271427798140815084034169653434281356529060515610
038724746752419724914728851951876177603990631918154935969142573829
302380599174260150198046072807425888009057806421470678568365785656
338239729246142010296146152617614519364755031241409747735429238059
394538101539399022057050895544770098919326745136868381407085803846
169827799344172318155802017764359811009556289049227727075265157824
222835493163861960684643523972006781486288845168780229529676435149
573829213356987312242330210331419777921629002024418020921562516960
647504934763113144043856990805350585452140508425576345518472750357
804059496574125015175714585873045446698119372434946188573028497356
006911782835641731623605076163791721258082496312759538189412034897
575233689748497197636350437620548135430682861191415259461455334525
570157035441735277310737083871112316497852841971475018468109174742
473758788227583387233545270090545227461330710911409615151054200398
718518731847880489126923207619130209809973078715604792225958398667
392062340184779690099386022186465940345377822601440065830490643893
965389712215079446408345218383213779107041014775796634782388744903
793302282823992375452236041880340947595909383710272661372000195110
454109616038270816408444479517122082044121588031831942382957743135
867534651686334730122674390415827816096757655171199165916263759273
367987785924112544601259922486666798121804189127793751363243857991
967757749657809098739748903448698503870821152072944093535461055814
421911472637335210298245402464829202997041096233963265921596553375
924609184005719436798858434670409813860875948603002141611918447853
024220020814115662616033225645371372274427616972053633976238238706
789316789391317333425337814007845796795004107119002580391396821294
845024502016089153442950827265810833625023181398117181302566483503
683526703813715290868565446515227647612539239801622847981028730229
917429632392114035834104917568016831142748103812234226602955434916
247476765977115678686657076425488306499793759579461231814507054959
964893286570014994207341245719371853408617364756667621993990345132
747976409547030740063235301602731902150960353190188523012981552802
175030848172139353077087351218166187396891806244744726040608419957
022232713089912007478836357317216269841017368239715268326289037615
958457306023405848549468261901914344716761828509774439440921409121
819495047098538360677664369891759817970112833851092541921263426137

```
1777985000301150696281357062876395879396564240289061419718685688914
1132194455731818074016563983794061594381452430193352117334879069361
6359576053282189408007675267248254521172818213651900751439633669339
2294224459369532903748583823506396041415937365805149325045538674 28
2050802873058138722656365780816157951018193451546882105540325635 90
1383756655255736569632808364545197344063828803859380367748890702 17
4048915596320796138962049584718060588704526537837336950162990044 10
7596143699190662195306126835354800798088930450517072801360282788 39
9932401321527021775199658263447214982785811187081793407781655300 41
2610771925163633647573490428415651139307153617683597579369043249 50
8639199854225904440326005898952687759110440583316520121187232425 69
9919659582920849008001763209343637445609489851583894011439420125 28
9183999575805043785090569162765631658537281566211783253895109447 29
3438525467526602923689517795045246446278774616788488519756458410 01
3564132079685421622379236538288457756654471167833923939698938014 977
3625918145462547688955108379232578743602867543698247774210022440 47
5149056387077077350706233945778671547603565556948042851234711073 087
4890451405089053734278811282967299382262951119775983891715155666 91
6759634753028320434743478907993572788754127763041774752843491869 50
8622613198703140773291196016437155071580277758403189347607190233 298
4686240989756710238252446451105201946329469401767572376110268354 11
8272017137211290598874389350856202327175305267615690337856821 967
6091889701616295769842701076894495238597619000208524611066059283 26
9226587727096344870418842342965071712464620698502224744364475917 13
6591549425396830772342130669565596877328825097632414427921526514 94
9926588492394271224497192552770676033706420970688139275561663310 74
6822184632226977121490344695573691954883132076996927330172882563 52
5631730154471153519565444663536635018399475970023767297890882617 35
6478383611344767791526330564378488084762748661914968656574741011 29
0713010683488020245209979138229312157458007629271946023043840098 14
2832355641238387202894273597413440055045552891291344349682514817 54
3673638136132341629379589929361327284156863288551811684133721906 44
9944684393051036803079993891575189751257200400930624322252900854 59
0038203185679243443180420966221356045113567676002151063057744934 10
1042395428140559201985688571178632248466548754495205695490237428 76
0608292635109202946679498824977659225937523148712999305496801268 71
0372687574174633715298338833496404229525874763279045123919843468 56
8121962953005194018083114743631689001486211683338804683589937918 67
2669313130237734439713389910318145043368517828010656704709323265 94
9733589133491199818627376634783970073341224934770444767765180534 77
5390091140077694554870998669416112698945431351739473520535575304 70
6189234816302441375823589734651600529193077840702805768782393900 83
9344849076137331215062129868471937764189549028035896562575394033 23
9494783878947603723528142271502544355888582979484764740618423213 31
0284837502203134060908163950950539968351154932433337246180217156 59
6750921764498041071175454426770205214430235638905283502713990882 51
9005663633336410244364541772045652794056616470372581806725557884 06
6968252019919030476285024172135225946072730648495021991837601326 06
4290693974528715722339037693727875468406600217233244745649689821 12
0907632261993183049170130157755983312230958375309503224518075346 33
4721838395709078314020414435087553239689203646224239483430669024 38
9283001520335522916152405710534682035036202037996579830605582541 83
2224037829870037960385468044579405776427864734030620801281185092 65
3200265361061315309489685553823755080434070308918716632833474431 356
0602087658345841241537617482000080212677758529826342167639911218 212
9027850271227022523803903381795796892394819155637405004464920492 89
2367906201300812694323487737253523659608806519295883629109899558 46
7244235286250708616848837131846257857744148118402417991532732951 77
3796370121455554827783961235214443137940902828351274517844058139 09
3550785706658722515427398145181915577855923498352581291161971203 49
```

```
37387895339152456908795153669529686797092340000923108173774 1738852
55195330275079503140300755614140839172043981535804407104031 5001870
98203695405828287398717368409945572949326420596075798094157 2697028
82549536632966959952545071324876284878139376838624789624004 7965120
20042563607788221974042241983920749087068337811383847947358 6453121
89621657332876499492912211429531725387803986601263075585824 6403157
41256789370553917184618024029486339923011376064862064192713 8718731
04382161596532757890284472456146112788020479371563613531950 2907830
74584537989531627721768578158228682898511953089573620847502 2052363
54820027791214154149609228771940482779106200037730836299104 6515603
69898968561499574094624437007609820303899902031006007138453 31591523
74357298105147957439178896673785339977251511265330931568706 8825676
86417708443452539721670696928640492190499505653405726323471 5699741
08497091904931507513394815219482519317826778607346557358412 2177332
49494878718793477140340917328747071903820436587370349510108 6843953
67925249397182128247303031558805726150640020473330524038024 917633
86426175202469420198263925568804304199222519391816262580494 4924993
75299174449050721179760867400883841305997789989521624028806 6155011
63884171780126054468792670484234828300630552604137741719522 218590
43820716520210290953990487621340562454173053101974852497163 6995174
32470456760444061400206812004126317838683510737421478710858 4087219
22158386588085480524697512595185146198024786449352825698620 0217495
93458588583267211438669138448742248853883230737586091781287 197805
42955852729368189438778846539518994604845992786876271543159 7898759
06125272181987117919870861244626600234396561391151574835005 1240998
03537292108972350511716773165354674600133628478448156597974 6795145
88711256477755804676930583327776294965732060926898995263565 2348178
41896413901667988219969128287616601096846874292425406977897 8019520
57208119291959140383842657315439918873315691817970464737764 1304749
85037751546481518470953739059658262320145754892013287098068 5186996
06860949869354836509699526177159796209143686532436292125743 5523475
63187259416109682802913190636159805149150440848875921737777 9462173
51010059509907546694262561506238324223642031081459773208351 1639951
97307934070806056586947418569308074488660310167499925902322 9166827
93109521633832489294706096203222263797396544914457058997956 9556732
30297787112018519697255296175101756353600153217005454772237 3017421
07837073954538720688538195550976283151127571431395495431439 5581916
18707905182616194779507892984510468976196167961623101159582 4118524
35003979580868497195733754716313429492742805027247685756982 9524698
71208855535490590669753314397936562613307357842316161839606 0466753
76505556642629334031305781039231585589901907059052428948175 4540647
08978798600713719255703269797355716521753818556172865971794 5483632
62872553030117937314073829502645595335734551888683486842263 5577288
78780462460003425656999470549969546283032792315285249305810 7416405
05135127371543201164652119969629106602557761671448740237198 8134553
91547464952504251987195769103537663203377001952881713994167 0653484
79121388433372425682759947913341465755524118015958765585707 6407224
43497164909660157788713594400291580320549884461894577086681 1659854
85620943464362137750040312745104752920999334387415529133296 9605207
97180977834944370362354178553448076340934905476287310221034 8585250
66607697333499094465909851102469034453305876807371487603846 1476711
26029973801847955908009290507608107870440756452245464655080 5234201
92044533651564777758075064013710834534240323526046907079167 4211944
54587496220993143456374478206894534646731666344194718778759 8556854
18133859417634042416238570369980324423987700423403070804343 1992119
58633793071613726602808215837319236277471124509988844151073 6281477
16708228980205958357773838485342658532670761983507327308563 446015
25416990301394340668454533430688042316306847759617969056003 70378311
67477543127700672956796989844653254901475618410941192277780 77848167
37340663899806738595192392308216466584092179673401584882850 69443864
```

```
7226653609101324079292543512097333689809646832877105859655931368670
7977745481869471289562080939357893150298101946930150254954933563 08
6747549421972188615254920501713828160275513887319766370881738045 29
6719788760954026232956712384815624943455406774705343539508573750 91
6354824236720894981237895461903382410712867169959080532767740149 43
4830648832483603144200028599027235756598799818205993599139716201 5
2819529399378733445680309093003923395719821168059742314766742136 76
0450289829533462642803751200934896773648656630098621265019263864 64
9500682721205530513258333012440491367675785838085624654728359547 57
7718275988086135568301745753318056017308454480416356151966238873 14
5036723928469017191434459822486277681784673048755682512077724449 88
8796638627783888866025232823510702500278657877905244991663409417 11
2860792993518034981259907946449123663172203502423894902459526783 20
7868904367176865810960735666858596685176029563815565374723525815 1
5861174184278218150556409179592479805770620296336192473915937592 12
5186845460536868909220744743374997307406055070045507388040547562 31
1250189369080727119835984811533583884437258612294517151676901464 95
9854710825656734154444465597261678250384267736871279506242483209 30
5221148388761378541244029432406143843764587009001940706646998368 25
3117803166576785155146872830883690603935581754524879903899827957 41
7405727219916190433921025022964867375639913962288865960601816396 44
5395400824409048226326340197260966102961871852468210738311355136 89
8412123846748760868803612742773149986603855878839713795996149442 9
1431668867339803097519744906159603074630022463324657602076350514 917
4653697477084593311877513785783273462415976071743182415944014256 67
9270658691351027988668568659099800255945631882722189876022924787 2488
6928001144962527258988371960306783504914237861471400746327060327 41
2191953218045575555111251377204002246990597792515966951474361409 76
5867058871920777976634686287054961556199798693780629489206128883 73
0161301600089086699864993294146389428500672513527500472638950789 491
5212569659946912530302192669276179478598448363263934791316537195 16
3817195904569676938546700776078665418734285496065653102876451211 46
7239245121545140107504694852454609672442098446767106869738633289 20
8696372425125684214443796647359391663446955600125361980404004096 05
8573527082975342761818801258290220873662151481040031577518431719 53
0517125999072014980128991175879273057018379783714159125140357799 53
9843517963908242612998625380484416655274494810098333214891039072 09
1025341708229832487162813659183673692087741663668103618753684456 47
0033982321981111856842826050106665974850452222811793132848054025 7
6778914501617505881180581069488560700564820969037004379354780404 804
2153937190812848618080755833422984586273879873132690330281511247 51
7345928640390009520196439787038803894765678019670245792461460170 20
0398521481117680120141968258659437701989592270669983506276957246 81
8209665641271308007866982948897845121223481350313581850065358286 97
2225202218486909133373790695655354273114413089296241439046901450 89
2954797312841519966209791635669705350195377923829588478810126154 68
1282515165329071590041642203965973386522013551927698653192827262 17
5964652698522859531529799653881748397486238984174336645361503916 14
2884656739290458101478395414218389891084741830266001845892361578 52
1026273923333582487317347654279106362883037001007411681229461753 898
3639631606244452870275042267454054606067483471412955858758473102 72
3470765467406448727857769725122042414124113537184026463181570336 74
4633352041298713449514122324222055981854058823444231807334624016 33
4790478341781903472302517071429720120869639351784565330599876349 67
8784198817630050322854847856834952089678360809337997394346978144 96
9934329965047850730639840405640606541955408529952176279498862300 12
1229215425874871748466010595838015215791009175267597045606699170 03
4889562833028943523868141026437964936317596285699432750136861067 20
0165798484638017024534793775075235792601660006180887276743519766 62
3617840948673392428372087093989511056478353179622077005503103987 70
```

```
332181974091098151544360128101117201200734391242768773573991471301
852760984061746960689714844314579557330036065437633854940792167886
053525951280712545305475541530356292088729596062295497216198389392
743477258590672865608137366309985876966879872654456650566988899281
924288852699066484468834039294451529617923917312556530975614694732
932041957544316231040523608496273429268622992370626063500697772374
564950930179591702689633315236287012141681175091750287778057521569
545442082988137945745416255791615646326761921145457379443408955586
250996719739708195111968011282334176560660564396085672533691505354
324673777309325120286402710513820041587630345803522443400548073580
995065707978331264947944391883368051139021101487973704312534143270
171175624210350164214905192184549927724346396468677333318461719451
601635439061014137332819268873519951103047498663914067469915378401
148630346944122997481578675176795193516547201281276102747517668716
510360524223262404664421518650546508482366127985892445259934250051
120781160722345003666738553509826044017001542632784670191055023984
305089319393968200686964151214676995818920736333757005171591388991
391581909786178478365124495826715265895968475027775768624495630577
738200697468170952332563514921554279330540805320016726722973788365
270135465130870551278602148966264504984354149566274968296548455079
968773013487376776588720009114274437469362883540130660491344062114
962364565453978472108776202919439201250731405801919776763577969797
437840526848545401881128007269220637504064963408548381520387095927
426635251840841047130386196170111767253613585806487765431680551569
544702334970065370907869407638293571882661712519463494749937926585
552496505217380889595640466917320900904830718021955483985307302692
303724103486387118591655364467151967463251586736347966162747576128
636810990366226222798170575254546810081499367353581304083043067792
442766127980949375224638557414247113414835149026669456562595917258
454572162887387668742210846890509973552698079613748983187687214834
729980687855372901217544477808240188596187867083788312953135161167
238607980470810847770320742272993362372907583639149271019198241535
576979839494423502699259775938621180347254547116398072990410220965
888779362424087554004536575226231446062297350460041811254119109724
428883595837710267942963304733030099133408884628021954937482610144
681982020868211025564492347498691365898251594691388582908357601164
373450780286683693266408145605974566464457885927039922974850025790
051246099145116954941066460148961564237043307111305124903670429798
339302876782033620667911071098794813633686708146677784257189065912
842395180306041144883478989272413106478805555066197873818497143888
491250286501705562939046785903045045135161033683512696040604607364
317668458638337529144935025468288758145173542063223400843291306898
497595480020534138200634532494713402531003368772604089876027177337
571829783702629992955006310655896015495086888929062159733217254731 6
686112589976255384101225612384766191913386012987170671963890330938
598411891729872586818747649718570367017232213969825412791533745388
859942586492094800048494076686995251982850861694417054440683867317
987148646915059231063581110952977296370945941100007601585022288879
332687466509055423164439223454737595895160582499992538777767956139
300327323319293068837034887868786756405577320591935045213257682850
583926398226617251579561931295486416035758951150552417356412893658
506730919263558435591547724994476169851464661475585842519789456519
628675924314651661909060406124893454390423306781566137498608844738
679593283521922933728270258683810324150127456186784470994877648078
691655658660585454999663657695961922933777991219766901558382063500 0
637303489053336338517875826132648747269731906240594692788218525 96
010144861288823745783348650576281938480679630055283108116676257359
161391281134438766054180919518745836838849073347706854410270296 75
695956530113000328792321846651386019410590887359221328973140713577
256895211021825662353787087871835004354060708002654324328213637684
```

052798234437142606528012755645349017016486713824511991621924070407
917734895941538136293025424318349963071585856237809319743208490650
873355353309338837222864969295756670190349752736462706941322589250
463535311289938802734555499273389825561073426329527342647786393027
361758377356847134033429296454754251031427748715291593268068806873
088482349868645334185673092397973565454826107229362325660532481833
795571078528433700514676251590468812386402125332158396052997200291
122808727226218677839091090339716289219920026319213308033895467392
961002694616867112893396535431012962449240793275054705062143969527
747959741353556545228270968615543113028139491316544876415892976516
683475501205308022188303579567066205743863569832465716643108451576
609995127488097462159329067720631836959158832027387616281010521054
777138453874696363689693131236684056947396989588400860717877598997
199835160477700553685286987322796269953737305039663813103313975175
117028415799042803289977485125108202780906285876179997542666367957
883913536655743988267567964913194842972667318805202974026365312346
03390546891368385461763128275339322094205666857657588863554009924
0772810791587918967281938483878661586723793094629539903661757986712
670373892194459680714979061475458619634105928522223615050606552597
895753995329587844601220109724818610272448391474393431835562139573
399054713508433665733909905940739848029996185504977535502818793851
197115749565875670021275764387140845731232196369997929620935851920
953135865106354193200891142793670522556147803465585548593923101303
587586502796484850848188429744731830099273042651595886824630862036
305073716535129343311513428628983758958492532136197360116354311350
999913509475167334613544220436826329945245431032345475768553681987
482311953391003289374814397847672626377620439346144875057324761587
011757788355538426335276762241311657387590277017241092110143017845
435409987274620132427693360447497654059738100364471372453776515977
371947545176880715766533964984779367265072528486886139180789749014
807421700430645415948367256083287928943908103038956357217794185249
571634632763692089798231696655510419008507829456639390242631728041
214230354441951420786078427124820633758110380509353558504792524992 3
241701515885236499695672722180236730935018398783424582365853881544
259940128402769038532254899119259282667271878723140806050675608113
240896015266395080779287934226706247271930943768464840757447211668
699420945364818741048505659975987829830774699824714018555718247 05
914895853004822120895235589145973814964070479939856189528044301 79
298927181559453258581360114665409966306009453564711479482124161371
053802447406317301152365647359814053224715879398095918223404201921
739501755879115086784172787190828056076044262483191795522905382926
160583803180366695056278253351109401474385272401651151291154862398
502927070497198220937057745182808120469349478989457366845606400787
525886081860846211413434451543803326311226421170251580820866210117
614333371361814645017266069708809900321437531729631770118681823 16
455594429487039317663479338946951628196458800892105779297652665321
879856596957369794675244627901079883166452172240725783942417823008
877811311358435021254612767932239610205270174362216991560948221098
184714753763773588754170488318570682406519559260434507987912913 06
186272271870382379312161055499627559647327536117045089748499179605
034913325319337781171165309786791245148027589806096381341423585185
583354415599893638300396324683053732697386875632406108768956235206
689348884155378112241230133476808068162794265703846202895975786158
276970520575397567098041092312876709459755482237498381117688972705283
326005203141070186579376697299915568070661157684084665875035555729
524088105916676204546324594083773981891311193738934961645225894970
346127605742613600741031994518922548078322326025961194019645697798
988135841602639913314385148595106535810337954056089399067603545067
285995525978213042506710281116949258442232648103557059933388050988
584303832454646938228538771855016338300276379717946729697257291589

1348860253612037138999177013996502644745354711704646775562048181 15
4426293614245033173527945506215058768559700259333766229006792304 45
2712383074300095332876407181178781292591211449540901635704326308 4
6205215631175717236959963935181136642666301348566122622123067150 71
7600110959778813087054182077927289183511573673759149840786610151 44
3934698852673891014220364769472447073735063489802833177466429469 11
2386173018245755393889976693421633496020160001595232232203180076 04
7331825105800418936319364263741233418315646735387552741857916767 50
1466187631228726232881351867756525138853906018707536470945563840 36
2264231885197896418117131867238456712203565152348812244361593428 74
0304134867735321811819902715044929174874929478999565979053641006 36
1255936387799558834652786022712898629409984747973771810405725532 83
6581006546932720926454705632821154823448194571949274059103657298 95
5274789168307997233246851244982110777822911574489821576314219863 41
1916909225917205524274583930951138397879455096892119149788591483 06
2602592927816240353535287891262743476927085579131469934957728333 69
3282794246019761712442173457041180703606935383311082169365712461 93
9012988069516169577373750958075347753649956253168713715988097010 11
6908840606241889051251268375144991992235540271548589084929287984 10
9019902680724531354118236183284112277646749721578109018233467652 22
8530446447830752192532637401788915162782395452387174324534013314 35
7984596018306781726112405000588584719028869122438470622497694088 5
8908944125621923455596100156374081815991431013389051690666157504 6
3875487132713521969428109687094641197709857727460392278718772703 07
4005599498995698079155010116003327238942767828012247886204242469 06
7297444997688939531719882876634178310197725368059359509027511153 13
3070029303775940268440940425278694985780051029458388290972889103 22
5828655097877735835528963078061056500324186090193431288734811541 85
8150785477547767191373334052546673262082889241224522008182379425 70
8473233805427402387267805589497452778417695884171848717728694354 56
7496218627402389790527737356825580675199884207810921672437004403 04
0211686781293209407493566897056674109543237560127577227632604306 34
1477697127204698220993845792820758356270031652071851807098721761 79
3828041719094063460382717169627058752088677166063185447995163135 87
8922820842929462559696133915980075775230865344306520986474462627 73
9440318332191660179666963669793106215622674310772371651238807721 71
9045791615251200535621533158886739692271517460275046663324914722 75
5692248335026428947278619221549125961757601034943516169147281435 83
6658569689241022672244233420349044284072342032381114487780302367 08
0031681736759216953926514536204738446498335141189850155944836838 67
2912963412657298461535363931912144092092770339678591471254562096 28
4329384659713585393585809786837960810524729740762783953533109171 55
9869015037801011194947370888999435853549250418108259715114608504 10
1680638534235591139571442373100862189429616198855196576630452384 60
6250447550748793969444597710188405632614116065335976734111765074 58
1056892785933248802057325220744226005968579874299366553402763652 77
3016161299872674934143857346336636171297464865479681818374762046 84
4162967658306815177476889640798268825715139290617474838225404130 01
6967686778045087638922218764919013870349088921117377674975493937 66
0285826399772337532726800724803068452327353121859761026192127553 20
8916283864371496467604251883980698919899532866445763824290943394 68
9457239556327263784446666232123203930242831648473231028557984913 7
4335691855811033260238767755429454846918781543101814167047120361 81
8263202266139788880601303616653008514610716592412533144477689657 60
4423865202528774563757448819905519979974447271329529612419841656 69
8929774958909538335639583059560466052696275420988627053958124112 89
4729549555305571649325754200857027185742984392635229648657325811 46
5596484831934612398897129141440117870119933105324167576388149731 43
9697575953953114229646257069235308059294855739709357403398033528 58
6612856942130434603283565477702658723038583549395789098172487734 84

```
046610352944938492918358200445946293178467568364915292084305794852
803056279125768990473567877421007576846005835053712621157594470074
608003120596718005216151553366136725577721389147941023910912543549
482976254124480683318796435811202958557354924315301241399203086481
916214777844207065460174032830492750242493693017697642217829814517
811182209882878018776347180983092823651810708792408730356936816152
718198708722583589944298174190266769277345694413397445036447997046
861705232519352090243060734085960388265854237368372484528823514682
630141995493850383808848259186724579794030027561914724699987083480
606222841755340173892726659526469320852630623519148642232806648800
767680027561127418896704235888165090567086045976165645166575511705
254555255306405795799844422116961691046727216014314573389986141954195
535639729490593218844806300746566278341117991918272811995101650468
788638203291480598721554159739883679940990591322804422181494396039
174908905521860580832342993510864811537611377188904218267487737766
631752097311907460692313323726969343942222412224984932038851220103
828885885525981344970391938907545592235313026345545706563757376761
242758972725558074079665161753609647052979733691382355812547649622
641771632205599851865280426342843632505731755803494446675175402544
304148489982377729632739300010939514406905409344542167035067082074
441544580447886983304075752894112575518448294819729986496130386828
114259238141424451883768274273499658076137484697097035599860688374
432674603069429340467641157515581935444489156228834807001220577578
434352386301671664544697683445701647971282077715859255157245939300
246528675554251496479135629224149385449069260894738698315222988899
044488427224415696882214377901348945440492696551969838250341743878
291756479638793668312988090150366015348565825559854297517981845678
365897739363401827356835988464757130703908711730053235636938554313
571530232295807030136213247587436608411376517175131521081923531898
121324028862993455501016088637360505178542009138293801361263273496
416700756502572811092096631130976115017064950907079533630318180943
181546777120706720544080094557244968021173975168995589045698972781
687614343367130472012646057350903090510693849800937639876629402360
364103225051727889277936029464070608864256886103017520247386682061
040153329610274840749433986058643031423633303712572216176781754397
023935589931188284896198936689306530935510474803151665144611134741
871819830138074989589448982575496259543169432371159267523330784851
788278689582598901592828230006846682803786517012893494411226835207
127385869571059321094724752586939733221141435534160382364269906947
327083758723463525610842110159064969702944961310616865895674265923
276377126283375719365896132088330578424785105617685375568500206409
749575747720992249574045230710359166628371038250386225355392560505
002132695936257913299247436338171524718617229275859275503779248648
612046813445909381118655188990291388337276490068187558298516249664
103584569041211783329729729462850238326216283882301869364121739086
758554325619245107966028438426261829293544450441761231292901047215
307886042253390868113406282962370712045908582654600087993519085676
611639033656228536998121181743182810187669868531889595762571536497
198555693811098877078604766878874177992555435323842339153868244681
163347048336536944413096221561558183639502798454251066650059342285
158563182845841392589208980052514102346399634097500185064174011145
353402799695586430482912030362691208465975464196492154305755320335
532686969535516213732429221970228158889402031601265330555068535059
114766740595725372016542532633363217012880544311047503245561800399
566123264101863036985945788003541324654169823380220623714982619319
872557621523654262748422581306423520521527538361844086703859423155
452358304415428786711483387262176833444646559193780763868541538641
338837166591028588778351035054767270746794680616977103225093262439
624778553577775298666833623822712055366399810759197142993513591285
355393472490187829772627433099552429828322063862608017168612841295
```

1052747869225709262425429231842774727862264438384668576376431745 23
4437525712845074336756535559087549110101922615532257019085712649 49
0813155998035513435060754586509982313960303198469089128650556731 68
4772150100785189302627592539655756997630197702660821564964775938 76
2973915771075277111702571877334152538907486340071971765275752066 20
9270468660338995873870301961401291363502586578820787128214258868 01
1157424565204025299806143834124121338462796005996155643896138717 10
6718540013415002409908909512408965749769488530228763036539179474 37
4369436026351840466405676578000394651747679034581563208596443060 14
8217570760886019873072944181787167375496307173366206913234174066 72
5824790570054449350389157871089134069032178431350265553154186488 82
1244873023970766775799797782023999950793512220262294300174660315 38
9659596999873427652055860578142428862287006827927765937063933521 31
1003886003001582693748180458049934297782238044786253933332058220 05
6175307536007991795767588842115229949890583363576953486653953910 13
5639503229963364980945755744721523142953756243221035969994041767 15
7077600614627209121554444076589400174370541804053750694709213340 19
4138793684840735534774817180933952555319822902606870527188424404 49
7564198637635720116635513313871753284726482151307916144099628584 45
0642974807667319243845787430935195898823311805351672381577369436 45
0732290248091069618736981907602603004343078562376840795795826120 74
3859154764478412586370977216595454842158997867200453033494605798 6
9312204514531405790417363024565170452634357185567525898900103800 78
1705835951198034012969266394704134067349458327413689932620117740
9110223489759916327983787323902303351300280479965998988277781993 02
0251732068326044282349907812020113462278575839831746861750401463 20
4263924511615637318788998867937395664356907701493528159186648205 12
9412473794200348333178961491657580174137125552926372007191833979 7
1893855308947986105085536535729879532448413222802091455529160380 82
6409754972075117604067214090915537646508825065908806866688916217 88
0541655425236675787486173661983641905379316050818294278891934987 22
0282022185720250246508112804467273664022030915720998486227604181 70
2604803249070203320435118032420952816438607927974673356667010131 50
4551236301982751900002729651151781461794182142239372584556776435 49
6067001326730759557757229158213798471712454942165221002172972054 76
6859293566549598717131930641334958060899166675887398721762867672 29
7511985806635659047681319243224412754203601407920365768369869019 76
2508853799518827777851168700752991827949510632352379815291748840 460
7559708211743942000556469540270575570355723911036512778935708826 77
2446400637675647238754141130634824616276832320783640365501874688 43
2703940573376207529679615256690008183298142748373993851284143509 284
8728358610579630108905668081426885292927094685278921008309576937 17
5926689786088811293868423107913060353145403594226052037634929845 91
6285550375551878051025867777443378171080387648198115779922076001 58
7148296166875803031358019924134130950563896970133284692318145643 54
3277201079857252565360765194475812985034728580624785195565429390 95
6452162854807170320194236686580333098864986859239460715029167629 61
9839717947618608862540247357528319948195186060974651525849080606 90
1256760823055768477440935339183173636116074252272465721021230872 873
2164800049780785769396482321698428604041501899713050952322271831 53
4441895623248251563074286089182149131085143643284894277961923986 36
6304081461143361287992269053031866648511794090771748393171368224 71
0159117567384687541267133395445641789406951194247289423857038961 94
6290641794219716008038659775941711000534189547260978636727084036 42
6143566455310237255208477338283997841820192792088563714913901631 71
9977343931753209798368379542419318047622420907980649551083400320 78
8635848867688684131457148210352572603311349942671930848235811791 05
1207620948619460567695466829279694584324277988720349305893133075 69
1965669102072369146482535668681319355114092065632263035159573545 06
8911433913559939729940375840177850912450597224599017301374626284 89

```
68996545053261612400612922654793688683215692214620900590263506 4446
78373085095334118774868764945214432976661308652282866844727368 3238
17869547812716730410519884840855278522645458317169667873052597 3094
49438809071352991771611179106704612363957924253181301930041989 2119
19367038042105417282451454845953003078821781457724325080290264 1700
66874954628659327541867624739193897970501396569189845543259467 7533
01609549577268176576361019948868851087077772746355787698366681 3992
08912228322948292464898719772297795967170980288694627182616710 9703
69767312812893695775767744316286059739876087440510955583547427 5085
55557258287083244094757749175151022223509026428972334454462690 3456
79398252927773127606554627894604051205342071003488236702700482 1801
63455661777828364345304502555695520215112866550446552776469127 9237
27821301685011832404399006838865279552379453131392075013525883 2292
02163194283520041974891723711810520903130074208203865060932666 9201
85913493323683078285324090384177917850045980290791374942113940 6443
55076602157556738635581328684886400744222852839142228730876341 5860
63868261773733010541057572219040321375900652208452536479508704 9317
50698924434892442610849825174687418598499788455424915457488836 6620
52971589936235305618898881568981910527258393816054258785200678 6010
37279322286721651235512109918062883437447299595674216446606336 334
83309225265117419460907720824331768122314335574351013476827873 73
12047804519327935144628705390391521169566601208918112036712370 6524
68637702306019714338814942112725319092043913029369244720187599 7828
11455256285880092327563263926928241013698203279925024946721284 12456
92446290061659298096233521708237620065952949795096142452192304 9924
74597309491744032258325017772993098564593283299324077148566773 8676
08102981745849508920919290152173204231618921733569054165462838 2326
88795059078354123775344306081551180388899517305134903824079660 6743
40667747246565635144118567824301798042211805851451920414199575 909758
06501985482527746650916513862600821335794104609600735964422002 2870
45567602998604410599899996236015234897278993519282747828241589 00126
75277449238326148762836336587494954239301638464467319839570809 1834
99473806046288412902019301655814370313954295486194273721197886 3727
86611811163917596722109300064561088578562031789157554896827516 91188
46061355062253154725285019581639929046003365343086485696046302 2155
82538079833861727709510138797729056507210515816352955090625337 8150
79900397129448734578584247575136115374430984064640986028630863 1669
79708260142445703981069399125483713062689598900142408518970248 3484
43938501082368579774767454212244048874902162966115861657879347 6126
30169839878018999822859778014428223418031964356863645775349037 24426
94426165375390310087342026258792208469609559528646697698901558 3444
96943944406190710174584307742023895301614106877121511090069879 3731
29531257356708020278953202830166339982001758907112999483764355 7970
41771685470209772429391346512449194489624472790038321269000511 3629
29296901938340849868000861096586729996444769608514553831688335 92997
13569943530670434487872208714416113895643854461518652586433586 5796
82850252446974193959699751726889333547289287953155083883042523 5598
88120263736607215651718215778712939142255110624021905360424985 6571
69107316722320877730153852246922972399807477727420434482825071 9543
88471347405055376838481589684558601371636149615703420538093221 7228
10636588940768428293746977459848522676320291975216649984674774 6754
99231170151091594087726830404186496868521535519356334267638096 5706
67578700560682857533418305400566846668456386489152091248226492 6546
55016642698758836213641708554052931549654889567572573680053088 6182
15510608248559771378278408294362283749293571408238570635965457 0964
44500213233676849396220860125301498797179273840695858042055126 209476
23990962199074007428554504110816770967880471251957701066460511 7
80981424761792927932285138855114993096829074541245823778465263 8297
49468019893971736062443479667387109250497646627976555427791682 3913
43893643247010484663585215842466619829903248645390806717417648 2568
```

```
3541273681173701031612483581704029541199943357224401180700915676726307062406810892323224681007035807225491209628793981066788837577611555029576819037015100827518149325030859365912559721395004483280816442609610764313971514501617475326767203138902277093487063137505859670960881848965783291186754949153125939447711354419373748777399805328234128958088323201240129951135754493836887986516239087157372828841080124783723613938214889994646324387081384131154104793671743470856853594520301116777405858831744490077072362022279777863331141809413684809851985210831531006919152391030520417201731215191712828624961893655356249535242515384164948480904603986617750371842390575155941257945110730500603123075863349228266456211816377871709089481417277740837364727856666871210731179457443528417145689617354927799782932153758540360368725894108716531552242868875848914580224065120722134381575531361446657995123890042500146530919802014904156555533607332252628153984305416687099147218501751417717148434988160817790143665960687841503996868025715471341284120420208818918029782399889917820515910765200470163693188462061748851553646973674113495154449682958262681800559615547729798983998399039560171327921989244308939970950396699662446245719644907666765096564248288217817687305235196242436337885513691518794726877399490000370340740078490706911791256690212319923839180793141267392268327080816007435257164969889471809430805625200338152005034150734500727434905752674719409978904292452813497866721743486414205529674116307581427307047552828270587218524077222719820930315157917505439024172456194171102616642916346784626658089220183184951066398097800909137269288458086063406438811955997216008258642333257664586669077386966015618903696067812564457981657592720751089540157762033438177775464059196422836257657823518420252810240551435106143891104921360577572580383825408981627859401585405471306974998264925726364102451414822387589292293748114086917113573087250028197363569051336471490406693055786952697592485426547421770590552738773294299974367725348132167106332936980537949861936589571069728817126332241581921244590368004467293753739485721867165595142936747986085336071592882848814270757530112480241519138162921101056788492021675400172436400078817050467477279189092264774782415422955162368360933299899112537721198188198833828445586336004749869693525095240023952029544427840172200360975150883219104471527380610014148000956632688370129555533686758767641368732218222986725754229131225618261140954038464913601579043894002182086416125366222358769906359495177924853456445189131604375306732495012612858544489016777132574746829313594361083768576684627021118124636460427951568425290532283318328406574458995719522013996359086528851119519959807565743451340814078944026378648951255407847888640113938689584123128853116831663456494229654296032876704053795526989094807632341541688004768381065613797127431628072782361018950655255974169658716400826849155649103525168134125162459596796735476611944674526373636474494144842154960920722937696359099650169475102710515117730923017079225400632181461986791458734400998498372489346539188198732367042792462719268785957435874734666263312020846872030821051036728314026001116882939080242741495823019334034935218776372073046222359963784537494497628882355850374685595594097477714658328350542997199731339676565226429076641084494086457723451934321049239470431601314174670355685819547177646285682220221873981708161945592011157585686575092136156217102698319787300737554104815125468080875099503578563026691560178033794636924052412211638842207835553563608777460304165331587912217865544327012318089345362290942217986576268224057549859817727451642718276755549569830044423379472514804171369589123100924205713005834435786792167392058079576580156339133087575334603815339306143253910707784209335509741302744511048196343581785042714637311902035500140834892179260559789524940054018519278863782351620022400387808626587107322600687191271506422264003337019371734937991178292901653807912460072296
```

```
39763039759946402985900848978575307928023943109250515705729790444a
43016694805152603291786184897032943726463015805039826085313494049a
28696082203863834536436534013099859253743844351234158061027242675a
97487130884125532174923321550108999880998232885145839963731419891a
43901609493043246497670724373826088183214028990704279005568261328a
86737968411830012361896322131607142281976361396724970778656235735a
45902145025020463619245368911726956776442327284046263585058260477a
94019253428124298992344873666118052004675755026094251423076008653a
81706562075822554259427441346247945359352981249724531628107173572a
65201790786647399519993456750530014499600315815126906244245445521a
37397254075450545033785207024622384213022086121936275815849178724a
98142002185046283978913103995975760147606389789896145730800743545
19936562598314572080723466274147687654743734511534415904694051550a
52372034710772766153191465364731430081640670020188094346268908083
31195022653606255391390079546817686628761690497181765365753968744a
01319376374062473999089692493796179231930912713916228743023603040a
51406955408277052428027287308281535217952009755001592541657539681a
65635029318417921492149521497743623957174057243254694633385928619173a
54820827262092741950212967324356365945520611572573280384015107264
06450514668055861698256945564449524682448433148065391391526193740a
84568764877602840703737920188138010199532816429669785243421539609a
62550154831944705845055701406936675601735932934461667136650076392a
33969642992782805781269771806609339456189681262696747438839955683a
76670456619604133077162848093062388033602429181817209878623186439a
64823530124519568009265981371428395601143359227418109647380365250a
74692035396520709391380749308584519296461836110510885032976120486a
81608715524069914626618079410731104640242962509924054949337648395a
02903713049262940371730108229408959382753145687175285254549656311a
66132184300814892510960383207950258468127740937818320575065598338a
63463458529552151338898280172539289415089623321631901997921936820a
16945018584314589443860227311587711831153880638812584416687520037a
11856918693482126566211380491034546178076399242906869825352698217a
69196148921067205210032332601134376543175594723369273489499616409a
34561423155106865581535391876112907306981464134184175935549854425a
77088751479258461077226016495882875174264017529446047267643336706a
75483304303029871413603878338706961314436289699377240811304846397a
61780894359186000846259351494718498908296349163363480484981818660a
65318108404131716591189821182771336232291756927599043357353685003a
02519867322784975060714648676664303842384335580617286748596725758a
60659317538773426828027267046985851741223417521039750995225960646a
05411711029395807358384120701647267091563439833182068770440689556a
77315325462848637903370046386649623975806197724469193394879154608a
36076870140133694754404830761827645746416837209996725401039165228a
88121777218088702191561111276672463687362849017957394715213749584
08118483462781577636343795335480233875994712188723684656245390557
82716540700900714917484760060350862354094604847422643043909785731
31569837293573559908115314765771056240164743627826853991847208003a
28024356825312562274475519865938759563496488475654587760716065058a
18716996645947306707389441496101265918383690397989010001538539474a
75789821629286152802939217459423983487473279725908627242157261944a
42020634417130374862600029908568915353793284013627853841644628140a
65653564681592719785394896717013279470902749492648952471436332083a
50882535430703038495008211717743041401115632945832376248700891479a
23628410634308108397875777736946758514771458202045246580403767
53222100584603312057895824079702256196472200135746602310506059901a
34053738503536286554360037046670378112790578516478988473410334112
87246049064859674665516122861290642141449886254247302762835165729a
63921396321949040688581450445705534720605313100438953184499169882a
88901361600387146359739309038244837405314379297699660140280080566a
06807926584701930050588572893501841241747429479106501146318978977a
```

330333183415918657879063443727170755202513150074381296886403218932
204739877902846949882744854176358086808354970486968637405737008030
322928517004086984152527872647289791871946683180927913411025814730
439863116791594137284181651232128058466378287376317425227033916072
555030985033941099725695533570568292835395340403320129977293330232
950259508401276588756753482862510148086608894214393499198071640669
194047245413682417233547080744002830162845317256390345936408918983
748647545965577391361120271036389166217730049538527667539356121497
604893243448194289239161327539608291322777245377092607084790878100
189447719161223257884293928375380925179486186581189817163099734098
656142786485274633847841909981282551340405962789642377885919583748
310928598678929715881356706540454613837557876333344044064341549983
774866496205003802979092147436638653263551647369562420827246097544
669417852479602139463097010825516892137936859390448817230687887923
896436653594042394835361481266030474818721693945604343268750001016
223369320109418879890571366004323498698392300213836176801049799267
491557782664789805062808943746584618160303715308108488798853011886
140268311922969565262258151906392911487441008319100384835704471008
493735132000126569794414544160469542072517869721770690297113588871
213198343200093789090982560430180106675056309022232618365933376
132828191587457234115820068068859752944074546478066002217078303 82
850039985120232841673479744786381754501680701973648262727343235989
813197211030429332157447769574270501736041850955946049832229481143
073901120548071900786452132378859251946887076129607201560307481602
316908284462254691730614157646466915743101387753052654593233680502
759392141544496889641711377125494056307816645262102591514400121915
312268685626990357235770225236609018988172242147153024876734430374
856818423433473943948522433145747724231499039817987999977849273061
344976960812245746428106830166307014524479471868205094486966637552
511159430467728840012587033456228454611697479133346863140131935301
521255569520149963410279991821117699012881957133781305516401423 05
136192894066176666128434801247112666698338353554725264444282609653
037287762056585012999939422495049723939010811293652752200483635088
500394398433677997074095488819623446529637760440335458460548802186
696612171628377346161981995462681751498678219497077035040528313538
612684251324318444496997578127409778304532165848019789178601143 12
147635368739016451949718329612651037252956176749323974507654372047
368324380953457674897811361083634907396386044132780625808725544720
460265173374337940067216612501981978461473203603423605808592029 52
625737291013627618208907465195446636800560762300619226365064290487
517376201988737248766056828086852925874835819671642027546552003889
266002328030390725487615587960013833287187017397953222771235656267
208829574381594659610489044053904337861143883100318638714229243978
446831846289936588299718333290229587852633902172720448201112997484
937666900585811548503724681580162022474924179266891042690103824878
661550123959541876761718121082819840773452292599425458611684819635
994156953861278464966310301827777849179685528345471048222402895683
685195411636786688067313652506556176201423848999684624559935665673
979631306762594437885614742903377475167723157583108466073630059740
146844516751982504016510570573729981026756346785968515680032554 29
679088788700034593008183311114021199856933198912697874535964858920
349380460386257073999277064439446156884561764819150459969150400406
743666430123298045990203889512794143419148954936431080070218487632
694728997711018942610452971682293032565854696802953284639004083253
676778517874991501139965558863998596206909583242723424686691226 27
242204066890031384674149575628259038067760434712409834660665811722
385465566528068175633932554499244019509697595179803495552624890835
300774833013327868512631778847273881285231050424552195130090212690
833436050578388270618653708825314792220303363844037697196147084505
329650530816839060312323664570166369224562768058918975971274333172

```
703235128002619126117646302568872730591582206739191788130952997231
938369090705461576882277414973909080313249760200223293527771544939
579623884029403974266493825636880620426461348164313655977405963423
458676834792226793151054687056017717449281333846576433783071392197
320659080553527915366580270289439015521983929848396924629174772905
122386639856396095502158225508058184913188596683385969029948215318
991851731890718979948729155005308538611661151957808566690889208413
392783140788858761117817806873892699296827748002216585838077662211
058558248032083001312242365084748236861553511824438331950772219728
392579036713809165897149797091523924289992329906651773377987799723
462600294861260598628399482314391583652422357428099347824622775597
780850204296343796584364401628012083850545812304047385102174923305
644413741222995298233478657762690069295286573776661911640185750654
630227906715992475494782294193310804468961727104211026455523303542
468786000844233006275689770160807508978169127196657006243669678724
716643203917419279257870516114543176663046511024418438368521010611
327241970901316813135772979040021277615745512719909962247089962277
519981646066953658684059030325130220979543796441668946154076088416
960111677554739217337908156089243836199429518619531191969827814688
247010063502452281221200948709163159251337549959608150908371537389
884845383246508598837054788721217101746452828869486757868591945114
376675902153484298268726033665657066522920112067610448984332182154
705762810928503381221634504022324996721965240531943128016703549235
357294483601343318980358170737016950266460177552101412360591290844
151262738637579435691738002463070408875279399500481124620984755817
556796222415885176755400330178398784284572881502314234380879019221
942100040827272353129379085176981050167692865845288128603675289733
893081332261489985274302924245415706595957305041282878494513198154
743849152608017107035753651998770118366350935746686585595573264864
397394716783508475899844180879231062130249693668185299615318741646
434876649875418210418545875239390199669917605378736925289391223352
913966743470608882866841968451782200108520861380064579910683725026
263259403975863597801058916098737244302844092810280514688647675229
747322377107723604513268993356219913195100330181742102530521827683
970191164782775652491725666476131377403024126628679497571509685711
574401666201059394099154721604796073572455948840779770739168593800
250020238843380591486087156205444340856505387244821433133437495877
658020791409715214130749238325523574255187870665998811602995917932
031508540792981401358621880465216319038209278958235355459365129610
389326682498911940965109403399560907528673813911261484176798906182
199050539021536750643714999048308508821277160683043333583714745477
277797915567952702464447418203262747019215094053620953669698777976
001419938956767796191533017444687187482411560280509589751094324839
647594627826473224884906453765346787507487533044159586615619946311
338020995805681869928631660553477836931998633220332629176531704828
516233434050485206337800776706666902813522529848277596743226800062
567037140781648966875848483058497474274381408736835885033119996144
525960552583036453992483219623996388822858845351312806373592008389
402895223385896699306686797559425716571399148881602893604112281866
341972756931700287977550174335569415677011941264408766812697619513
350495934623282260069557156342035955988579371562575659791663752534
282958144379238845568392719511098757590474842598021640443769588502
604952677163590958023225430862401314336778620523565387952190904017
668350574800115182339983701084039412372709826525506246538217944990
403747660906025767784754618015462180660226521884134241862327280905
214003545019402701240844008677869098658661475316270957271153546105
773552246253316586319256387405927530193113606392681717068943971013
585386302594365655602034059285928439563877622779766032988579487991
238095244254892727787101862118358854405906563422077660040688974033
145737410328120586960044504940173340423597795773580188382490590000
```

```
12458725926260803475886317957281045327028615851139962002098908463
82839187152713248277537783473133548252872416262734992468833538905
22589825558401824876148629912338200634785541104346201566129784715
889200927601139799040838529334383986286458198694943320081772538952
38775671456065094202752530590785126734331347979026422945233136533
69745286926432272075855012734606501573691554545056945021797362282
406774228105295234385100752247903982310436831833714133088956414401
23785646853485493495069531256610467441878950381844306637843045989
24140405368310795536359802766321378493584082004362068150720713730
02000874571630058711643193945421959899372571701177925345379683586
2806547378410445911246868940722249045688717102490307111364807616
849845183008620548684089544188804213030429817712254721026801015956
827169639022739913110704729284456983132680830260110232587186631
774912438225993939415075013442740561792422752117788705008204510517
14834131295817595988816290601205116093293813031651338401303364712
8462030419016559992685859622291336965642704021127481399160962198
44701847469753016237947342625061910993762595492743256061748988667
63052431522252386720980298997299966685768657352721785770865227083
20093735670077326633850773973712118329030293580279417921440142464
31250210012654222007804783345249758831111145222497110304985489777
58291725896336110679290256689982145887864312857651850689262986546
2155007924213483283852952889034287450255798142838485225375424388
93391405521846349544772035686845524827785953389648770983120642
83856466159478819391135613073065664100286352548133678476032381705
33394605199146352368029476114268204891133710622141125343516818613
30735565130625423892095937180225989702419030239340523934285419770
02428584716693892511633440866641907478051717410543354756931477100
70368565267781370428807484965524990408394339495665337844091298836
6261060285546713817289034911808074713033544046943488195938494756
97041021821100948592921431787142456144665254758758284763741545474
929825974281460687599369961587582364958189551869983266068970123
0989630321447837721189515875420532778570689048855885411760779993
85652827789536297090860921798308433999578983256556655655670458857
26907298662177751909828841263532413063671828496222815687075577289
36371961698550805791246424533743772245700587242607210332754417507
59127873328214680943463427708616472195629989678188577180751106365
2643304131679827976722047064045164539527669955410360208505154755
7944059400958826033742313384938020880106358601145846464271154126718
66584775707032164802406968074653843537436981342266235733661266358
72229161730778356307737574721838953440350873045680382320805699653
1436662774970754199013965986513390789815729162517698850802791239
41815244599542276574687809371599343649404300845934422697469490901
96373442048865221936986519174869125323943121874896127961733219335
40278098731471884021451514871603806303683405108369815699120575757
68872272824449826738797822764876710340192795135995754846331515115
30080997864255321717062725208433477540277072789972245591839798800
294699397897555292126096804123399497974955169341206898184158747605
67566622058823123916339247804349845042492443084554745561409966375
63018159894363187310123829145666822568437917572446085891301829729
56445223748912490029573013319856968215722732905423445028642064802
05980068845374513535131885576624541520000933725004992892433851520
97822893740362604168608826308403739966532239757758182810404044533
19542193972633557330699948861458698082098147752901386870273617112
78073913568761733038716752426312138110322395430600885407634988869
77193722989917223532599077766320593465221368313330454813455517289
59989197784428410071265935630553788131472846036452363291000606215
62746406755409068398368527605532159551822416156033914049216968198
69399371599813809186246698317742747178972272563098108111439312355
63528366210338826778673114993363845189546259663826995844021862686
50118619907843746693971950749487241048271544669131345944503964343
```

```
6801640908235270621179256405696703978251008392041192663107158029 37
0182644203489103016041362415517692748216854834676640870091174854 56
7574506228886071812861950500627614149747315247789608052134130527 77
8944678204745914578656226519287747742956817655709044650117739610 8
9810431589344735913573747828844695873269559438846760168647041891 42
4260814560019702639376165947986279125341159916252711579311628286 08
4284109164300504363795387977110473074842688840993496589010974884 66
7768389065612525481705182403235942777040132451245836656632764214 86
1064738810067955660932573305726922713879360197740463058227702888 03
7637396460074709299304049469149557567426399931101195571778018547 32
4400325240138377029990473893976773931658441825947245236391039679 92
5695503310017049426524797816199110073550166161693531405317976842 27
5328510487499178347872159047742215316584117006576459966573587893 32
5014817978407418880357358814253209621530450474161910644372878437 73
1424218470743573712000955508693746655235688921388648997493038183 46
2542340003272040002184837670066988733809591190437330704087281903 10
3595064279159492277331964721718944632499537978066724270683654132 96
5988157018730031969780565331669704559133510102135238979139829736 21
9657365768555981175948779677329794408563331626559454221102111786 22
9973114083747682258307437908079507107615615109033143509868992905 9
0930998015408977554466559129193074283657774609701030567516236370 977
7561186932690360036767489066915977522243936591745268918736280338 45
8132968950099332438465376199484083717383094012863647933204829026 42
2573567788673356059318320694597135231659259811624390100517842610 77
1123390609625455248631413337103659572100206374819106874204922708 16
0472565512042025850429097836317570454753224800187698877586582965 89
2826558615896454805438203098444956987929436495803761599576780705 18
2002169101374658298581197684459666124191661813554497852306986366 84
2652996903432851697472141526059435521559065628983153184432876414 54
2699927384995400582899106577474741973442411726352822003721650463 9
4723957917125052764085532343881180459312833589603521245633541501 19
8284919022656148184132904306253015669319331949467427636383735601 90
6336686616267704687557292534530613199272463549291378826888014563 43
6843925180086627401610749973374793597040038583385189859714320571 54
6515241076938777988342009007809302331727091506882120183429710734 02
0960457865229595072706540125039834089401338781024340069469119313 97
6564426632618983536691882413579693191451276408980476523351186923 35
3972593969930992572998668381547851360783844176095532003185768842 03
5913567840984764495260174579251448673121062833887072832958936472 41
9751070270557844463806684065445443221168348553921349484154567199 40
1121106800775351159027241277369987524704434156684345437334151678 39
4673460865202089295357820307194678282263126696308259540969876509 73
9940014233171782199361444572854668771198552628507946239286587118 26
6581597068431770333981982876505110868104051787690507752102727555 80
6725537113251801245001611500386026655769511292153177497548192830 13
3750124003276100134564622196346641323787905695779623586516807721 80
1285999129383924717138121188209734619879001317397464313355831223 09
5551568179440158831047939592328130974604267360587285091258343665 24
0396911112118737836776526665115550114288084697174389774831989387 20
4800729044260854114481174192142658592658338847683371049857428626 71
7756548224095790351944103582118987507288996081938145204727330489 02
2554045487388018331350838194709661333460780477358223844363806016 01
6712205358721047596484451701800092205332494083842353461489502817 52
8086543588728854741618376612300819093278093034369424802180645075 87
7702841899596203734213431793343309017836071662599176102380652068 40
7929476765535072682046550486594498083402954663413103250931261189 89
9565540209946114939692218799131738451876872826373531304161238877 431022
1676886703044054082613229959091034280663070119737981519360908002 68
6021776252451151646544471679739761518377913366853731745667982171 14
6590625172356933294886981468394524640913603914100416384977628716 13
```

```
39699539474505169787000463523348213665724737466509152039267115 6183
62466224045721941269234270388504527470613815711018152332031469 6810
20393037167794877501840033788035027559363116354000956709793650 7847
55744736939126757118857975499311866934880287899640646409031872 3726
20392233455149647678892552847926177765829958592989826651113237 4295
93878898590748248865589493366986562401161004840137437322686502 9594
20227656809461792340646563103580938326124336520435184148898048 1208
07201573294788201529455259801864110989565037603820365457344682 8238
57122578922489340387292171669613683670040335358542494851654672 3741
76053326533899467799886593584873184153045130550797512246061778 4544
82872471159942812253179419920242914695749774933574033489792956 4887
14289542998456867382991122973941371501362878184500351476663754 8505
50375349734867442109283880832057713836702671620074979866954338 9519
74087144281008884201694149363406820370100493660118131804897924 5710
23923483702946288272154049390568794026098744739188574698212852 6667
65449653202532684366338299288075594757868040272082253957725129 0177
65695784072382846240509242683138979684342015243270477139697871 2180
43867281164363906653656172170449599200639935572444776350290266 4522
13192336425081821211461817866064085041173206410810878239905848 9697
01901015255591255414809594289953581213133433939532434131377051 6051
03495318853141395639362094883819716670521582596454726029667672 4927
57210826258932277664006760365095993552315006184638672545487914 1494
44070962209756343142957569047759006696612233859920857323834763 2193
30822884336299216153601880713112888816951379845488577640080086 42501
80618495732246545245374153648330250644045049836401565909345925 467
31779325468068222321257061668437423756224027355428123669958229 4708
89111423207683192557555562727523595939552335473310676327792148 4206
49363419589981185807750948974917833847926210853352946336322609 2183
86429716841010446852392353457379021997496387399212122108034312 9488
45642983546047602028976173180768028540871865670740100366299612 2759
31627658979500584451565584496177962734622568921337962044222938 5706
20419753828283184402939299753481164803787268885916803749903844 7908
37867278389594993773455264644408781648344425192947134866938367 6806
02724834823059001768590766905483859898197503192528415789984008 2954
91527573145546125881172076461342828032713198588324841356756461 7561
33835649342561351345412860839732316277839503041201162553005576 0757
01455480189684571827381044792782691287953229307231054492807731 4444
76601055505866597643332470470374511809895354939496944531113652 5935
07744352216219658318739070202179034644277157187888311155866469 9252
21536798051474928522257127243163111263540134245447899223961037 9713
38858729923084521343999177894719371481976512941348563855851009 3927
32221752904451063929562242734206335691389200214793665268625423 7678
04645959181655826779756968234308890746035374995047793359202105 4937
23663216268660344610473206937760148103162230741770805902577357 50698
40951342049516081668064555188005137666095519040317294119154534 9557
66083079448811482127941967799229178425225353064032350558475978 7448
72889599022394728454408427270289860569060057952982493668306683 7398
58506364109955446941494864350413431930955590198099973934085881 1534
36198037675482624630505622314427990242941847955027985735360472 9785
59388523717952357080387448962392944313322004678822643225916313 0403
09578307052701443652908334294076767721262848730617355563862096 0647
46132545894048684754619326744380764222864152961504341209625054 0300
59556472147579542163810666257134465105163451361978907241993420 4693
64894746533336761182063460544142709956341288926084551315961279 4964
82457295019407452064693642659950597065604140648024840669127861 0125
81817170642712322794277998279074274033472248538706251464935416 4155
68082902797172225748649782491217139298509027535431085278519837 2280
84890857216468921559248951078679206014143808416824113367633152 7667
90630755674450626413369780300831865523476461800281914299315635 2334
58537890121458173646954349980606204158395437388629675806717641 2059
```

```
9253328266062609132257274454513977345663881597353267814730412 43784
7333287943655304829997734145609238976912981671591438521825711 73151
0038505006895997362289928981035641536906691404872027890087674 65389
8500701790224820844491915766061653243414555145159065537411188 78294
5185315757124097570907233885613614467858815877726830845590424 10104
5366888951777857038868690934320171136457225141012912096376413 79081
3996487479807162623005137503136288987070056092784655589361106 70983
7741756364859020894439963234759211702382278780561813735625589 35092
3975103404759776437130138006542629168108388052608499508637710 25427
5787846614328730311303020109241519284544487150230626211095692 10229
4911801017102315614762530961825794499532326435311038325706062 14372
6325244144794832005828668456130887090282183486032320243552398 623759
0998166400428295996032279073960985386651595146524084825075237 87476
4072855672893021943602176924832242643468955495853212379366206 71918
9117829277478416096667338760311108799131353359851880201874476 41676
1934532005250326831731199873482323018425738543097945018475740 04945
1285402326951770705246471779573910955561025367056492992647711 16173
6840923395528621284569193811636074115281602312942592661222246 50320
8364591401786120030720382262484725841773457715418552857378071 45698
1463942666428986532046025741875460116929718126557172271253068 08351
1793549860361891774696228600627766814935570680095588135522584 61648
4148089838398043325558596882332633074447922170426606282606633 155960
8165038294978952308830449061227016819383434046583748931507457 50723
5865502346654548606083695663445105241307319747379723831211087 70332
3413477908683123440910396101348096812867114638752939914540011 39796
5718313702076671955548495281577918060839922458175437585128548 02655
8418068465759842023143558167527477628063579972361564133495811 44079
6663217101971879110785746429492971428226147746500694160508928 30302
9576312720637695285701296756252901520750641230076837716644079 20286
7541067084511998140083762216616607045042998946968587508146982 83732
8372393751733157349003452980100503198392620684449732428493376 23732
8411601390504274802583377444201970082945754182447934325710167 71695
8533438746337147101202151141135199680793782183360285212127246 06224
8133196068254550090184597874928128545455809270008602271122092 90010
0979740026367673228339208756311900035558513820399606876296313 49926
3973244562738723111249634427474487294867322363008515562084044 59725
9241883320236963743732369656906435659484985378426707326569366 53955
0740660596387094242483135201280001901725514229856590667129551 32205
2732876925956564093906092990728800644859459939154289363750326 50171
0012626708787051966169118836989024496812092808201191892256335 15728
7616917274764482210308384540387628155609580823922670039567535 28173
3809814880871880267549296255848966800859562004221806087551928 71515
7660045753815847732022287079712258142403898961602465742837519 35949
3772557918853547680903869235169007435045716819197965723955318 13997
1686268479402912914655672520799226022040100217903444007577569 9193
2552533396322877777531714668452701300323269589855351387383334 42529
0825393473678604504126692157835368485337174688451673290286246 96162
3377075127366079635084006526438425979036036202577679903702643 39422
9840654085049874864833935474877870000005469988759344454734557 47774
7785612626776977595248451333971096721860524039926537331892174 20240
0756771966131115757325801946604899456955718419284578663577743 0944621
9243141494385838502604011580951790743998630450074120884339677 67852
7480456856802057405954804062012381041725960582495478434627037 39079
5038697772814852130110665290630717677165746695668250652879862 28741
3970580096715646061503385552105823031141873302744124480793057 17984
1339338874343658378791238168569278903163392983354008440384310 48950
5409797840700172869996085694335663742424569153416468570622857 45508
3638654337685089946264989105928598075181750652653636270500332 93320
8544510602208230985217936883888495849500759824331489370585843 73845
2286824553077774354230524419373986112150706745872364095836234 73291
```

975320951731947369598548039413115592110080726173449152275620774656
870173372209840162995608937199116231422806123499552475025364788035
544518561352143801588879611248752127261370445815291977755673287668
422643017190085140764096376843909699412983364684492712510853075195
993655548418487437382697821368396158646646481027754845520960397433
711760333328606931328622684333519056043118294494185411559312111708
410347157284607124570706016573386723793192145758311083188987447359
799307320316219592135662790426497581083354850014177843063443055714
385985608731402811854574833050489319697320842537037862089239089894
436934301656075212210981424110263403150364236166958808403720162040
373304878708433479186329430967760026713039934892373226706445668792
340126952876928820063102291656296038768452883392134756698308628235
078722567019612096237496563444766470613281554887172780407587431516
795177117324537751384914925689061202182920231960903057584939878930
359118127613033933032104957748129372747200086232419760305922322167
727833363964260503397982859408640951949752030071331457498005556729
790970156896115635456302606550663665343580922649791502649959212480
621962505770307477548497089754452993094351930367914064175364825670
633416790602041225117328766299970583381062474604895611285424270257
933330628988477218779385445929919922949764461258660311545759977930
193280024243216873022547921909467835940569761940897766035691333262
328347912525258717535047386241203721751498758124745649052389813222
023380467384677909559849843040211605878401861127084471853521407151
278720248392856798882973200115871056448605759412653610236055945438
726904246755285529830987009964268668535851115260752145061176594795
032251253200960973602368909178040174142810272484343441226448303272
776049947453101743146159989062342004213056872101017575510453024326
566011251511270827409658805442106673600696913223990518840374062 26
382892737526054805693882093754682035893109838542899959585427167360
381239144529131118922959604971987265219953271170997986671946858935
713748275217034704122922042599732932608759155939469850134136614381
566271704068557445120261619676675697567755607380271375639380033424
488280831471953456901602305142843074489050935843205499098909591523
104776663664827244460255329751300772156107105279867924544202998233
145983118571614641507784712414351543339723065369679384424131624513
079885047916470094556037385725744915198511770495253450078117211673
159416211415112239337197718695534565980123885780516898037345792748
572923683389746222018471364229009965317585238381710578061223324396
717609706380684944097174411504698970115342571214529730398949660091
503844778546332291464814468620936312474399843820072687148510525787
008862698613585683681770692731547902266321474161445533203153227630
714854697234542830930647528395978520205921426113702111416263544041
042404686891744169119545354589902817107349223234789185945901596651
097471346332759114095233034389185506783734974380243486421946035264
787171863570626099690744365005850337289877693529508194301709504251
559475799065389226625172481157212928699057724723032091149943686004
644715336008520052163518631909454078853328588895260816269911019223
674774008490789522943178275873340804505400678446958053896350591 60
381951477381349929416940810089634922340186600532246499423053756973
724265570491776651721253483788552766942632781523054579247466191662
707038049727293558060517357108747298884499966109284954519232334623
163956242195731615718088098029881671873412277771985624528419358268
759166160118482262582772316724344261865564499940453622936963560670
990301891207730845005545389041401157572743806633167678371 2133307
890414088409423798422823571781681351468712277564863816829186047198
012688993616953950722595482376275274985400399221987702501940396756
540507443423973190671805195990120463073414245856554810287312599571
396686817816569508990771141685192120426040403292008335668906240452
203195458756720827519185261630629602871330108960002224055059755 2892
750089631974588409553198317887543029103218023953889249275091155158

2044095999508741115316204949772950188785457657567701195378736437 46
25389715938881600406903876279761088361568114503916801853146031891 9
76340695030044613046222371890877543800184643522739839794830078627 7
40640308138328889609873396179188168585981917551015934859963354984
23151589458273646202543121457439628768342181307569364238741998790 9
39108270139609534261248631688130274104135550993008235112403523023 9
27604667833861505742882785779435977841961592940660942043635726599 5
72420586318887584064993857758707634860901255595637349641939827351 5
81410500375339093512771831223841469537697846562489791792712166320 8
07582836747418975733692348187789666727220943402133189772448388864 4
53893216754761781331285576579043143886942667910370925985470088199 1
79069244936529969554475513131547585432472264638738120299239437388 6
02593066420742386507417422801154139450415513871246950884262980114 1
03941134972207185075558352305651674965378309696153524368872203380 1
17221255772604285810160652900959962085372993879037406961793194460 0
60661723105235411586736424408398047144702233316402051554407782039 7
87170555198528169580887907938122269024084390317658345984278868689 6
88849445150587658585246865730586358801409250134358340229850125832 1
23033766997755275734733674485900970843750026048895680867308935618 3
61703058973852538021340100411341947141809883760605271101056517849 7
89108598396509968164378200025021886983192529076340114962292489875 7
68163082135062983062118702313633806675357828995571493542907699477 6
39249751560096021921895771196654062850378835993482613372077909920 9
65421209407563390903200689577077470945835247243921777946307455290
83498436360025032996645059292464779551567391807168001506277424479 7
04364756823816870593314177694674306201397595359108004809576628837 3
83218708380022500004081362318111366738174185488132675101261295309 3
26082840088903576508314276358345077332377437973431117015371210037 3
32667324451221997933360883006327816464467570472471755942073733292 1
71189732819532543054019478487040199198754693061916138200336165651 0
98083636148034547884389730168860161997860700961837326369023129212 9
26432883883713274063331775308460861689892402352158352988606515531 3
30936630284843950629176953284467014241580390924961568546003448872 8
86734645277117373789997360856956695550097429309623464041035666885 5
32922283314855935229333411336940393717186570886278170375043399972 8
01466518033338858998498312725902365153327113695211199584249973278 8
89650084252454930555885464942167662703581274797299757161887497216 6
79012966678086137745496613319997933336928453435884766189940386007 21
81323618452752815025697526857999780799298657821123549818516929831 9
68307101185312600663680759673338487081491867868940118156125942350
84438967874430414369229514532291044151525499152506427836177813987 6
60496260090803180730434823088045576860384526666912862670855219036 72
92893747525774228547114323526521423483301637859164311518993303734 3
48168408215365312697284908953279416728462581353458880868565799565 0
39281627527931629928585909090990121310160430006965747363329384357 8
41537114939017548612586520896364573647580225602962005266679305547 3
06785054115030384242140639930960688300712624678975426446261491958 5
98772089640117364483647522412865265221956530840271910502191083398 8
78462318765070093993772697167304643116003677384569621728477736960 8
55370436290560141442307154744529605310833374048095070515978878877 1
12646849694796153698799488981672296585733639082079721301101509698 8
76687927733943494104715986465661808388398026671592974162303351102 4
88820051386859748575055397484092544186586856075640258417918251630 8
94813277201239093830788408370970756214276141656641662560114916835 9
44404989709367611173189633519030375027524747336111294567664697080 3
46766982979174307796861354165713710897164168809783018983420740412 4
68163079996603543135630507990424070301068232548250471177626227495 2
03734803327631266810259852849966713449510866456940811570565057697 2
17736469425777079917520577984007356789879503384920464986877663228 2
93021755590424655566145048315544340655423374955892263154856185856 0

941877211154154964366217136596348498970616008549575631782904357451
518362829927704874146263310015271662603505756722727544633995157147
115610003276282055377467212851377198081357030798657373497147357653
273407248571664074209420221307914803773608092725354150413380468118
756936499199811941002689412730034309844075068837299799409682309282
861297762507351973945712465928453175882490987071215592250099447943
793627161198731017571817675387666516229855846012693132885045722273
795949817989442325013516449245544684121262604625518249031981067160
859328629156762086572725209446919477064712024078908325975191575967
875186360629271004083615345223032871294985046101545633827105752256
393350309736364954281224608215627413006585694059625451225984564496
877951067520082183682447563127486263104094222590719252493211833105
070136952015140116302494953499269659366034702040568633763377674053
522739493434652662130736137244021750986284676041857569140527377295
908875459831427746016957227471284131591417179015517671916474795965
774680768825848395278810081099155039012881116365202923704380532565
955966842391334858912745886541708713003797573736928013462247086330
946825629765769657690607300757229386697765150820823933796779784180
785374713805450041186874266932224892545405487069969216812372627243
893083612736166472787124441376681828905890521382672563625189124381
524279434751935553187005035901113432218043793940442034758584486604
746523253620424589750536573122475953609686622440612958644842228999
293688439861352107808087679455172498370075157230223119794267475334
415758804853230152803462507071751817540490324731594063778458881396
316441059887069821206008729489152480269678107160497068534135329964
993796866660527624679459945316887987194105647429428587382700112888
416731775223660707128961882585264608764596225629107514348697920507
451839391643924091603253614786351871031901384966258942296629624317
679261239676233722713879174767519022579342893535489619901757819427
817933593161744595360665638332847918392927659955066495641924641441
093907302519384576926654307356251850184875331866110170943318627558
698235413075997723787451461876379861991523921319469302737645796131
131902538970790061114326787943118411465433173122941399914171444069
211255769235155386410860736931599720005933289721089672594870918820
784498976268490001636997503708595166526264717906435880491631963574
761396746980084683339351849619422971681256835435096326011848789480
232032492998090724425651232754775131804537735483747148393467844960
670663959742050559401405116978505143354351297571864505637850605670
594828257352703204372921169282393677893880038489311179880361269267
975794325820437048309783433076724567126394849681198367222750283468
625309169479256013739355858190501186263793894102085801231150741100
856381949971113216269700627741316417464326504654111974238383994615
519978581513509145089890491602188152983532939107247663596488857093
183995115015246426189985287506422006388028981905723747804925005396
767479408894855251022203967810710565254719717629379964488192929081
394527868110446577482435519439270750408387096604813314892331516777
642423979397525725333381939527170171649364294678205998288599551076
468306636450267307973300922124472278208289671031033436209170447793
647906845854796002505789440932648047289887511639497799029421754840
107880555939207095629443316909000537632663845173865973553908681399
848999068860340073833326860832636730188972413482306523753846177202
535377407930801061076110586075844156469892409324733017532341832575
663499892399052517016767139851876803691791999632561102921719396739
474049764749194970869989666433612162354356649131158805002583692888
493278034648565214378341232978854151569124473069754197927080182190
606224687794896482595684623819177663672704358438514383120475500099
073849819232104517612816108082534660658707228886863057945546339979
626612693371105637207047878301532343313721688689585307394131820925
496857472082526374324951498085042624871011207681259226553229833431
433870440641859980968616888256284601569219695540703766744240807212

```
415449110535233627154321006911075188830206090038301832340876173910
340465572492450679569677202013015271255154843532031976103471121682
463134153185657194557259656281095976916781662492760201745173016210
171899080518790938074486615586976854658726340491763168556468951 86
224524704448452494462442547739254307863306235368554149243177070148
615302747175121962209233799786020804874696407210659122918909947864
384659594379400597553138859648572757227205664169384270473354984 42
642226264638782352890201340490629761599022549027778123680771890806
673284015186359100679251184185241529290862764073570691275197566315
664305284947017637218114598604716331761684059509022549514939834826
213632332612770140600746959879293826382062969078708081619864963059
370145876479459371195955151917159825983722953228733923206659457347
582468256328381216542416160115829329207476150007135358018337732406
522077220741014001415853637643528123923533226069299417835361683 42
429945501119004317612062990951152662205911116341700214553410575344
823491043504578875830473592064008848261791512318970295049266757215
779515377620747907317871305028416770948343717593250658517838540960
937042936927319009580929192523972706385220709376283832256831539608
354504523509117804427489392569964055919890683912294044082979139704
157743879423810221539233954058355041136217686560128829922435760952
981212209263729851756570813992122093694690694715052328077459229864
228802826691663732038393091735460239003176018105496752302588882031
353514553890163755138161669782250564807700621959942869416952786829
888747875314021213999847388311978639357787459177248686031267162 68
083600027669552432991228440207446373531463385311809602295643152796
469593931153310461126252911880488436452716624901692931313843640555
528055220215938629825067679738446746997160462117301788465036921382
010072482321852304396886407266624343463175549936680725326876465707
431296054646965742014788858724642343068491804990866789126274322 830
983449564193022689464187915082860849298593546532013424279174874714
925827478475925183917605261506165976617674102070800222106231073571
455306554016445689334764347519281791798470631383914982038685414204
365746374385926388279117330639041778218904168557232721953604043659
584725034467565938077563496127307805179119531944892597650712879431
631194138505992994323917823527258765690107451538624193759842327481
821986397431559810766094479394385509857150183427508632159811107088
894614481063158821943948622105813974976406334645423878564061388 32
786046277033452607057425996733921274136364848877618643820395683167
724710909811550720988487009280479082886280724014562943571340699146
780953798412303895325877144722435482447280121115649022645292269273
333505508019364513909123534641768229486581271163414884464871389404
383837963115598489640878724792941967896017048828204682167661834998
248072593851853562748543710045361375358075475937360984404829982513
746149857434320496948153908318865262555301372780855322652968254330
610329116915712261880921349298791035451283992973173936168724209013
741053794421961379256273372512851988002560065639065266863278940456
902283648518061542594724961253689548843469558035037758290683977922
534665895505296126402786481442640429141437663771124986133839163750
355519479692496764183860762888540604234449177947266985151630853471
023288898337446012779578241258449183459392834716926530762282560489
169993590442687461014301090026056871037561995809163191668707351331
265555170951651809993839524437888179706960083078503068038846262126
599257807712607758088000343978346133099725900596031834762781505339
333500886524280041528529854448308039533374501320717398934951676 3527
663268951219615973031083195747610274987427625787481248856771095790
059275024912592107834128753027900551374524760746339353860959092224
433102444982052983235321721183797190916628906109390500402623348585
681650527343821854679856425946898029924257891484989803170295732026
503142823893677165131142822540680034288951042652723736343632704397
471385992955719142741004177216651478786480623578609519875716526435
```

```
0673599768531817026536271897216064812956543255780841609191199958610
6598379839906828921815633766300414496379518772181497754460656574151
2259061168841693645807319848617450164353491704003767367703282833981
7242787641863756003199001521726127959422377791151153872534748019889
2958076662795090705642519676967452131089086547914473182218421069455
5481777570442946865351605123203642820288539716842931782850510041755
9907372745740296984221711084119710170795111922643843072817024975275
3651155030905669860202871615617910147902158628559745993928794208178
8934508872981175579247329676942683585199304164996729022318615637788
8929649449115734862814477723127202692621186149838937817470626354699
4173222422816015319757146885533485406721211391333621919638585559444
5669434487492837730460747918788271139283493151635521021154616782922
9011988481625595829761615193054472850510333374151147662706878119811
0924586600307766117237265754996908922780108687324358788964095236666
4475180760795087874878456630271175092829135106672040865553916873800
6527994779872704403685680310190579525132472810774268278737087538855
2652742896724891252959605683983088711526501068467551331864007459155
4855298189914172568472931033297155171244619221937580068291749655699
4980416611800606616464183021434819849515433410228578600407557372222
2238263516872415418726122793533915949498447451325311939388867638383
7299514048230140846017143241234098266699060282194266877460203500444
0856123473844899148670537831106303374206327358401816044149500068577
0068602918124555728616694557231042861274115796768553677812820691899
6725651764175339883558926200047249897831017586848064161068770319055
3857416689621099578993119813635548462355616448959107906215750655222
4094061165808378122492852393444197072202397264214858448508938338166
9487532001102786055974904168835980821272209948898437879036421232644
0091237221369375013243249397480461757175731855614933696964757151455
0104101822114628193043231994877912406352433297838870214697558857033
5875533723043579421541476259481567781342894500795694092175275053911
0049791960226514967893653504178581648874766362889231645062157921244
4770538774859296836593400152643478845657191993324559344747610766922
3801393691187871716970381638032555734548073847310096280476653805766
4196281861646903149895890734516862554120703955650768180338003214255
3221309691462771351775020632499932798692285967842629892223273127399
9231512974400728690824045658370106295594598274543768728374587326200
5496052258545386919704724279552587010566654992761697912303802892422
5273634308378553387991228231787191445009537391142448942066246505200
4621332640292300772913090613314960019648548585268885040905245458233
8589668607420776816339540959709061513335189217283037426642612107955
3539208449765707588007000095596434977729583434296039999871266002122
4097868896277429934161333976076403628042380740795090911304861283100
8580230989940197766032520746293226652854423152310019772226581464211
9527121136849861051608049650790319832306790086301406419563194373133
6871401703250785327825122298468480476805350515330998641446625688044
2100729187467442563040546221798728087879880207052481262870753252188
5865713368444173839714169097599502956152112739516214002809051675455
3599113540600983565669403060724635737457613486804942537541670810666
3760134210758669077825519577011850491490032552386637367282042181211
2935257942452259304563670629610158896763695599769235175277540740633
9227552568933579899916376600628768132428070200197286316393839244333
3217631852747067885307662056060553927218608887926980194155380443222
7113032983994800933219991935009539701205400399221178090680151631955
3897296930294266511051546471845408357660135323206922894752600757955
3871400188580115006871345631406727263581820410866203448650592835566
9072215718916968173875350899806818840774541815848328749749452617777
4253232726567168802243584886673830900090954141764274617441401881599
9833260656922812248536053406121367821895038208833436964566820512111
4569821742618691084916111435009659495669250245317287045960227981922
6588355857776578508629064108719323151674050884832933715755136229933
```

```
033943427222548430156049886855791326901606872143815448408977297208
602502628701747276974283685649028705897972966701240278445898679710
664095590357943526213458731841714638552958004262630991469108253447
820609322621516227356451330937306062659179975281426567504161762843
877771060000359860447614824102146700651903709097383892418097894798
461757188017987790284840595432602930190803132976336394556332258558
770351367923434782799263460182262438370102967548913435809360623380
902755114498389478002427010165183332957961091843734524486807631302
188286788843341271074049726514005617854916224579379741839098791108
130123850566489421629280175527794355887808000386601631852093081783
082274078138572116470482132066006921750489241752776322641184837413
419129597916244498637897439894016470738203095489630576708782990809
401246189862394193080258971036787409195732090329992909314232739957
535697486049779146232868572555981070210686958161799239008272615848 2
849283957362159297333847526015440800373743988763791120663127552602
092181720397891422937879315489160642267221512029110069862081935008
363686334565576071736599725219738489354261513607795970706514904731
982703838692779734358733380432587338628275322504303466841489618448
932121985294953211274128520690952131318086924570579281027302615871
732515191519471733322350739193894503969786652754959285515576129264689
991099049843462164548229398062136925784305825712134946988100956747
091236149854217578024724627853880746576420884272849742808541946620
112434603058955936156263633871923949560350557759331102115780258988
105423394810420976173620332215330746048986906569049968263991201122
582419527512826930139512055387016777719827080791185197916197707609
703114942545924758511271024341987507926156400969050821382267456354
989971178534219303844611894440692912149442561379361727597599003260 0
676914795063425471654159792500736762870796854315265259790347625207
073168634843918593836663606040637241663490554890170418743591571307
184427914253679593195656333137245480717221530840577578044642233828 7
344842208586263304966647279305733273065924363013158109900042232647
729042237048674390922852921826213632000100184063717479760803609387
198049949150747306862108255562842227778036758190735480274224716924
723592375923146544819585527602375980384666369570929250941793871867
286191114353127649697923373002877570185544188682535850699517644792
776807683681428714172373583022830558557116393991920068864708077184
062678483975885046049219895478016442622654673710993630273145011669
046843793757263898085131003495724944050302455168768584797126682003
810537665528479347783666581357206216433346853797800852904302040576
808988727016807068496860145118610901835855313273010050047682399552
936429749124650294987550593110868145788324134909376809665624527918
269467237278464004372598804121988816235541287990043633095102879634
331407334004753000179701254601483650493008494901200969865165240028
819632296926630584158784085479024428264032337281544885248335283818
743466204081089610341604115401265677552729048583277509799530992879
174990120449267690732709694378652036193733452896920831021696598061
392499563015838939492372724701023034043089877400363483390828284790
160818481379696288247127175166094071649916920880206373833905913548
025528539005356957388598543379747435356287278076085312006658441737
541521686044599886117586432989965034184430301870334938203136191107
893466791446327152738154649707736426281792653268174678306140819431
470398738656598431555876530410721889565429664806683085327935403287
780657764375056472374828496301651859633457313973324239035162619758
179556237486947156334795625940342228337756781257397894335701955353
516514412719356618008778697611776883752571349922591714286585806510
590053832670154712530284054579150977509670966903989418574325290108 69862
669825107051887615026997013411359730637519674585325333164843552568
926501955625879631454890139292498632887047290888182571584825305574
161858235070984409521145667240641148166986651468335205449792177903
276406716728685044542163729176578894602990230713681129075662646521
```

```
99909281492079920807312301803707405533232358635502721650196591519
17802729773386816211460846764385355770686205506782006758174821422
24893026420567107557488677620715503502239198106990193054319560998
69101098227233634759362936235840800647808165376021900225027354549
39130069700150023782163906086244349733611710484583729397115541575
70861125878253183425719225400091382360224611746895432426076056752
30157798269352962204191872414250464432416597986587672613334434985
21024850813319299695162529915816548527096428413701289945691444149
62044008016660489640567587899080732800176041149296058718101665410
54524903598638874820740553553164649796630765503922627313456771546
84600526554374833059965898011997233692788191902333129871935023833
57125575312525409602678780851890719326863008332753406113674180199
20700940462658897838935021757920286007633661094554212734780646417
26381681630833344331042514773445622555256144639360200478639272096
65500478785025786898471315116448531105967786869425898874680804942
96173657461559550439944946381276474294791667692175499588922969493
22132787305587090917959882098151914406382037955918608625387577261
01700765423921244283192314571055232886960117631439273699977324386
07420643073327539794545173579030142898755531991350214229919828562
75198285065111161287382218616531225288793952233051488054149456463
22187894543988939390381307700087959825792143108023745646017468115
92358748205234734474686102525852265695725665778062213000602144581
70428991769484184396246967115600476223593394760880632921448035538
28710084533380776324120208522743619657154171935295257357191051774
63182096405579138460868824339111464086077912901837980559848676684
35546133327133863532136964710194237460337299823466189799232943197
31416929943430743559231496454102320575850734768809793842855931138
84197446806328488290759704281963057303575922495127602968783967338
25715323292151374961877386404455508115485957573102901081844909930
83264405496434267123468762300735764666207111490105183645211787539
38309832432676930517976051879436760598930195785242452797452795681
18960119890643321783058835087967723372958920263733197734509417496
87214267523407582386387768029461056471666499151525487666105031123
22750995060157292642294147855136838872615113143344063898462237254
91421240486221836012030346150461461587879208351822117430681101712
33262846777123571125829460873819278164584781941479554810155497985
83427459268100095620021214480643914102290447954780580366456707812
11640796918523194565125098239694927626071481257567181208331616020
90113583520737073135289779341810319953603068392690561022097585385
99252548341494977785927129604110468486015003634064162402063484298
25854080219471132156831838966973698627826218013928376431630481985
13450569096031258337359726711160565935756560316813755203550470730
42252063888999601330295519985594838526466219552186693485276514349
68779823011091934039190913761659144906512091607789757699244449190
18979259582423453835653808985434414709329754243014700239125882649
36265973911712363586168823622565884469701768584701472034782214933
92956324483168044464223721866743396681028148542643319766980391131
07286354232427548668341451508645330767690069130662648575539689184
25319361025433593884678624790866746844008485371610939752803778290
70940977606739781265510390326033224104125368606036864495017827303
78352749731920499599173381644650860804849161114368569302142133802
12826509606993149076410650518286331501355415857687187281311665220
49586716259933074318230520070935220373189289759691203281389970003
75314083830204997007940434617457713410287718306849390256704805660
18330118203958527937982849042206777676765313453907158370735208769
28495577648716839499404043101085178850087034027295701830839685377
05191465033388435660725912401066184035940140460881369523490353777
36291744607877425437700245445266366624745864605215219326398476506
94881926925786685075223726703782092140886840132936289401189709047
17335590858398244633492906655168232068477450851227916530119164041
```

600902042702503351745796539824994799281975488578920208057182198276
970716844784544403680671294954906109715985855018680832060135111428
644063503018103495665038542806291950124129083446264494050456936516
660955625647759268485737475806060004583547601174164475903988600397
461771614590514092069698890596238437986869489478901054433288592038
261039004895755469406519439106413635422557915617894533706049905433
722474532449348781024377392557504531001760560948489030682617047194
652379940122222508073220913247844152502849880980593720656550409857
317340683013158971806996356035872133943911595606296207304016875051
802825904781612068876466760337382804153294607723464874635122163732
130270989503489990930410051501629658999550047014413632739095726193
974809981426105640132449792693761414886575398937194268162134360997
817429716521648035723304401596066218092645191830723120786494525597
809327789032966503697349874176032645169043711469323144778766626862
764507653586681999025414612033462105457705325525873610717044999424
650844481189709435942872816858183209717062984115401278802978406374
367171376207873833193995413068702323632219636085915786281366686858
296797698931363645863165959683348454566631716429837708511500037723
860959032331903395622820331922124674525244710939030807333622831407
309360390294079927419931607926639015854720274262995073064166333411
751295271889504218114408893546754603721212530028833772298281544622
894544281016684959989553625263667917390680273625474096540385765629
672473875260383346390216756595335505115358725525145173143805347697
502507033112801737632119325474334076036381065206863217481975386 17
158370585453466785673411351482377767775345876966751328947100291738
086607216929468553114582705135983868816264438060018141926263542047
190310455893895343070197791301338020742937112115546401304606205365
865553318282664657807656172258474619563574176583106862559480128677
529018243617397120974445092733092391575207758575816362376714743808
213779174746605001470055331028783884759301238634059186376034046965
686720970151291902603551485629769337885099340804531433262416620341
401382042815279493864507548488764632729412866717171780717524322485
562807776431868383059667541917999686680796124735878641898606803864
195229631438024890007333970479858206253288928111315490262101177103
076604220879348716432281016376294068614277070840123666967227030120
044226516584262830383378612358147456068674904194565775722043459353
498070316522105841552123207781142547882187818493372934396789 64733
489337870529236918819988047722344214986742621289259026212809424267
470408493247962172399967599523118253754651821700593990369456585466
378323447951852564948841197616875988714706148803695941809514812985
840825514379252655285628426622873729882585626811639361022177278230
929449619105493557624580935613688619763663828493091128294519162457
024826451644279218564873600168684837952572568289665713366731855051
373290841855366701777152394957424060508383782519366468219311098552
026392389473535548104878407176785651062986490521878131938552104545
926339863178734302624679160738049510570138878323709263378116430936
872571488699360662289617087021791008907509443315890984698226353558
348261495327711937451420175952537347122253291499161666331375025561
824789531158568855263929960464755050409058157440869591918826316002
623320393467199783705590617652599206715071179816543197652547613614
235397411297290394679109262788187483902488261365795250334889363482
480435954163034689226040140006645122362473100581374516809069978073
171300207367592322580318929860613453792589069640797564987487459686
346878447743345688996021606158402125006441144542887187852696438 4944
103165345782890234344192586334650478202008074508156845156599152248
269290053329292393534967341101584297345780647210221921708808521 6380
185558932981042594876159907626853181422322424197679372125951713646
105051644930672870514641759124454290891340013365144936847542316765
970744717833386505452254589705433510969977516389599979474317665143
270856830419130249869521280853966349318982064979088840349265006856

48603469591757660884277596798710765717266687974574721360627850819
04506417542767831322371433571313406724599956361884629842205648734
14622846025214726433355409487280447497190141286481213783703817312
45878730955998855489346232306816636922505033129818818928775020064
49625402132974557088354211980186516657101309535010729600306094438
59882198212505781897033140910810707575413346194572567201345659049
46332335104255508674523558941447285505322125476625516085790467086
10066560991352830916142340279940698630000178533828401691931821644
17201240635272637990300131078466320099514687624821067866394827501
85025339777577171355184847392743941369320063898006173548553752839
51196634280206153911106110478500899124613554195888213215516471419
79112880917674669363915639566425778360052204298401157251681800457
95318703202515357136689129946871689861939987938405360090773098300
59966428177057588550985796959285753094574501442604818483627428760
12300984251652298369686445654255204535571566637392991684244873041
07724063943839981315241016190250177033345208610297433408830177279
75298127374856242418514396940419393704349561226466787365608099786
03672870854073398724204896366714674241802740422508139467570055514
37547057325861407362700813417848879567988470619162563452991063990
93019673525498503145827817907379024775847590349364256214629553239
62170458675860534013568567084249509945672134389089347433816142236
84019563023894150929323939811197847974792293348005510079358076374
73840644316745646712057218483906911184593934319943340622073407795
69235470617983895706309904354177619385956556522941634196108403715
20652359026344719365122879150394813289323427675244253817440321604
79701143874597638097536404984186783172757347881611323892557729389
38919425439743346985340998944082458136232605785345523575170078338
84322942947891669040490930229128525484910451825565617122590970956
56397806299494185201447465224341812955953595008276926061591296381
62546024873469054359871892664177210644073306130565019596277486507
17599572726461548123564029176571520034224596070192744207622351601
64309537397179841664358457561428015561127044948373499492648265574
86036485222456989789327607104588210666846047061723042223442426839
72530977982363671615580149778579058450696920774842189360359246348
13348246762374911535986819745214861027876525360411218449382037190
85993177588043508587677324307285825394530014372095476168262822230
97430312361795546395843734046058882606562297860961986953594734826
38695773164000211409903216525337561384909794780668567025842418032
75476523051518566719625691666791385689213227570121414359956535862
23511680232053174260935222801944184082485539870670674057604966914
26402484645513098707732174621460073286599834006658163252476738789
24000605150660126633204406522119507841478417143050498569709664964
91844709480052791372602692338287011986565168478654277000624531361
39544144221139838011705979798519013148454376246468112317652736895
16201312599204109385214819988148909789956433699477813726014812677
56504999335089159780640101912363733441296597531326488286035025854
34444741749894997010865524201241556863722684100188522611836954893
45760572029359554938414980291681490449788445665003799536720464750
62871893841912355978359436701193605245100581430735556592922535454
56980968681988244627585273310700439881987553469990977723760533384
89448342423687486309375738685082862230367192130462356654779770305
14219779426462825642248887271506643859395299243357020698046539581
22506689031574146606991437601950334194657317013007086286472878492
14925463624909416021722197330512565076554055525324619021321775192
13820980310085784322538763070870005422040035836592499227074248585
80007949776497821601774966828801895797192178827318640622339798882
92016211940176807176505147747826880912826761274191735751985667185
94922701238372806503228904879528690192151907269046841860086482067
08930262868568039663990216108571514469587466306945386538526681419
52156516231215740071245571224971512127848519799078543959778415825

6554964942914708945178695354957369638169277298394787184737198986 57
17540857129767016965917750109467591238443152399614106237875640020 3
71121281773682090436017914269664426585248585716464503939306429031 5
67296195788171482637145716451024043300586685580518820993526993893 7
93803076455777292181125462903931497755571598880740001490734562638 4
18041766961616455932411465562144602230878834053482281489033332964 9
08071943785388515762428135902889902836827483834403454432661312510 9
37507318802181519958832564800775503889302598231880179838923992162 3
89689122263446639928499727225801598192649546476291906273017152107 9
15728140549941354640052731624461332490216406205876399507882908849 8
02890777976908136840151330002029581489405003953870784952922942120 4
87065085051156817518864693727584040917763036469020976475709067364 6
10018518382098608605243527160741148726171244859732523793755405233 2
28473825741606306644692681192335459073159949615230406803366230765 63
60180272476243832911433685584759340952924454953745400121404747967 0
49012455859928707940523626357176415693707657017356987564510179876 4
07886562088554625273699548063035425695725174554050190122727010850 2
82583036545774919638509612675417699158240408441619294350973910501 8
59234934508740134037643993687393656528058038372530314827538661452 5
14113072865270929814869783234631939084960700856891462072043385592
60038117026791555464756324509625364918630849806948599862110069899 0
95761073115777100309249995511347555486636475892689952926632753091 9
53932934693165445052752985478158814128691349275868260494612243446 7
62036909781850337242145109405967439781151402968533269871475459599
47334952751461632794481120847241688139205876042226992360070027883 7
33631982359964657487607641065684550127095961229614687430956367813 8
29363024999375153580690030290555077565430746286831177708182126410 5
28856519658279423589229044782019030660470681390468371963750412420
78392391757831874143335190980059498982505295818014918380908633173 7
61546983916002304043286953634639362675939558170129333430787581774 6
11095523715474997915563366123894550479702274862910753795238910957 5
27497415259251676134040832355266944969298497579696964604702243165 2
81335324286227480146287980114036967578609273192071433179784592514 3
51173246808302839043136850381412245192249969429055095005221221208
45285208275745191150686745600798036725250231736456801296716362993 7
16561854034625133417727686028880009419005428700851982879247524425 6
53443913882176217347218781946792891182697334363303064178650375486 3
58585341552588633158529682265956724908593721692350750577356544366 7
37970980466654238436018544197204740532119258765262307897389045404 5
62583128820234607103211199866145149536619078176180590990797962053 1
04538008614372904160690506593781582267438889148227171462415563055 1
09825784734273973293386496295777698360742288297550262403966613219 7
29370433835295893186608538903496258893024020933683302126275662553 9
58826315234929169063640456363281733406268079315615281702611982145 3
43747049152594796015540104459379582360808975374300190122189519125 2
51415689444568373404018783224122472650024246856328932028449614363 9
58948334950464576978023443409998203602199672319808822067023517913 0
78060003951072861535413950804650160124365077933144715095913649731 2
12004140364308306553385696128160769225624181967053484993863023160 1
20843714880738444300287007698108982148213512376605893165125013074 2
34204753532683485275147300902828376583888568853915782989588372271 7
69473394917851324807205206610845742147098527028815268382243380918 3
17979446161191356481192077605323901610083559897985923576302690812 4
99979474604787042719585252957953874938005194430528610728548700286 4
07021569984600504130428886520172859385283718965774057306701237936 7
57214155873235974165173539778766118420258686915368939634860263462
53224181001950090429852263498549169664185109985831161239849860424 2
01729759298690562492642648181299369177317548622536585512393889556 7768
63117714777099344634976395911028256188884828004614779015730658269 5
48695011356233543978785294946501896831326657950179293910790703931 6

```
1888104519406245545262403028387304604836783334799102894789075355968
8174378074856423245998489563953429579799382544124608150655572922512
4941691212746054749611565820700645757506362207266379857847490086 20
0312061885919904875810904853494194250476185826370656349949931 19407
0540550302775205758757903749338453015070866349856440422578623 22356
1925600939525837404919770569409830018397654080050084550223533 68073
7717572445953300728628386219050647411784995993094476412108735 05657
3897900186127390131055223871863144415390635039693880588692658 11096
8799791655972346660854735958445892959466443823447011509024644 59837
7309758624819178301453403545292408387257474202719225138997658 96497
6309573977113172363409513287260110219410569040476883192805455 61084
7367485521450143905064216920646040861783251711369275016786067 27567
7739419940353910688455838434347011267543795690427828636625701 83485
9286684638841293074618509391956934200892014381057621453042525 67016
5835626878979021145507559931827570215257341281276346717910022 99133
7032683937875246692081894878663043328974469405907224224474807 82948
8136011601120098801970660145419826925876524647187375032890335 13056
0808991085377472971309993310691174435011770702230787771476484 74286
5090866141993639249611796605251989257927746311579647821951623 72077
0051595187869694380966921790964014080788359844355826611830560 9383
5501477187451602240918456462506837308485967473354147638796718 6261
4024489893543923110719469578059926573810874500734375400162959 94717
0341120971648844169452130020558812200012140884508509317113424 63668
2133795176132185132229150777006616463208935288945947253383339 47820
0385172811828029052882909344140988682533433485515436723269135 45074
2814420178249933794831741748092196658162946570414464789367527 06677
3380297316385973098386216717059673866749926805885923186318994 59540
0046604201040507885026735271986737895812404840493991514141224 72460
0781487900917854589157336373675174072334618487943180070023235 97187
4349650645236732140719510207918349412382591078826015981624963 44673
5984317035245363140821192982323980547969910374124387884336957 9173
4892743660348731227993654565925979488044432051542467511391290 45822
1905298438497172956110839822850890992400271260329765022451543 51855
7158444688885272140472407862029956875487468241146979869117031 37730
4409152219893199641264068421384082592530408821041735855111398 11223
5084586539619595387846160360586702524837097251321593464179940 20162
5341956236450936507649571166523163999242131030124847391237636 55015
2293540639445055504763041776850999426460760338186864163388965 46645
5302156839306232717806361040658089746183965986913162470259326 2587
1635640490108937237595578075483171928497180261119804639411842 18068
9969126811794300445056319381041421157541470964842968952328762 14505
7258451489910420649694653277910266421477100924463779458875662 02797
7811884839185139349582803780818211522830530659097088458648804 67200
7254174162314200506512787585941722494557751562177614940010278 39471
6417686434605616007816037578941164525364880854607372183665073 41406
9106936423446354137561514804361758575670587606104380212888081 12257
5976505453082421272111664317031538613957579524722536982783999 78527
3044080645562203370896759402981324672920993526030605826769946 04260
8810206069796099401682605545653693401361960057276968379419536 03165
3317518218727171727103063774358645451235531341024968370726497 42876
8742732175747446322386112448686118714781789592516862625845189 09675
5674924753598534085760624407534227786371244436828059005968043 19318
5213497578390196601993234424809499038784508100756418948695731 8098
2346791267204068035959698264226344063680716848710373915053397 97386
8947502423327036034864260430464396024465541905125802108178687 33732
0096840483744822074669211538360923877652631337533874451248119 30772
2940680305157108097223453117844905169073170238863513155208504 73767
5269304118622822090174402725368178234397535825927269520443746 47332
6395199434797160252674768155900638026620083393524381511763502 47759
7737309844101987728237249818845821897715934814200985910859791 65506
```

```
11988550844491450112849555488295192827103807475390095297494129864 3
196566650878885555053953333565184827510133396872778976780061392 46
95344738977077911495402572734074987720676344715987132516781405079 4
24982643660814027301506667745039242546310768331908469250967333315 9
469574760338703834400812064593076601405858774679439451851611213 57
87100003092633284839854976100655719690124346890220970056691829090 2
72198470903091600702910180622972757569009639179359881511408154779 73
95081514817657855643218038001134785590840165492836181317056267218 7
48726697450688681395106437436978017372476240306130516606594577502 6
36456728643527590755267950937521574936319073909298562717599131997
67988733495382258565413301087830445961468973313245974657600880938 1
13738895104568019673401334081154213174957320483212164976545660228 5
41830271084108227822326251804684178611922429371095607435591371863 4
45013736696441937194102951788907561257962427297194150448912274781 5
39228864819305918491555737117849994136024971957972953122313488900 4
40869592415528254950559143680254317192063731652762439603533903769 4
30916512673761033827994417777696167838993853788654383800083558849 9
28068717792007184407632911756631524222547032698558217102557806413 6
66091346896755047300315655030474956491196807103378651178706974091 8
47339366027727565546804817696267585064968896917087351534937861116 1
05286343537567587494676527537106578483608773459064709856280520757 2
31779494805202427801411184072820434282134673806904296010651327425 4
31528475323098005012090625543881386077207561831857530516492807263
80702210971353857263833280129873530934276505003191939035182563996 5
30024940478949155970459740638945908857542839514073549544634533290 1
48403434237057425642972411394189372878354234519022851287914093354 4
13073629764791233596166178621535948295041138632555821001726567904 2
57762961486573648410407704759261345450141665165411539246464157254 7
62704340493437066791176691834829228284942750377209154209275392574 4
74542855302216082313694934495490512634089589516543198071654726072 1
27080713191448183691933015638200606044298095622895426672128715932
61887068681039462446912052966685578263179892063846293118086258967 2
63202237967988110649719809483110774775679196279169876617712263383 1
23617341295938875182874603770718953345874627199509100400245768969 0
55651869680164020361350075626699625677421664897964928721426659624 3
80420913066713038364952265172427750219757210397243733808452520792 3
81005587047233800636482832543083326346968756062687479000016646152 5
15616233750853810167688899095267466036516441097744336958157918515 2
92811296668518285117870896633397531230716529113574344767019207605 6
03873571000098927119755980651623231845680822694203906660547899777 71
12698426871997862645866055712884958711859969384939057573656829007 8
37328538477968322148451092481501090635680445546825663828506597175 6
79915034489352378504735755595962440996204113410746220863124434644 2
15194892901282582109420917283300659508859594887962135632338095388 5
16711669494157788962101120804893364781536194098619012529420212355 1
62952964388874442264325827231535971198199809895344381079623831191 6
50228902466929638440892226734542517747096469188661219069302501213
45733885083056709090652176696355538496106263933222770075608686350 3
56794075906120275778225655252326079956128315530666836458219406610
36131534886810154715453847525107451097323411519633519206818278665 9
33784134448502964088154591272993196113091339902038456132483546714 05
09814079045081634932422053448768972383058734482765882519866103093 7
10935113360450871862543020917471816856785205333676565918493264105 3
80173533375935629827675576245632053862193016383553465798793043335 90
31370708614265765160244828444129290364829930674957571322544906067 6
28546979364664889860691593884041874096694502835984673925784130676
82024530058763395742992072105873225031381760286739825984212004551 6
11929920789220448763982151977621118671771803602284257161229118097 6
09299962757652462902611339138109329185680780828688834752921721958 7
62030424838329799703693833959554051723014233006190900760431985214 6
```

```
232876600121198397839456531186741009347825595052468894568233957496
174582202266967861293352590635154175463329465957229928754817520199
574873563623029657862121402121096512378012440580111120213092268982
656315236303125117245144458478194150933909364451675873514922270972
338056794535836444284493702593310762091465037458365833111899727544
151511595663860622457261601815097262801451413250585776175996892329
97616084599986469940788799867179062665226461864677561196422421260
252571583893643216673876735852350862923345835841145257919359477443
535502038831127989928783752659679297252970297328291475493544880173
368054920420540417415217361300367424792564139244460652050841900480
031134456782977739802427036881313829480053016865108970981955029822
7591160006348761216048579113481274492904800189835910154738014811583
529188374834086145350929806678432647208067218416161220232603304854
59497620485710658731622670458373940316486070476295855614579869478
022840191851293461725167764263896798391314798012970108887561496285
339229066327799641374835497976742302013625460036072522645273288403
0657793758552625013661672235809802367672688868432438196855787485654
037563674736063035153772157226660827454250621294223520622462745544
721201464682539046590062437838620315997659359271874186304688549569
543760657121461618511513877309626102419637238584708632829044753104
366535186436669160838483190822957317006039429994315769978280532813
839599047025791985381871046390917443036107072496693162060004073667
307004778378952034628001009807910866379942616602131761775097931324
036741319873969976646945414922100108586915148331790372358535758866
895906927765602810349083072343682125022380757404957571429840540416
50363099441048934392836819587281863580348820377892810363051
700941692535418188190674273743292345050244763808712872399630748152
71902242240307180676737856979630613884802347230924612734527909826
678665443652928124212179544058442822716635262495782684337418238456
382990961624005574611635297224610752367587416715978145234368299927
414374222602409343124195131439127191269409146684442225261525961704
090631477387902822215642376311275534526406978026987198047612488492
81633596742704244580977393249918080296908144343794746822220542671
057920970103677018239016228329676408918804203352683484344131312253
308975781993086726303712413014435850822491513173672093967232806262
813410281172113428370807972877031136735756113467692832589776481307
379279398602351556135280769545931765566451422765304482812000052639
378891853502358910435103833750091268298131699163930238645684392782
610197307280129318415249983696174915459052890043128598097862110282
180628336527151415959762198394807419105799191438100026438290807571
099416180040803520077042982188180528909763538006312990301333845475
536563264574635374487327418115799940026061061323714030872000011794
561354018661968984898724484629711094954953353548821708558861744368
680923281501595648819019983849500152688632931687272061342320885557
609009372881216100810727252772900582670431078825429238970690554171
338970702773839735237096874502062018983838235368731696641643253290
994610620439218921784622012102789861415462328797796646306948357690
539990523695860731999736381105808339355326638315278258636186165125
622601126484083340072484125503338786739484236797851323081081053508
445826760488680586672799069254463942227063878063483857350231705852
682818704819063452184301938609887864454314153452335287362055466461
740673015185573421975046839780728757462173888345472019597921256789
095861515343377889854541689799026039038789395080742775709070198232
686675116850297147521407378750274207835096752827653138808450183782
449916938439074827763702982332532495148917847026015056924381016597
300418004478678944699371601741111441422983934050137039862938500119
886769063452555858572982311021474589612113571998832584478741722782
019553385785333762555216531011602909625478493519485859104998755609
039430696560030913165507306035661931218285951422459554140626156349
334740694969276566399529735494287170348525326939019857305623835346
```

513719006772902048411213103070514268707682379326541754460664580120
883585892976667007599630276151583071423019732704812374560966722070
501104497112020557412716355303072019454679673017435298519581603636
216867285450885390591226355800813079145199904787334711424346483697
922994509826191456207675481225905643480597174389345266276391782998
015802533702938182953594537610880688289507698213390136691823413240
479449075161772712104752375228330241165234445428047922197789538675
534571764287544365314397357330318704335678175203303144311742667104
127383964950541426996023042284600084060739917312658630441112432398
031719388164361055464458380755527464098479452540936167399108167728
338093563082605288741967314948005906466092921574365054213222717529
514342409363789579252567090403021496152028520551213609922887217559
344358950665798037524019396594101871012885382445997299071689295432
788609539113143228997658666354584230689676416075157229746149351095
862119360976113382190766083788713634345018723944715738035644778 32
314848087701048783955230679261706140318411094308559424218435372720
396332431140699562114726160106404853553878079113536947601535742641
083367466375888024717039365754448442029796351351850990499606322201
815601237483672261167907282100165143788294312180046183447833547532
904536095422002593790216462691017399709438039923236560213250880784
809977206116993393695147559364014313987938955842375396278622594059
684844834363867275431732447195608539246933921547654842329235268144
760564612484007228225998443124513463080900371318296588981636842164
671437137119670122552379790517530467391959793330158294470193884872
262760286582812127583271673751915524499834454864117450368699461583
324838248217352745614599622387551359651265919824132824055373535566
629108180331181166861237618897638291921960716620233858996696630338
123233665395667345895621826934080104388459058485263751873355585082
796420083533778872557478322836851368484840407753343312730217351 0246
706337172747106487089749970000868549870904218310720544410975960 1743
565043454299042435348717173386572719542000235052610383402766898 8945
615386719547745949010981431156574544343049582809208706055820852 561
566558925270395112202807506431622607941629612140079131810201616 609
714147710572707876978635900466623949771119007131446169389300092 014
664131850829828356403818497268937926606565559834833805671679531 695
368451910535973307995535364802185109557386009308729258158686747 991
740856173779817889364971361190710898225936703537982303144773209 122
890977815060604882342209646740988738345868483993149990118353889 4584
614980587936035134622628495093796826801641685633966783958155163 466
091812131057332891163903552747000742803772996840623159640168846 455
942193546158144832066771463540502620397385666076916322572611414 188
461557172639732131692293222966843342498997445155204136764025375 974
539180108550070213837906215175435948450943348406091085196213570 634
867868089343154823648322099689206931301837303942032135846486947 095
810741300208544188201592116882935434539188885804281148977453523 69
739268511753144848776152627684463589827086411007502309126683995 309
238732814466450309289156739928421667969278413968722663405765386 301
469155770011809087486338308455708272368142884401018790795645062 197
718821816242587193077279712147020840629264825868804726753637487 667
640843571182860885638894474949858108656478839989186614810438820 049
948273507829159927359325686153478424233955347690795447655337919 244
383244433550520558780287812002983686985015822941148091565467819 564
090434839735528560578646689424902265120275984438114896796818006 959
274708574464481276634284306334137180824001638851765354793081987 707
791853662382845488886743193785672608359625453433499947731853585 608
256903215059538514776057390680523876091231516355989805092168659 574
757251428747811449505938786609687700543714196346306334207974496 096
373296148533906812428740019326523173138366336011835692951839796 060
965869977053828030964913030593286861197769796651831284785135023 071
331731695074051015799524386460161793374626310000414274020505448 7682

278142423674847815586665169099254514552000674202204085539034864248
825287219862928546633281587585756077339476502470944038950902 24123
160797241254008271757287532222291452325000508127300142932998161948
236776447113688629980523640922069053040607320769429927296183919405
165200432378707104876875875471499379926791997882190793456846749566
749374182076812563536882505276156836173118251333689049118411539624
576567987652816851852108072428531323595198102591476087006325988030
735221133291997882742906214752711825407196698151542617227938978707
318226230651765358144797241254744460839740878500244114689568873076
404759601355071143505659960221279252567853997474802979631871685912
068702578506795462648215852761776730415010331860978963741354746220
803307695850537638619224310912311873449127974867907067418532 17292
683740932830852263161168297608373541787081685063382634229665113468
884510635989985744948079525001003335586006845940767701463968 0414
464107013360573385948586198263120427992540965067552279635201850128
192385450715234663929326413558855234521349346013162067093062729076
965297770108957462083032666331482739109155096879224794488579825828
205732182565781915281324956753167709842654271824776115101866820959
011230517674651421781908084447630816020836190880911983449528843763
280831143502583695598820265657024198284108220889272240841974347401
399758359031863456671608982911710831526950555434741331628586294540
502912330370664818550072033386697336379810769742538944410761 68732
394648792268895586631943169267798755848558738286873269542479456397
174703451253222001845684473225232601538871198024614190623339 97496
890617074331138960016343441107280187530114440168832777424975958379
403947778835156646751270857162601877005659413497709415542888830093
551303828187366838775463127482223300313052326766466842719507033147
985622120800970446972655023779769613650473753949939483077456926542
101845676267067421304216772166963844224715636893461676271091001529
163264283413553817663949654695619550459202224292322827878113582315
672925473610207328842050854570557767244590444881757264648649601031
703849184093819220268354628477974496865298045360479228670162052747
561529435521100809500845198578464407517193301374140773094258474998
684892995436845155152919276052250284025709974378512062491192280459
262607033249792953796657262152407159882784465534633158266662619267
791081020332009328662273578378155901365219188080601385279176240781
582755186536271191465827480233087759826386046161452214001683229574
338774240833559549452950898685409724541759107027293264888234283358
469348350435198206898399913825110154613334473253392693622531637836
944884610929899846685212578365793275893909628593940212447612164988
805504958698111803387129064256719117008292713203169570276509662814
546560272430625960996734994418328864461390304296179253480160958273
178109401345441830740423272383052783274087458840586826076123359704
548125286593673907227632353558638947702718508957556762104744968158
806273628141907023418279801904275387322997905820325427909 52414526
397343245682846652013221898014884010700769807318472471447494992460
092351517812667987292836590941720054801749082800084099419356506044
333555395448980731721400590880320993524598212321530107831756352891
940298632288940023908327312183055751877094578344481274542371828581
185984044041057221209992055464904385510191083069424775142993363572
822377679065984098046515763113771680980853448189329463405128625625
810915186631367804065535788977994417544290461707655909196172335281
280643862318627953659399575427438515307653304013607366929364552434
339399148571348200135033445709826232111606977345560274760883038723
030506581437632830873682731594175393720007130056364215391353166291
503713543967816276277491319976775308361243891797032539325467938313
776989141617158352943919351790115081694627273305615047561269990774
658062025023041358552310680223074248426915212816585595537419387271
594835599322049044774666557581945790488725144808731807706923944747
431931041360459050645306301748111415456506945324836010266296212454

```
77614208574005609652746437689949840716052432513650278778911355381
94229800554295932178283905320875508976732123708745028312591486 9083
50448500932061288649025132349179992031059027922511683727689 9908422
44978301693555912274930828125654344482349876885373530756170 3021964
33287571007338059495517892866190132377393133060903571074498 1834871
33250545309960413231771998467266919081298093735176285133317 8068879
89757225552663279533302563543820177083272883385030036767215 985502
55106066802472260953804420168957881186861015620788220210194 7964047
76808108087111725092256360487951048717533894938161711112811 0308773
97849684279061504943135778697201039357680097614421029532741 5877434
86449459607604283487804752699371929467363894281842301458190 4732174
20135425993924523071465444794078374323701449226809458152558 9283434
41394205308328443529536304556315248022873548013172822423247 6083122
02414770961757014903686771052718313671330809638595109206320 4850957
26058359051748397812512113792044824316001678107117346853630 8045746
35931843096999698388196231042406158729675953574394928648330 87791207
01728725750580257872495270636152772744713743912897932184953 045628
63211026986327602302957821289379539504989758562962556628997 4058061
43228763685962179946546909782276882966952078205710395906872 7462190
43420949429709911113411895266929679437599622587097188146417 2698031
02880047598790844592309194459286640658326122178216887747071 669226
54843497013146988087005620487765927746124410871694613679499 002490
81388652890738287414831880643743475791628629972220200691849 4470560
25949929893807319761477818955156694159199596642497543749795 3838348
73135139164528724251025351850639919915730755058931316228751 3233617
80380697176683309182575668570820278948591639435154251331502 8022030
59029910560593018637205118445883378829902725653531268616051 4571532
10589501414527280378871220594944619909771305048401595434401 5964940
74990529726006071781155180986358936937575457419051652626618 1583666
77319309061885278262913626908406850436978181182890616795341 0999527
61156931811693935952251357610830276142326352796455075971048 536566
17220182680025943904934450333140573043530451437796084149344 9302697
81849921240142524943323003995989794164067153388611976196817 2252856
46503256228327421800527188462318376303518593696049418690781 5977121
57193941631957770968018564554935445201366780266281139539341 0408069
89986698763532415809695257037873119361249867503267573598362 5898431
89889598839056286921185933362544099374161724019542776558131 6708804
37043201145228993175477475314468237222731315914730925747104 3338386
41140970798262908385278852285639087193920490301139946595731 2746051
59078957833210743167427099485411407187333974522118875342686 00599
29865085048787946910907794347200777844302830661104238874488 4610082
65382802984021606833170925935798568025880161399309510681227 9707275
55849999812967113165426304136471072504527535805753066071120 13058254
82172570076519292437444005932567811061735566362158561029992 3491165
98795167860880531628128755552032086501618169902372895728129 7742126
47366180811672232925547375070256679867657004848787270152865 7855284
70169848268595955504840904460730019666484970065265885660087 0767936
34469220002669555368104087810186770361007861315403987499868 2587881
35831508294813195938760498662913021513031416477174754277579 6743576
34241837800694157155338063743165571288237140592871520780394 6430925
29752988252392426258332166464950351641280080158816688523681 2221523
86258675916088789350765900895516159274157120991156719956796 9602207
06644037519549333512316148428046435367561793113249306349614 8410707
33135543681536424660422220338487589002751842546830811418608 3409685
66254955914622288102352827689775372254468712846966202052982 6964252
82328253453602534017578927616924297424882521457514260327733 9605028
51799843334096873469144697618993564058875611984751091099885 2534673
05505962701163191414672305251513961885516405509122518270898 7435134
45994276447540953654831976262128660761077983060238280665666 4184611
50612168125550355225127019545225630364232773265800151624759 9255652
```

```
238254688125940536361157339558675134710949526883703211494119292798
400986266255942261351727610166803760065246775201816858401172220626
742269662416200903734098180835143370881521935507151274942633955869
137292685897181485433144313802004903683299928103154380150374817726
177149752231086458483314959463489377745730035751009753005657851925
147632594092732312615263975372396875889218559507950826913350464608
635046455949010597958752690599106607466738124949930696952771624859
919274260080088449789769783966465416377847642313692327726968053379
652345765962949942763968110317027578424483983877349030554675828821
048735033179781892810745928029746191486936784113197622213640367202
215568355826210580439315062000939650809803343201710765403059051405
473448320833643319407779355377740417646996447497202055014503618480
806527982006389055498780347367665765071153373893804069673575285421
944287739847660457290591842997700250289865527431944493455457310521
265829822248903145317010652078893447262457924392288329177671004406
195544670238791811192507381638124639738649243646601301941614986364
5175498021596962372670291070202830828627521248858940796800775136297
076267330759834622457924291329222165166346732070616172330633141241
788279771760126848386096937135488902477648681770988353071178521266
502027683348357304902022150257264305939232437125899153043778629706
100983890686389716376347286759784185490928923782527528546313608941
877604475194996578853070927950657680787679252781929908534286082318
376647592595816391115753102047615172505301277604038014391726694905
0161005628570784053842677849122735551603547105737366807521898155274
15266441234593303658670987041668802463441109631320190281992706807
424414279699487580890618168242937128996909922190393898182772627499
195528181230233825359769963300713702436696183256231971672391178855
920033669389901294227996698961737919055403167327902653570229195606
717552715958774585852700230463569417141213699699937904803817864568
976563385902753774811334272286780803313487813899384852066151170197
618926580970479299192790449075074008225743235184901891825055543622
635377318724704640510656285845719897301804989928002231605312868836
988847664092394534403723994455686197506548896374658129096538698653
509855306964206079272865482544136582150480241853365936646089150948
529430279250798576699582323068499207794960304019886188423598182228
76454108231207389067989649922087201347233371268568927133531550872
974356181961269438938239830757970329241652100250915543088897803102
193463232348190839408200870249781735550586952922006385115367975561
177079673235438067412514418494488932849835925859652449934282220248
670112239324866784521294937525746839474713360307510427760823419481
210899364533950384802792754587621681943345621696059539811101256256
192751186935458279444725862967722152810122309126072315383384751958
026272160827578784295323078705273766359912158752804316485683008133
451348963098788083350853560853082739964704136440820413602717194053
741947893004172777063361761041581224196526912218316408294182778238
186346219678714876819046872036146234033463552594624901491743806127
735581241756272756256785270101281739122526783743131869437085983916
431971144268732166954768143435994782809094643563890670135290517693
802652839654093401787303819952009118903694044890907374061028173309
072704516551272959565845861395892350552025495617431141679056591144
540040618217049194516932904873828161489192907001585306677795925318
931454990532685661753096785433460991136492692917991951878061329048
070744298300076740674987503021235223295263987843315706928804135781
181414035949799319537994174695618132313828355351273763389279424236
991870955678299670646211602834338706273536116137728510939570469505
316699144920030262030701115733460221611811140962110232612794507975
295250672404896420599499981853111922352067098764890474448320358479
3797777089881705580551585659003575710913010838303952048489536753897
939591091456852124426746346606832559883018855007288449457734297537
482901467306072968621959950015077980431874128687419726672324680382
```

```
292664678731205133980449871604074834930661166953440637702586485725
190429781616742138846290477226455101790621990101064723991166245178
996306841376747614290773613910599890516479131402990129874001851155
667329000905959816848642302428138855182045092494134942564516176438
024831981644782441753695904089262540896713293137672417971184497881
886641208500538501296658073585911445560234972394110401145615524276
081418037006925282208168327641105143640153574972994893742231607297
932846688596890779529479049622774184931407468245530910024807546119
711207497840562123258456736068054598680926652675729422881406450993
468358928198550973009991035626310461940730356061009786589359486291
483333890852349035226016112673114461448858248543699354752553 54685
970970012005981056167495030237194603079884905713479666526183524310
329605348660554276444504687869498531934453062179768591989190527250
913975087304947225328826392861338131243081633610044989140357151600
108759995996203276239973756006689326216632041271564414718553992163
745132571787569920045331793734215392554957301422189144445148817659
003488084946814467645187706960097211812601131936079883761087506199
583298067505343866438560342279149210478085670871212686110499475512
364196862159005355189208960454911801837087351213302113927947949498
157160689254439283935066231954258302225993904646108419126977283911
625329700899993604983544205589250763685161205226873792836010453863
705169396833464702268162108047704317588916650909690797846215121153
405850167527433048957831728319550911254913959967434804307493724232
093563415951628524296500143613754935532420224733819454036035016
471292095100620300291035916923877115341640257130903323748011829648
153048860911523951995822740111528817863718246154591014153507 3223
798908102585207913123261833783663973086725348955880826566395904698
494586645327943651353041329623353826462598103372361852919348742087
087405531674261212633731882937687221636803492356352005450381560894
384404796918033652023351402976755746861433330502083878381115 41268
925151166116930274296000426113671057593007503033986321195030803 3923
935794087416870532450153670572923374346617729991193485193153195573
999537691263697163367331179764771187970106981594852818917269069689
185229060502882341966125844911596623504305364360140344435869935105
208842215655791117254130937401610698894316863165155788868478425711
435293078457378268381913669237471873765036114019931274788955613142
720362250679969666727376373195224645165671555403406941284959423784
512097673759017436858546715560385708069006797916099574890855980086
291652161786679405219875252687974230877054749630423028293035258427
612765298373725206252502198535700584024882472881571439333319693786
895013767534951472748248357170591103305164106549341448296256259053
053650440190380040202119684943642618549571897538200290157298536474
141922268166590819374268930277118424017749134844691016489250387237
200483465001201970901451290948959111337775558325485971778735550736
760415422782088532065005677332973421264736659144908426144499208939
625766086675034809435890690348945624497298523623631671179811318218
118983491109144324354709971575748761833386808392758242188258317972
873376426584738419959429063460865025641920729018922939376884454225
160114951712985769514509191963635889005402162681622563351387321791
540055706200344549047466943494357183394855708615827904260730245673
126120480774370352283234599595601698227400702221585367923961094 60
875597226270351571618216645216224885958641204904458419056864801654
356011243206126652245916123841780807997824843552246179405853478344
642156931094236963465846509937007349026356989916371199938940366672
127024866305173877943713902426423067540464969108152829509788010760
020123834434144501356761975935634590123911593375276828652878589244
329251305094989943506488296096665835247177063141529760216668828967
756246178265376697131371999006207021132131686344793232850535267949
560065946849113093180985273308269830327604880888179661619217295908
933911350111207115745768701357561706784912286958407024018892419839
```

```
84694794786018629451800458079873067062979401705718930043938659822 2
49512044190847007469602478155305053419398528919926120190091041563 4
14109989470424396920286616102997332160924507834779693074782697807 0
54446274758827131476923967144193749605293642754698007741800521845 8
00027965465030002406063168830518507664008071804743220370390040582 6
44659377538081714755380260079649446238210566352321573300736714252 5
98481768236735021089086290403614913836427450332857483415640643206 6
89194496748278940363347410411129740448869877217655103148076248706 3
50678154665044564166122386658729677939128805009272305309637101886
53216210569718046969660534235379944536063368695069147708862131474 1
05827259534365885139645919276910688196786785926689805446422570520 3
82487012622034850487050663011061956622881219469864836947195537553 3
96360211212003066978167067709647235781146125189170132398210799728 4
37951364368656117137375844300761437122867678000357008602057620137 6
88963049414699694740839791173586156581516385421636279068010225866 1
87136039197017311433859312927496714361446797801995139148220631826 0
22118689076010896956529129986875746214049984071404754056779056886 8
86478172452832666182598182252014731626403289192851128238981205542 2
15979937190486214375637288455209946833473378417669478051400007976
42251172020185731690589739215012381335881183622775348102737103649 3
20849205941783873183945781220868472822775659980082173517596711993 9
95926440743943861132541162237267684977016981507948315745510965331
07287006255223441387685229836826541958706343081757454391640275712 4
97594259399588239095670698447309993059551743615911540093159194922 4
95647179189322539829596326491932212978677307304157801795209314 1
38469653003432072705420005956668467709137380709334350231022160765 5
84543084872951310472287109963386350969948308495605329871456560640 9
23182381332187880085000847147047359852126560582850894450924764761709
11061238375632689015816899808061554801395860245107500012238157368 4
66567152911442874911742506294132126303656049544904557092778615881 6
15668302099056258476911368027652260606490409134672599970331855604 7
65827513122924021916965187859077190738293006235986798266744820382 7
05340629601026097181086850897613679624080024505538177234291701351 8
80426389261714612890720102590760421166092922287019966075188507202 1
92487218937162291933192717123531809249959229011922950964196115379 0
05918800450090414502557094671717525809365458569362245856635466191 5
22005939376186862182146602389800418405283481475116055290537915813 3
42230543743333883189960419679081119392198891347151801832111774780 47
44698092576953907403875358960590991079429935275665412376629320033 6
24012886186572176411148614710928703561099867922619713934920579870 2
66661185237530091959236153546524607046867492021247646752375995171 5
61815218544376399845606352889083661392298630705036852971258516986 3
31442385980621275692344988529172596300772508654188499298386837924 5
11373779522606687743092259125029143791856528769124110571365532855 27
92653328350731609735879459481256348199351469187008885214318633657 4
07874409562713903481807177941295408148322774702712456013713249801 4
90169161162967363763421079348217777934743825904799583197158802620 1
39937881200061591076306502008382680044033662630424658388948479810
47086088166318448901898087115495005334664654828434918409044756679 2
98136111504752642423929686318136169801325888194399321084257991617
30699429030399715260431669479858677053524358781682644296534046909 4
51594538884599209756255160388159169316566468194318952235653786003 4
25886548387822425301616483121234856316260457457447991941258992523
14994103400282037557235851183385681775101953078269804845441140370 1
32411875899878064183237927374138272928000431842170701713686340646 1
10327457473967999314353635135939771359471462109072191210558903082 2
44705087968192714399413322756685729668311188854483475867783232347 5
37554741753400810503523635715400480002760553821741030745589029987 2
94721707719511659677626716491035056707037998233980937584860808525 78
53805887716180013763260840151420016659197591559855161156432249636 97
```

```
1610232015439245778410930840142487666704772241120532075597790544495
5882575957815705876544333341729663693118388086691641836589628003382
3065173172027256657887306730547714590104963012916483881832716635142
2899830412480643860570873277797636794984448056106688129005501332788
6047499465992178967865322463786920743009274664119565132865552741068
9264535287970432290608980899346185812290542975063008754354855175622
1851856390051984501949666770786659460018771676589622478442779685579
2924178545808906250563334856870633046840571325251540583000082050636
4150151722566366387352243636275974681799746274291711960186211618742
2746333695101851479720293519591311026246612966336658336043111567980
7500204107398356941643638789014664311899162907027183524788322999447
4808288082142012209328738457458021827257092445656873815999656693481
9715065611105852340717662005463133871634358175578548151571126935579
0187529814444877818386510196147537951313173854855241856141555503648
3433467554627735431341641637527101352057735283621006290522700717857
0501601175909342079423978043578816110011339035332835423279876848098
6769792142956398542570374870885134903653510265162623898267334690486
2537709102717188687480399038216113064153409720024204447786248248796
9346281689064257424301427927471119667514370936550654375464469559285
3671784361516330821674589086848973834011929218270034979373496826566
2635996939236290420059701346477824818216741757146524225106143669279
0223054364748522778253118977548255577700282959990655236086555632879
2175074705216024764257288446600245590158652930419881049851680444986
6103034712788372757241790020774595938303601112745003549644632237540
2979312639141596209281563349075715562463765525563752233563459000278
3615588938679804011876211077465058628373071615023932619245933044121
7902216959227426875524359429261780244424898126940443276913755881366
4686624281946944181357322024090773078187959524816980838323709448334
3046164569627705508644267061164306103156214215165708154243512007196
5173547455618755482137247149869554903457085895053511432394021582246
2373179918552318749680683707121469217975229527765496855904282484768
0820504383454436761380315223777505023893408898888176380547193911891
7368390549117035241713907428198953496019307209748054434747336641489
8546993266910483005198534414859328350139417445761472321193035677351
6795279651808809212447277537320225205325479639953602596799644642299
8460643275461300472320428319136575898090629884651088063978772133628
1586946906446637709059991166556720577545205312655215945252856009305
0834200984975785933726471757737222554577493056737322675994547442309
2454280531965388848263110066019704801436833586079727042714324410
1918596247937428199108107646741156645852669096974672371970082211147
8899498225050495065962096324623162602588340771283899196083119800139
8252927316814808346717602224727913127914975549582470411000628077251
5848247625179837368358886765101313189469685489077951472474164533245
9020729886518696550927776924292199661353104048077394395637556388717
7300689058980462020283427661504434311928581884257353550708003792286
7781520000347001739755791472630932078727070832281626781492610376277
7650037037939641990477124122525343068517075691123610546586992144845
7862637590609807988652926828464980090413512656417860018872927015033
8174283187137586592654274730054724720857382664639920533749403999635
5445028721624010184184325452614504979432962002087201782575427776900
0056668128331687727441937700516295116386048758318237549131641401398
1246864700943821876443121389237939296349015970042250369644899205575
5834348198896493977398056279967628299816707061439856639145113334523
9761167205780188989534450690953850517607639437966600185732608911171
2837055912519200832211049553238652831066348481108278056662596066195
9186542470162512017013181749677241815842906240116076747500922316594
0884273500977129685035860543207495195758116118555389127067860366392
8285636957219434773389466127632991488961139409524522733965166444439
9561764463800359924230737648590182772338070147910565712633827227070
2618103745988957619840029870168680163746592595880895765706206811808
```

```
015788245235470512319493403010897860778603558016256474963668952266
644297034750973757293655215957205018380011300642483695274978980488
669946312648370712620175835288983470645632987962527812423814972340
269923862245553799485227566164182094297284938152661553100382476458
941010631066791559388078731554225148099356347040083852563338713734
005789665118282648417522204303350688567673456703495553783291284572
140874341844662966193453239340102186602517225820244369698286892594
482292569580501473842032234552749162420536991531864251835132393123
249926171102090679208777081297331878703254358945819299157922805909
629313615804547525082634659987958292821403960671715491841579732776
669604006761432161372979962879694295586470962389699846152883945119
399866680529478562520047926648025531104334482182225375355413143198
789842444080937262640971866322596396451409108638268364912283333873
204238924585565961231467157323664576166407856925815730177064163262
773408491365608012574138431829515250315688191614312101638047668216
237140317556271005829767655973993498456511642712547906884468423062
617054728444476329963710651294135427650142263767469742212676631934
514820117451729366819512348454810184828070389799766164332625967071
557641203485357545067392252078078608458853099304390756822309941363
722218810657474734407718365747078729497306288848468685885509892782
698668673193115049382377532058098194315835426983040375068899137050
166988926840535519245647408292679246585580218926399943633169156591
990968485444353997200992769810704737366421869025712126960845786029
245897142634405548617343956352245226657966744084913628165235 83507
583693164682894041678369725524738575238764987591524330562429 07916
161697923747260277756820967581833760682148768889887355900659817808
858652732469911268246578940497843423605889910792877680494259713368
582031968619868504750040513473006793223432926837954984462333505247
263460082690068256150580587416497636645560901697038354464898578971
378176342974766943836393242204257238379652358937833265993732944425
338544431136852105145832185085344227497507375521585479886496527046
750978569634379985928566957606237908255374065993918733488900683707
746155489726616930391741958038525420236316279949081363400652774248
626138944550673590388362049435427415319751126340575592208112892245
179435878060940908912240877214600264516575734886351840493636414562
851412076587021198616414722225541119232174474798899174912396282969
229904038628541701808467475656375497288352641939168352265091911716
387965696975529401617570874128981880416544974424432650797686219165
253206781232677694837429664274422877198570625317789009163812625149
268598835683397805868852899607572612324059082101032768468528289163
892886962291219615830219522140590952205228972377626878139597405380
951219895952727502300495102059121377517497365540730407591887400539
984780761585134218807192758602383601471109843561136961831373268637
572744207189509549635640736905992260249409020151463495614064290221
465656120894680521708255829751858096181033709257858600807838724053
485637997824320509476609337576351781081771721920388391768477538656
929309953682828387716996256693672496583637164397842074160549709960
664332221742829985099655056667494594601913868672654443173397918383
359490266481059102340343599790310442218904226364686067510808226902
546675198447458959068566636994156846733845847585095636870779635931
230765940437498221894244750281518093959819874655731869032409794891
587446913819980388681581918671230448377568580829491220952824741513
947607156627309898236541080144691954182331228197046695559611558555
558315806826102545264309773230627583922003237786026453847280259878
024119832286203580916875388115444827489547270015957334586400959176
457689844417700579912331788126521625196597494202791280207150219806
882664821063030190168241191958249015513411969179182395145052466745
793017346324703628358480990854643293434708610295462057784 99327479
680655603815266584241552414020100328059312423538009786819459704872
307836886309617020651519710643599949209993665076698458740 98782049
```

80640092746745834314104223160821853440623459815908708177850488861
88076935347619189602143388378631740462974922741162798239194917 6987
36236162054797635694462437600831699566411895725532404025593363 2552
56840140231223146417761662232610746582277194030504753182900525 6782
86119800467786361796944009774017135335045063508793226200874308 2216
99646658676912952816629185237161253242263525276924708417951978 0977
66815332733428522923216581020264856817984465476940137002477436 8463
83495383813674721416742804313646881524370710670450070487649927 7719
36088402624574832544329630553349177467551235750843585736505027 6037
63749933947741180234829655157373241432111527755175995719456924 8002
81214257652224094905123953705898275419286162064497690080051672 8105
52043485522695502759535414010803257607794755217382999163944188 3871
60943799454569368152118768894453778958210606427690604129993591 7059
13374264018843127518832934810663329392267909394999477588807244 4790
29792535905463174010381493970797634841266157383566709345257924 2745
78960762472142228323509197054000044625077648492785138988357188 3509
82013055342590379193536315223548885866324766764288077779308887 9772
30434364036100687516427741220226139668328077893192782831808017 3077
35060748152484249186724266164920442291822597138240224595523248 1491
15768607770745469519296793787891155746845011974262916666793378 4946
41156038274345558900700550222560515902758161648765329583195718 1357
69435720399299808221473778306201846089261721074472828622198590 9857
16938933609829022428242439689621174428745278333132623768866435 8206
85652662538606769966708097524491429427265459755066092677031564 9639
64401146832516886069987849984107652527278128521589795908956597 8155
83997375515051171785190558431115800729804986743339345409911690 2870
68434677418337742981763021220816227692197448107582973633218543 6139
24559326523471790705810474687354824552977769856892079562945220 4538
28662741920457455359753783399530454822592043518197267149264847 6022
04843973694833692230263902003159471170574057396373805127004419 9045
11231506374588102661348578762009022865409928241283772367672335 8777
58644948698520724860512059860666813723635977900580542664078404 6475
17048534148182759136249435263534902946088197659060016587137596 1754
05197969008855720693723613481394247208117409043862656836403896 5745
12705874912449954240863294279551686606512866368895705695022204 0105
70307968554223702470323438644259531872847429722147172523996999 2359
55808240678681724403658322319672710300723700188475128326853384 6058
34278510272647405624823640633476677570828300683362955179358138 1740
55054693283242616034216539333747842236296193124820173836444254 9652
13133161882613006637459749913737857632874140797844793458879589 4649
89455274197287491029563632473047956693085789648079076313688043 4588
75781161717334732314995804433882253168862429307299348551379327 3497
37185100257796798321678291010493154216629361392140993885253203 2510
83276297326084041385706294131133013332097212236675607213832676 9687
59337121436284932472540613991437390017909655616950524523664399 9228
91575648298692358153737232565173629326972342367008436161796738 7854
12338336605625473318531815294700259227642817480718686124832829 3644
51204862886214768184751400687476326953459144613694264270032683 1223
61926144791063805896407722430206150826425937453033717241764377 3503
12429148019752390957197447489447153307997401303495068279209810 8358
63986342997346393046896092447257385173993687206821498042655312 4365
07173091861222523205627160040265570027872774381741924351552055 1967
11728092850514120703459095006984213956799693632645159723833516 2410
92181531597086428473648179525991050955471880427769219470454917 61767
50371643819086697233951943546187685076622930252542954588322796 5676774
36304067879294412760668121253522202311854862065044395394508331 7510
36583670328033948358812184531326517795025651828998076043311448 6012
05065275015400456837046219154066798804242099576049931925485478 9084
96535053485400384303175668826105555495951848785734176553929482 0411
37537515972816222666334348169402368480231249335660860924198839 9182

```
16617589253010571401660681312694306034967362830291568237557 8192779
44328956919802459497451946431232736194957476648834132462543 9972015
07598284347233931957944820680890605891133985312271080186567 7278159
41198540489775460548919763872856096638049540086111470844426 4789236
58831637835663924878976950720266270305790423051402224097192 0328906
18982219280978572634516485630007662024212820434855630908521 7888747
71177834621604616055143076378705922646900867158759744494378 3484827
16114874286991483216478467463043615225653938297974670781183 7307823
03118242340838652157653475203542413752791242398251756248254 4830046
65359400831234540528967787410513615012364025900456553133102 4775186
86159099530602215036676918568220383470112931858156087273329 526603
52655011232499265932569003251529317531669596146622381145742 4745620
72830343255348702757698771108729113549593472195838335632826 5835481
18449639775922378878880443698468515367475136504896256293112 7022799
54288498999637590462367888264316742631006301794525460633268 122579
20178183049663533451603167736701744684265267167371652582296 3934774
84332059523630266272206877024372030776327349519981568291827 9050247
37048173521868736678248576506066689507158944956232274185656 0196805
75436514545216349960217441291710542209291780186152456224507 2350571
44146713040303935895220734635199786488490136414609947071319 4173872
78558076410191673085356151661972454654527568488744897738507 2760144
28708114938324971883354015861254854286094345601016500901599 34047
88762709630398110242252039676657399265439920376967110949173 3084464
55807434034349782575542771003734773935340022527843803197324 0162990
81604351321157773545428348123214910963071254659426910055878 3295870
88894690246363266305833672598691227243130374580523377669244 2176024
61473544109723669598080211324536318322294375562731168218744 3627167
15686590240179593292832033078534081437656749140858860369736 7284457
94345813589865015967905882725387953723136965007335638464756 0701395
23497003811083281051092852569969885229199115846194565350796 0480232
66663068860863398722083599188993123787894994525263971843821 8462630
38806708103221148931893646089799623357918658262048904237262 6132345
67054286538678238194926812289610452338739878810489315027691 6978748
29278797548288026222797741735202538719874817121221558965948 797112
62632466281900340288320655475681131737680249891788724425138 822908
63760549701234863602048510962352788936018317083393455622844 7900858
65865752430453023665277458314451397463341174279958675419182 932860
71397475418290309318288160077681341065573711653229177055981 0560144
55829108039258537954184315111810045848990282610189618735389 02625164
89433474345425411051909808825844379750160614581898377507792 7239640
20771463359458926436962146209482092574281590456026920569349 8774464
95566802957565063634502667112303664053713637921468253905430 6930004
84241489982948550537979741889209218747397160936460563377944 4175212
42717732268677829514875455730110253920155091648218454830151 3687443
99874561010323693261734825836121129624573517277967744624770 6082603
26944708335640980871306498911061971877169031312164039670001 4314465
69976556871940038135907748198345848806954826394348215678235 9633050
69488974072392523827676064298876086778199572198473722797860 4150425
08847141789436625691408065834826907194407427499841510891228 1119130
51921248519779604700512055673188776606559225818143178499501 0340698
24226993383036307258992809336341329081507735767554427542729 2178784
56766310453681820132627582430723942730885784466671185314857 0372638
76145044348610469304575822631580971170407984117565544101324 2991082
10963724984197149968002661122224116276547846529152125615752 4919064
17181227853263841666063789634548139137325074333493295520879 255060
08166301047291111732174134344707817208507691606498536416819 4690932
94115562580636748311490050345521695704033767928104814918678 0494952
28553682243152738419076533074725564798419465386919987607622 9865675
35375501126911294187359741098240305101962773791078584852883 0286786
57739657260218052770073631746703273419143890887257693034172 7435167
```

```
5871220562305040774970000598540138301228241583094504827045811370576
2083436166356803525303591679096364635737449892093079594664704643437
8705119711135476651047196248719976522180212032436360552826775 08656
7225502496127629156726493403603008861377930383929846716416489 57948
8318065478362585224371184607435896594532850844025007063152319 43932
2570570843838480860847638666083129044921942701551557202047438 76552
2812521067258028872378353516790737504820421646968848385273970 33244
8486614200670309698669312732739408701914957907335607884443759 31681
0927104473376190517167994194272434899991858007977170192147145 1672
8325421693706238292023784831018312180368392385300272093929380 04636
7845219553744437335918195293510339234188363467182012275749517 60590
2586857937581469684328052733987145139420083146474738663417013 31961
6310226035264260276997130153799810893166441152384485344748952 29162
2690657249069698253574060008098937455425457779015498346865358 370136
0571082540844756238918313800795420469206625270898219317009169 09072
7485969825754617575499149518372366424535263757761246618159752 4071
6700394012764052654196207653579568198464121712717213496780050 69572
4242270423461355737612051130500028600623319658976344020079418 09250
5566013594610679654763882671693840917637986748068881672301651 22403
5326870341296838588030431636797141697764690884964925186326612 96859
1180822955482147562935296574161245030192512494487074275141867 33602
3032000044550571665918474560664157354343100659668155381470814 28842
6255251445265942387566188509938359824723546002667897548291349 03449
2964879853003825625245656122403458073519193241049256885530198 51993
6587460251292075604715527804249973643853625123837505139962066 72398
2725049942735651643310921578194708211735465984446363850183844 21738
5318968173250707498941286446645627833018858866846722891553357 37178
6834665540805979509123550839391770627765675290146648139911114 5590
8505374671018603253138842315776138822079615535244028504588072 953
6438207464133491945339107175026541072414391771558556437702870 09009
6055190569551238343860086542896430499524187245422137029233789 33585
0124750754636322055183980391580018756549748460340546829537543 21052
7463467025064988757478463846259812624029303666021729082334668 22021
6863300462928431012001548886389956510305458151264885171704042 02635
4946457685237540198744587664460443926706736302674268383523828 154786
6960018703286109803203714309296852601233888418932413143901788 6514
8972948510053826103399072492225697267155022131197150157960576 05882
2428532750813495343802529556524012489621643938806157633764802 70364
4020310318993813503292271315782228052884490075472056133192888 82880
3014279339989326803600338663534172241390925557258851397467393 79162
8216102164450629590369524745302822355476931054914819701890239 30952
0713248153648723430439342435631099433019840100502550758530041 24402
6424404238852222961599880856377431052505561793050579151058220 09885
5971504054512479560760526065239625626403320783976706326971766 75193
9183220087628503005281370336304047346934863287370120644040024 70611
5757563633464623420853885742508683124515921662253906650358139 97403
0963612615594551042278559045530973039515490936081514712383830 79236
9123398753158274112833101190030245140997046534650992438331645 4555
3507470392604605420891679058896372202895689938469660378116652 60491
0479215974103839335746861356601844496800854923767707389405633 70648
5266640835053035097500300002323480752684936297327429055905074 04654
8621141241993097114343923825086638683773884446649889153390677 42210
3442500858291195555674136959664247133857088717736420668105656 46068
3440456891341897683042820378672595329790210932878884230294478 60515
0348615654650970837114357976522543012008451452657191301406323 94775
1158789994277034996344818127462233913944985799843820216513074 76512
3255158161058565389111275241653391801717517014033651096508769 159776
3642668327681138520946677832207600159909731640949142678571043 42911
4283672657133251116203355968957711225479495800158251719864324 450778
6834358999521566560895963241705461238602956978680793633944069 28720
```

376886394929430093467194237112999426715828062697160220315956501993
55045614346198036240489901929791683861761269984848386282521983462
627527439077416913115243508183815484999134998491189086713357234145
821513518709753433329334117864125477183428843611192361118601572534
777303771790420145052442500265304533746747274247241307543686827352
169577782366362285722212574261406469716907023631855265253313948844
694441132213592102575917442576444162610089185819590932896831241250
228165687230474970558548269994285045342605886436467743207740385575
616021926481784189348684966755846588426831056443052330222059182322
152064555441692356070265021628312423464389928978048600969082699053
279818843835162931189526210965761526917261483318518079648547878377
329541028454635344624104968256047365511103211895261707578934075839
542344502189862405580135523875392369629974295680022459156685866735
116577937186653136884598757172652058475012367316313714039208273519
864111222750401779687189258415722478537026569434975944609321990644
909193761157561302152111919485022347414906313907930352869681136780
555168187461228166402877961343639825035282054824236296951282564871
514808762270954716466222090180028479351832719291469522113247516803
455353655516803305535415205342268509850239079036255961460125341186
620023175689111404158688062280100065457672718651307305533966871734
025274557382377184299173532291004018364304403105539935807300996129
198738790173676961039006875822473662052973295051266221645515160507
868398803968551187435672901354339735081260092916888939360994352719
263469977561373424313959939568202542844037193024591245658870410995
186249481278788320000640519675128652587144832353429917159173489812
944332976619902302015829663106848405115634609610429609663957887706
302738189464090292668777748730774187018836037942339601237177137111
089620207694470166037648120368885118373052507931871273607429834890
659100246165987442406773985385123045805649367146087741830695765359
376060070140563539965112977327928143047192408525284083055149052762
390902387607305809083414362841005473369032911371185910961720062140
515362032283392724115631615663121491405709907816731408865621123093
133793593152561354112016603854068537694209929282007279649086871907
709333610759194144357805433648744948304015733064821442793312043041
402545493349813951084093101581187178133190305351591215151871660170
828589990442877499124799183769709956819940701848737544179754789456
774995863112301218656691586033447688219662075519620082899559891646
650879573688004972612383079308233800299882392112854092096532624688
573867371667377250403117533636571281982530565990980506311202687190
513926870358844994522021549940557342357801066085620771804795540756
364826167454620040272635625473404309409600892469591212915887300034
688459434748067129844564800901495585933490983717144505405660941246
161613444251643650041227385387113342843152411110904767157363458840
394511553181219999287909451721595473113489943827370301380015649485
196344498868180656430972022772819365759901891195670286825652242981
807938563226873658675315649182570075051501412915978208833453802244
364270016511477591482805286338088397274022913414196169881437128915
062105654412511982119083442119508926880645326353618948879380716965
982927814705873414181255589115550047842154593816326090114093190107
698494750964786910513821952533492122918434408593194256574468782605
040203040030922234016547348203587939611198597395568059622589850874
055574232428253467135677408319415331770395881123007211984334542176
026009760420037872880045971380596690167954585763966112755038535284
531651406765401303840761709625648550598589161914676175732457246920
104104106585120047428768416175965754771526509497675245561255554505
123582180694036306406574744027339582364542759795435237727125059712
768384844596059418208947458150121997261837528851267232385185863240
996347343810351080008318852064973304981747077677541238533413099593
588428841587593014099043291112822844025317941849587372262283674842
863128018910431550093261006611312919720433495099623804188091585360

```
8226945176000122291267758946648087598375009614588965407571011670549
3735522319404512385707715996100064652086243427133070170680596371847
2342524029821212312904945859666684608019769972775272392438883052000
7707845984092170120817759072476198408867146778599969312163140944033
3144691597871039102584860272811071172157926631753042114849434133601
9309934307256841825441958907531859441945868781290695399968355739844
1358405575533897147626545690390998797678149653224958079865257885400
4811778708505875308526706434726155986137991784617755855199953889066
6485938693122754314684637214344502743848547045371071570429844589388
9795774925121055291297996601691708730353870330537394715804505098884
9224885896156574770327203221252601245834379306106529366230400119240
3031657721463708645357304933947540573918691422318192791590273256615
1555477509790490583103778938577212125282106409182837252073670930914
8128093520367031010893678711428951658930386122807508694551260224477
7765241015129543149753495247855456799175726592637194197468497232222
4840655414619962436246319300902968832839402512952271976431896328255
8326791008565624698976302482368533815362639415880757778523255185214
8506182984431709490110012154054518684462670305157443294459626276523
7810832333067900253549913176647627358575911164437335402155247207900
8084596545257852461811964709576204891001076068977961606606780887827
3846459522294658235679679146792836532220887277276500023647462394666
6777432360025874812745828607798783671599693309632751208644649310417
1989092592839356915170156428218300714471140630875384642145490779734
9928374575418472826829221215846258752919261432811574092883864434436
7428063860599275328495734009681328839506608447814898971506719787598
8639535216163749766961168422687005373529620665043779416170019545147
3572648489174993965286055550645012638057910849665021790963244871055
9335985888892094994413999111842605240478408240015041720825944396493
6304050422432334423412231898366259071657411322119024917339873698977
9087191906692587929184368375037768708134990084596374623095803562598
5310651551231093349921946547070691747261511176249697840416431496349
7149156813881662405609098234901645504865834821322106776934780797466
6687398105150897237571460947381775184673851619229096048670992721503
5135742014713465416808025570378438242131536081317049732945431018124
1630670915230530947431530240913050787997736866709720467029894653478
1438192571666402154608067457973335844781305655662386139896418584866
5055676785938059526425552216892696925251769564723000760023964516180
9663235926092583895177938403990526765175046776855866893072337676228
8726582289415350117696979596156241813580090718922442177232872003875
1710603651464829833033053722495645744798162292303589962916773140904
7518400658041704030819834394381654891068288206327462119824833333898
5666989529290283506745773869455592449308447108296011158850439402580
6950398954693995075967060331038807206468560963210090329813694337119
0665316827971131126381553798474363341054424592149284481660789722422
2569223936776987105775043118705183040841330039336437281082996533734
4118435440548056188153151907911779005947207137322565568901763171369
0530039449867117867765760616135419482026995648769679190253527415105
1805318047904833299524707387105836691625235149737958838129078224417
1818484141920140211151236625757297330152442395812863869939865665967
2257084747162982259672706700530278389166151345353868004947749187295
6611414023828964951927673145521410419075872805809654542738818448510
1646721648412860818118848555428125653196041688466411776004696192798
4695385084891010095957760848958110087780931833420011471107536645178
2495401231276452531676859891101361387098255505779044828323184614801
2386799414649794199831986564301242122420097592584329965443684001329
2175343954831853021121004716189080692564921648315239461646650153757
8187636285721141963760634590380804485144448268896301485547363567177
6361126655204304580328615087831950798810064706561069059209256965510
8470085543421756824262446773998395864395309621857601668502351345733
9696192916600677121304593927898088510740
```

```
64783706764394460094322902404672837646877470672089842241557979318 2
085269902422050852871413657941189781978550458739575252874600365 22
7032362527644841776506081179839508952757538706673058464281051722 44
2379835925436488986759190110848908531258565256849093459137873410 96
7007559078717216941505715952074434206547730785720194909655059112 3
0715608365355174411382300982288148106012175954828424074282604636 97
0643405576271180674187884886305881954050094623354931598350794951 23
3441213572825926780315452329051660143981142902824654494029523710 57
5220410408719852639985642562637879054893209384738109277312745761 67
3514839306932765056656216510271885039240621199732384491744898198 45
1671836974498021578014537918841419203669268865841251760108777421 86
1457740409016994907180539710867021896549013771064734514113106185 15
3490595314547451367557354141683472166332112542241507184695989545 01
3350691631304662536485450402893329655830311197227044692712768445 18
7936986647962370897083791325104578842326704877073426002470128939 97
3636738197291834119862214376692896693353279645140506039664013467 40
6763717868287393048276432252090231302919414755942393487176956896 30
3242017126782671993250398151667836940800966020396142600876263161 77
1576560678605542088477315621474647508369308685473326069773361088 07
0161145554162831138340683635597954777594929953511691429683515415 90
6531174851453049638023821127319513900885682425770076130904374910 97
0801102693328707280843897428148571623498332863755675513226515384 94
0228560697895043137447601216535598183368946742757422540139789594 92
1292535499754244350393113660773385319495920506535804026006236988 8683
5741773118120204487114491206080091414041785500406260051423898099
8588060625998593878167626604123664222145716678870393861741585904 24
6386451079530432107075752610454004327624195871577617842744430767 39
0248190189460719058147553505810236686282098810508614533179815210 3
6092562555849947227097303806193235909114715979964546453672331695 86
1621557701291316057096724507588063294889700033601732003149633967 03
1283239940699152200059759581699408488928790287870957148779104899 68
7207013326691020490833378624454127436386003821688457435357348569 54
2376006133559198136987649211296354545802945038845082300601904759 16
9844999749285356787888774055789934978442999037941842816192291664 88
1251892815947864416565586115774593040323582606542663128269069150 64
6614399822782040111367023605190530930620673903027388120691661208 67
7753644625729204191297007926545074607753128883188253096178396199 7
6095544128713730709131929297474923988857945269066393153737183766 05
8626386851040743922546146929987675748887570048359865472133548351 94
2068203056630757241848187405753097137190801667442371350489590528 68
9763169365100418441841929310641217624349642807246220795077792707 77
5940850184783641036810814303717300047893228768691588740232752954 81
2243787785994434661532103076203193244832730028661543436125006078 98
0064933067707551723849745218389985563254644198317277638544589402 03
6258928886700741919812629633704564459022096778867130780728851177 95
6255457236175317918344958866223470113919646489934196257032419095 54
5535429631166351171922664204771691513008825332423423185330638239 30
7999728026830156448174462806677078649241502501406976087151060958 02
6576018636858767595225548486759593091602771905617183110210000787 22 86
0230717868791671907153620932990534336850969225074549573359857226 91
1277559013741008525376269032264904762459823903709196087975491852 30
6887396856358873495848827637613568436402782134965320221620253900 15
6466223865132120426626370651407902029875583269543604100466557213 39
2831449323695411323766206922383077364896405814413329795519439965 06
0432690833736648864366002864911095590720327738217854893016167919 13
5899344261282022544397548225777539028780812390065881804375509744 77
6027130929007956101760871082136862656685640334033271152619979995 46
0671735858910215594898711891888924143102785122441590804005135088 11
6050053100242120192777976866867663927803620821921365378905531498 92
6604922873241925739059632046527108073528121647193134510770658201 26
```

```
146202387118614701500297074020714669628944080966815463011972005892
139093486574986557942005145573823194506046418349843916344149336874
974083843470409843492905516335636517364373976372182533421553977413
672765462638260850745558046710307654857561724562075113548871979562
115712816459467797240519847340294435121810186101346010377134517919
667029853915051226235427124100136665962748256585723817758172617743
576696198559931832068173283954483120554592538025706153304808341658
339480391021438997926469047391929437430101677948822478509958248181
032809407218526229428099485630965041460314445048781473327478920724
966271710204105098592639718944019328825551301234179887443761681365
121823531255731233958949203005402012671985157412756120153017742912
720774273943596504093954199364540141234099602139402224484085184848
332198425908425390987985210295900294077244141016654654873469273559
754615434107642381639206540366015300938502511115562648931999753448
906547614815812522512560562588699082706447409727111737291630760622
385173708214865980055492241775543703894118300654208876792251773889
701873225593493196204540543301757747213965759405417290481304027124
161138680548085002881786941092037720337981316067910743433191577004
656678794592800171659235707152641398092653464837560166236907653569
706215111504303824258984350202313113673906982067031660684473530 7
```

```
583015672082761540668287130050327141843913535004779722515537359244
728732574153753034991764158809964303915977315363170014666626388293
853296681159260307074378349493172572938180715062357327537218262699
363827186676101717472859495469571984543837804288399873151347405781
758477788929828485913453395015232269804934144342422970791501955236
630855032892744166646745265231744759082831248357714395442890453243
198733285098707110075679310504066200452348876962098360272571530489
223643411849427732055843261582846021572700419118345460000821815892
759556031386747432076375702119198906335226109059381935386962787213
405894643099310931672292594209303201545120120763563505739424400930
996671175227949248011618198871456715695135664054208776797620624692
641499503081730220478263264516467787690279893935775180730662032566
38131002463926570385617762052327895680297674766851344475176447245
258312970910015419521622604037955642485535376402196633191303376813
994426710033175743066999659383239210167996108136633285159070243224
754758290857648393765178360991123343959468337960098971368228065945
297746445434491643173749089986134365682572488751513214961238202166
634996724654336285229483325007883063795521355200859851420246309871
193611776095011817233800545962407163665886877579801876043281920402
050253852643518492029157968922663004568581766187961393023545599084
674216906566065194005675222985195908703895665391305994099122387586
849251535109322009586765511539614926945349453553111487418703067073
073762823238740258107067911870389654994048383178902997420247467498
50399181995446060857478335866524706462320848454898971455020321161
465955189844357428877919531709989618645143272900186033120652878863
339256503373888835809085824505643946671712631127365222765179414834
143168394494582644825056228069501372167870578802531885005058960874
295673127425382893414113514109836327988541505149679002485724238672
673691805875561532039795567263543170253025356630663739776208741968
670619931753671598812221121442088237038107076777137595628290593806
216583216869007668154133453036117384683393955177937753994627960160
229296331274113950352838281635891327772556109315981976442552798787
255896457661970963237823053375945420082077055629090442377252837573
508110404005631183219485475786109505631350714719614587105169761852
958639357577092454115019934309488848458957387958631624974178527859
314927506517174185974951125191171329025534462400496594781211334835
400306189304792243001524693835214136727845826651384912903320503135
720125618140067375400439688289943969454204245655089655523052486747
853094181946744093819014628042007498122032218374100529786239402123
482749095802769942777980523749211359552468218896391058644776634658
206540551964125977106681895478628696309809514172738463509286202100
179292347089449182457456884662285001382686177468576630059727671080
981917841346689830781965726813325022350128152664463746536716331509
770497790390182291755774507379557600646472213912092042930350652273
596118082236730584599004846823045812492169295795979120402568657598
537278951193786709755763040094704613335373137328107644115322695237
231053757431339894280068109412834869385111700642631717926249166481
519887437530404753051229303412313771533780044996881132502543423829
769257101712805310629329135477605614556555577295296310833587113574
218521962546305790097450301791395338537202838383239650271982402202
2884685171939123593896573534344443819679868819189721380028084324365
903979898299755255265615100220791561306614562154462171214057159004
161290443862220708341310270604207500352937586907696060053295778993
676737283529788896975737390416300581051474296568095446584870871370
744803629574941622608524867574871007900646598476066221465086785553
748183212367324172473733298481214527618141705964259555975594769606
995921153687692024947208930205658770613534441244809663175348980256
935441467110686655310403885738942699158158756763313819416869526837
528039867984146524504664791372878904688702621258671326016012736382
858092395684523604537881677282127733939035939845511624058786015972
```

```
76248532048166905332479550176615643982969584422851540720035355848 0
17027253600062735079725626303442469530778041400123293088106075610 6
98012462715867510892239757787460800745960633967273353183291189837 4
43157016649303149662672749261751953151603875524586891897488598734 3
62869409520294764043715208622720763505066119447488826420181246030 0
48771669584623398837341306176101609748786682829611382330297295156 3
38924745092073603521322416520944787761417122440642423615464237072 0
87873466242632382915250343181396110116401495042638298129120834888 0
70808981602199112421020838869134413189776691083142281971168431725 6
00291474423358236927082282264412301430763699575773213231494563412 5
77176283917753067282193150231606004724272324959231091658796290040 4
20713684678936787440279951287778600960995535894058529636240692497 7
51058664333171194664767596385525638909430488305635373767165788578 7
23760177719890735662545712570540441956715637897726069879961064788 2
53784093038603628787504421542371009616316624765362854161080890516 8
72406550803190230004017803943396322944785249229900733594571663880 3
66263073321918430478722765563747678291014042289307696589220627143 7
80034915619313894349706642240871938046498744248781757898708967916 5
23433170032356287773769757067581355416925280059666607008154991790 3
36824550365239265036906616000049864305571014890758622337098697718 9
53526042096938054346908609455727182229742010562096312124179984332 5
33500807733350813406486850640985010608920520531299184703356433714
85177203169805163289932325637879370186698277615276970018995170835
74322471135890957365596762968093597079255653225536712805691294393
97101040575857784053021474516957436568252363204276545214234897672 9
97439821411470654136684575184061145938834687277769163383683648282 4
00348196343335023033500463853464675263916228631053087378287659422 8
14968855442640474691968677362309956353461165720854247166821648487 4
15481064636243284935952936714519250567359380304742649123510906071
80320123568411458024951115512621963797012475390562551325560691738 6
56823814350189411424073299343439657569833820464063310376046192195 9
18934490813427463539652391263499003790727701346020616000226695789 1
05496485245899856736687949924275202998210699185867968581654456692 8
48585717285841721092008785623469765937938392570702019926475390892 8
89013733452199564621505753260890438463794706731391495497828832937 7
78002785768002952807663557611142275880570750154843408978672645222 9
78234079171241419381763762928687630808099444654399393603699874824 3
46263605946396855756866103372419331114415320210853696113220147564 4
73069564423746707245367073912183108195116678622040810851794780200 7
35093707763447336033875946175726301526942979316363022552714231483 1
86748170973585136294709286120747883861969348796664758389962477200 1
02995132146027158546024448966421221018009046327711407983308806796 3
72049941292045862743003752046672458094800325052547336016026975212 3
02487081084281712349580018872832330022365973481994663331514084285 4
41305544347567240486202195467460713049718531839405815030153857950 0
90295573440105359256967862742256519024358624320743426402313037473 0
29131934876486802191357504366417890064419628516656568992339692946 5
52066186055658453133668272930811922036790119942198672271664386906 0
09858620346054350841563351387986603381154869032503655430371938047 0
73837702207332307653733437738733560715276748578064457377113890820 9
75252911499301624672047017898602073711678312902000442274565947669 0
20170469690295003649877134480554656968693956682984661715078610055 9
29005426866433432048068638851242206756156956337693873247686925836 5
58407532302089092742646082960497740161465300326429663395381801349 6
22678976170631461332376257105593078414866795790744223535206396888 6
79370284201744269594389931713485918638923481728917183786473363067 3
09614209062405014786722018752755443636168764684772232899551632461 4
83755503626571622969289668994198030721114726060428993967138443607 4
73159161221978627865749760935700062923020625830409564520035082737 4
82265251149987596444806371038070147639889456160614657923505883662 9
```

22055434798009324056508210445366685280414911736346013907686145544
80155856712910544185455287125488166334703235761423542425515894884
41324633669260150500045315473100817938420558068464966794526369229997
14975741473146608603811345877753161450258157246920652604399775653
63812082511501191569960074017833402396153657918703062253462162367
66651356918051661880618520979438193955134451187167983889049167845
44252455209870873270685702170080101269086833598864521918463446453
04984795533084824305893835261447682842565158072835566926439525338
65315826289402790942428164497052495418043444304900683410344967152
56968555072772639637194748920680512967073891592292519062502379239
33092940502201980245187738278575772814117708739847916124235206421
92689555003081834944917337044646648361575408141183983208664587485
58416986689214006572642252426337188153114735046252773147547817241
08005288334574225855353549172048755497616188163967768275344392997
62224895923875632435846796865590670848584721108281114357022078993
64745383757880497290627847082773970129334417597072112812713802623
11013400749806850797799545270762507667687136202492594832463740108
30880461064658290759526028790923829951266096092344302353715533097
36914396289589871128219225883431346460831839618442741205134837459
79874505024032469524561824871158612953691028833226605855065822368
72666068319614334122287656201303248152467799790473637822125551114
92014931591401196241290059499686713528807571016493295039649014910
85368056073153850898889912580862504252144841235906730341468623610
73387577082752124294540240338848195514917862209189881878839238148
55367061471167274498951577560809064620351553918770306381683405886
75323428552514728558274170145104464867893332311953805781313414629
29896526243392499514824883885939472111980735663343246928006228732
62445711657586846266599736519659168630343446300949091923507276471
18171742484423982173937175549813406502766409943123583958714357418
93672324327935366755066379849637299634073709368140965546683527086
03764756326775419963375592405664765668711614822743198383326325079
99118535672351184524691892992412203972384823272636050331049578421
14999953128372903238326732288490455829184537721009500474998089712
67956436356445053958610121375156504797864294429071094145543754528
32423706660073181231282839505010273244221738217965446604881341973
01653092577280862321125944446651279642730606257565730528844326986
62226909325544992601830227042810266650598915702548740923532102487
73278901975056558518899116240618078411740203395814008146097569761
31406844620836895480370505787296596491915878712673844700657237862
48494202314343531372675192639740747077258528701088309769074334278
88295869058655918115979633626610693548636174452795679402535733760
45518320866498033274647381249698668811983441470674402479988027011
09740430718797012314771054914781662185771072599416335419429222617
71835042442352489704886041150421428030699939638048386036751011160
12183303779356412951371357359032584613793810608902455409578186056
98761901539116670544523135972995765831170852667151677163979162871
98001109089714168621595263935204155740002653813981841678589483242
37090451290494216013588643705602726289748339701456592272886853082
35755249846671813652479366044789597299591207177215780018012814618
19702914412088636604956212491431289342585380920741293154904146385
04800627931503880453860587802683184235936766632249900910802226938
67621730187396907775657480732772610513426312309811004657251746938
72558311169212088508246385534970266653789561632256791145495589090
01729153020257937951710936574435154228004022680932157639599067034
19009892672434795461177027032423161547084339407735444982529658978
69548574239932297854670534439274741045413360689618577193799899889
17473479172702345400827273712762177324039521414382698459601365289
81850031003314707562612108088128949193723341963920386230308443592
90039912771657901903805497597050946124271553611645450529965406985
20414465919812538818628797149098535021491358428431183695503364735

```
942005678788489698798507164626052538205482973148451641230102013272
258827202153745791855671116203651774391168979762885321245443987789
052834587442212357927404373251260178150205303999493994666637047731
547186061271709867303604704417012726122314812130719816423875768758
450416219288823886740436119936638865495697945988973505154940532222
421793262650316246130750773660043698872006695545964481196760629082
2504338443298426986372616108849150888008858850949141169082408392033
053290829239865397569612862716048237219071895598482361972871195859
292077492542771629766817566371219781746217647243027820267130982284
872971018571588816786644138260701157713952618255677855861982551624
654657006467994156250961532410705570434880313365416634365004196400
413426506250857219808853972416504639790142021076365831716462144629
0645114590485322504749277513532976871025225580211806484658259376401
982309427038004996609882034348566209119639207813985067290731084507
743202660009747262902537071117179820742714926117922574382686669915
514453952397224763490560267858499847290796572907057114037838312785
431251367137408485149933749565036311848218842992263229983599960566
232790379541347046043723698508474054764962358534455985677299638431
420649231249345584762692020156530890250772469068830541115623193746
818278520853515723639746791023602107590770299190318893254378628612
583729052339407024366960420879008767860420568989010568979360058702
497050682960777944294910461413551696902039024082711710071624866823
751543151013170541349119841434083476187763613791379361552772295744
192870430399604226886189978746023107721971708785112127944130612305
725385797176689860715816809874258542596991164933922340331060480458
807373393395480939291841132775613550168620737377642724727458998241
126007533615453614460514058389426073333756522445006677236071285663
622380824877199732491605446311323302532388134278671307298854954061
446161337593077796083397652173707242714182944872306415415134195710
830818303831887091855551524146739270092361954659303458731061732460
307724538278650105122543453299505937099453029219091804604626838884
599286882096816507895254546495071997866337256090768055235109540046
825056085094884698670513390425639678465338824259111117061677713833
689668754144984288023585427339131668554352942518548857268933870723
56596372046638858659932640159222946148417148410967972075288306285
420513711378456036277834753357916796173004294303538308645019822087
7062642225807935392633461244326924074350914140330433856564405098872
872128728840309702599936374017664035264423452439271348652492711982
649189302975979097532936656828439355820700648606620972597792666950
323765332411630480073784111114445245533901571269578051720292878370
235717719679646508004500413278237001591925010746999201533266504069
243164941939898282676795970634714300442230960898263261217334018327
800463378158511995655858876442818435043595108438852468531378181110
818048695197227403830245108012185034279999377407772728086572184142
278729677153133074123905724264064216136322816205038219385394336378
760478745807621422258481233466645263535297389698827945984355531433
993402452549557519474808215831708176625278483380406973882363075864
28961453030401552885401618691144154765871289300055800230413095620
654176803253332472934591700065330045566846969098246525977593850182
048531232336071546436409299103654411738056373092637256602077111961
068021382196728162018089116629941766843279647423121196326442252201
791544776814976460357306094700519572096095869811980124580981771718
115953171404840553935925105372573522948062426846997075888696209665
993695697735918100860645199351369601232438864742416680025976775194
408868413679939573832762965866230339672402485409925738978071657214
655468963163448929719459094100476799982923077302999582326845566827
616990418886424194766643720535579953465695607711410727928550104418
627748877915863267681887808751461838893498587937187515028676675012
848942318794103002257756603086355878333422494483219722570020138080
467540721641741916973805578987554542260095498691076751097565231958
```

452854951254533137971593423921816280739283239872688106910836116058
842379403055431581584742466304382776568069535427387097440123727633
999758565702619368019531336558244664140725012912612482469515935717
631802950906978737290105822514728518274887205199823488320746676010
649849880364994797377095273780291924841902589398596010430179098419
851164850270930706516283878624829378214117581291605783709045964502
797630680395586515812153563772951192766869600978967791217302753950
882540331838444590741843758720139771113896501952440783977805029915
048941286862487937264686565773216393693006380294282892791514568 9795
181529915471057800878374052249626074763227854451963839354066281409
576288838020924418536377382784395693086931493772915482639028747087
305699723288475731558572237570719170945642038157756693053544391287
508221993801052450298591753099892615459425179728586054882918019560
495768942684066173161861550458718958005987236080235623552133520343
060068903488915901526697628154577619366190896785860659384245436513
360202688638916276860644549394073401541133396245820782406408581884
432025781410335894730992666728241928328884242536445035709726894492
275040450941726887089918272350431273052078827623241187844724118260
133323228842122940596229736218745347670341001706560066604378875917
822186915357104648961582258044940105430900887200460418458805115992
891073498947036209771520832027700262818048051741841999061387810488
064887243085369212744373380224206588985065698067945563925673262244
111952407301843655821818625264658065663796238009033845412060 05800
139980863235752544119337804003992674563942102720869588152306 68802
426742881329508203222477172177852759826928706943861220740191 54786
524302959810384121154900489550598297485974912475762194090426439903
276402359160883608471520451100851787603946664290788305914424753818
777713661123702281519501953580904434779038141147327614840976041785
219250262410891807947133118746986676917520732020538852609029752781
348624778708553444865640346729022944447281667465187631339880914247
152634248414000048800494391180111299501563921689139067594717839529
236420862361780645908590954446600938027145943381704069740444245283
864801286362919365281022243150447590046820881776551711124298582210
731326101965301770787289169093195233892714761304523529362806904961
165926199346868180552342463318799829172783394715995351269521792238
078695303648940589031314955863335969904115006609494983102159681 8366
796391067807362611087452485454783460977012347325181427363661049862
908692456197634548939088003314769468211542703382827390495078948787
476737625534583104754559635158510292926741634570498702234666915253
043275373769653321807220932512722519888000043177182396219937253 7655
503399257620634632723726133908265912686833824345289852251537724198
880378677531710597796810665726588078987415741841471227953876706055
801177584618417206930140398633850515787401807926847455466429012290
440739217896555301559495018629521691048435972273704347121367877180
907337403039206770893530512040325752700045989612027406592149482049
677516908304086070787905529945169844243619792287691751593450114793
249928839097313839638288026301792157953420820108819910522320509603
833354505401415427637138838522610114572626700438679066343473600840
063381097704451442409938296140926597837271683182950256099202357290
010903319161204801495845179886247904676072893939404828710482066580
716937979235668525922786226051103221685915246578973360063733518277
368664505973907804777257300065998434152433204548516284377775402028
457023239646857918798376338486654615576095669248484021440541627330
981234175507435304434416429979345287534393620549380140290723502458
297701136265235833891006616689984600960712639035893529272001582721
574312065191597977374814665314015852052708864012461226224625152705
075252092226421971538803796570394869762481092586596239653673730287
862994838428990476727971877433911336834191671177567689627602607164
843911607739090261778882303794959907088106476276753539759225252547
836475967085792134991697459918210000607877474249369769839296615803 45

```
110583155127074087021453313619281954261003354047948206175846149144
903123437005426137511247843785784656078405455351992118581112555428
149552355941601404745366596869911078431620395331286053098306503674
850737600011073486217027069695024761709297607613648476984066685133
413704571022710774482739565272643596179170684103429856495885089893
411687256384043525423915716086632482593374670883675076507020873414
549944539909332127130089064201493609522129993285559239704106725418
502623683532314592732150208408334593571010145301308910008818947148
081292693719998159843188184680376283315576442716655172157190799 53
468178920745427552014965012139208889515829833451578094052507629669
897294456789910698272634805436531001225005240124461415755130292741
378627966346247046720054505089322435061296323030622051847523885698
470892744953223596150329209185266065451422514892298143381882742108
722165726695675538240306467582640444707618139759581612486663834 3303
573945773390934021097162892151464887823407472956291190731051118315
761535218564780036671749188561626932385625313887962238348886364 9910
996541877382398407803132404563460628731584952108853135076222875028
381678761318066803706815790233366791073776254145311595711277560695
192094329438264596608738690884732919831248301460011795531820254845
968638790622921244201880229117836672746072288730734454277811557131
388038718879991319179572304119050383094159772043614231955761263291
008956911487970665368233806282982128185598738916288354573000906246
332683080151465298906235658627228036203672667963278392938873197115 3838
937769064548603057847264218028492769503493509026684879771153201402
076348413817343233959783375248424674365421477367813565495902226568
005501856161853555367569483449051372522760941101351732950743410745
969395079988300080374352894981127609704243056855608210784852092804
291459491718819060062047266076340025729180541286165037101023494179
521707807582435121154437713297130153879228377783519620170212378193
718652208956160647346992797830836330618207123068927659891064542060
433132712788549873875291180243977945042320080778541505642932480485
509670110527383250128411097708755930136705900739271613797972949317
549228986254066407436600166230704610322302353691760892423722438401
623019736959313458986351903667490008280548074778534018429717308732
376019514312262483936030442197351244364097053667346466566611713702
674268317227728642452715904723587393178765697322305301709335718289
620161342908492566152180670434650304367530821891272344745739944376
937669272200868839076560932705061600975053808776045674346952108078
016226164554061140614965530584371079589321112947501097888109505182
017859332996354888136787173090214403979206021744187284467359238345
263313111390667956747312017695279049305984195992099117796525505105
119490278487890856381225527489966749563111199762690727424847817308
805228918285982347681021989033891759867693003145374608089828128 53
264102786855371524165674861076773783110347138623313486747333124552
565121364660265711983764617830087665576137546153143762438616931098
642768250970619571377743454903718795882065191945146128438727872904
788596040899890656935987486008835923496291822286584276212376289467
183907799935901564986920437561050411806444293307846180659302341499
099425526544801738433187758186969174369082400477384237652031190677
061204138065435002199155284622326374696813486427452533005015175444
182047470187959854825330263285524874201699778396850912386170113840
901863179144774059397035947256273457800169123941250452545220231811
183233484528106029927622886582632592508463096363771277121844360840
664934732856774593608146428979960808059944526672896734392694118 8425
912133674832735151428279873343374199956613588582544717987349462371
459703232560132926587159730253787951814512003419232079209935776701
648177616619431583730418911306886266008495622396283036401185920803
901295230200507955037736782814661374675999725438743390348141291 4330
487583182795307068510822955334688560315196487141563366205180730273
682021382587399173926891097257191723882612850608090617511104582 8934
```

```
39307290210369575263881467370996817441707665102359787004492183496 9
39444876992047588288864213955403021704441172365940707473468814605 7
49213515720219267915958354363734178101100898702763120580774175538 7
51459723264467354850209080123204328152383298265382542596944501725 7
20513422432369199940860105557415566621602399498546318737571877676 0
55896736122361108155161111958972696364757692306063283451337937823 9
69055645935769488392260124695034814919799624056980632405264212249 9
76183468646128945191771128887785974249891045162233963893438931010 5
33709915313517788364000755880806726249980578207352430014936054109 6
97740432278612114038811497968714761853432773753197555084090953754 2
35285619828009451673997821023647303462360122611500154138519351814 8
57128551018095682204435853551232569614286821626632797780839634074 2
04643784290930204468101707031460386023416555458470694606978221288 0
21686750211178893113285021445635304166037858378029067877681628635 1
90063025161961437559456963700520073405910634391039581504887808067 3
78136329433777158085209096632408107768078767723532859978380188929
11561602833350391681760071420277909306400810373076087278788703296 0
97327369920168936474840642258486197096992516903755386034456104558 6
31060662895724710423505700531917870684758851263557871215997380273 1
56105105710252548260800034669455727563388024588264407391111454018 2
44522745689616708884483572553985957579077090692374677843692455388 8
30051399891262245643334840572814743160204175956205975090606411585 9
78647522016204337188040953290786539890339000944792980896254261021 5
82711339394693682727639468829511068896830431053609599851460071100 3
96992511017273115121138327015959309255539113969262266848896845860 7
71838672246128064298465994991319922777903007805579568052558313188 0
83379436664895619384282173671578365127840311650885609063842243427
38573408319605610030395643497231477066603283762579088708668005628 7
85899869485318990890643905579060272924428741662208788111343241175 0
77576986236986948094382804735383674785760725547177608189303122804 5
63701884569488330162256909945771909810723213716223576771960954900 0
76033794093176474534175457747686675795443942806667715912116605998 2
79498653616783594827889707733498382234203167269800110605866068215 0
95618819072095103889156170303449627805116304332447526710485228470 6
93331703999948414736861267707122821347121952965354790554020811334 4
84676929107849095889950471444986644932444369178416544553139075931
37000634600065111694993553942027315218487180961736235428201888124 87
71303818057869959246043661541021913612792316615725222613150323184 5
29320326520299858890400513420311556049242512121964924779972831885 3
25328652762053100859519409616715329707656271682966480653692505530 4
30957626849624204213690666207286883659934597954677668212362769757 2
22982688582516839973357531876654479725387016063355144337296484439 8
73330202814536980556183579911843544523920601782951841864157046725 3
31980125608154488418938140292916587865349913701064050796443484201 3
37189910346682868567363941458991367486119818835534766101230169116 2
13819598262596882620892242357146208898393752841481281303014046288 8
46412352635710442628926489586838587983588926740022469545753312549 2
36397457577976950159261239095999177840084910145512102724997521968 7
54486267946219158217614879009704681936092260710316189692502047889 8
39501552321953661776824960654484246121318443288950992629256923106 0
17198163025178668464039374393086140092215558270901825358644238187 0
22652970541477446394780947647053796676262484635988031101925847233 9
54536703057179594253024607154166328949523992364440829108536832264 3
09217180922711073306155169406869535169260286429841112147467580169 8
69920638513860206023514466101955063946708309209555326653794263459
25524850549828489377377534463184183692823362979014937153703605837 9
15441240228758175515602340308207490662260210265322601991708403357 5
64397322162129062298748613474308819785829194679489864536026455176 9
25157684844804995586865257578097835270321180459811953300213486364 4
76163231541947983377802326227887116623269835650811840640180633661 1
```

```
8055049009559225083541921724644146775236750637155328940537149781 26
1834603234381943438857798770605378474121477072833264234862223288 17
2717264797503212504182758853139541965836000518492224367534045445 16
5784966206537538184768159918067434650179160058409078997743311260 83
3492791455264043149855751136637204313267929220013397086230273017 07
8861742590671321205806873362592099126805348556671270147232586179 67
0417923056818119402948015297490403242486118747069672686020716626 92
7527562434155531479235718423267476645788543469782588267265802005 97
0821251182452112214249169755094098825464928153227826898221841889 30
4828486698906722041212254579873483504674123403719581402574335274 79
9582450761577855334791133160079655910878923922832100001314583960 03
5749311312101565369511531882483186445117233965302116174601907786 83
1088708013571847368867067708574776272694598342719244173584579274 61
9034695371278891924429081513658537142811730159437859568736726537 74
3335958702699922062849699813460184395463552993513272028599202358 64
2989657702630111705691506177731878180988280137972240385087894963 64
5219372235279477349651304771171380844357710827015483286911519326 60
2664418889292332141273606183521626256125191646926085523195416152 230
0720705533735441347903770229247733045869734419858320275560938982 78
0552072580467316890666481510171699150423776860750390454077607664 83
8451603348801604668317879582051097995608729551612865987321522681 25
3819657542541279326195958780148812918069080874886255982356059494 09
8469166868981108108444119843718408700001812469809200471177439274 40
5616033843950881835959658740170664126207167672087583026182348638 63
0426942352342702351040424501423041251403976501049982865782620980 88
6773267229071465209265661690219167765722842460304584842767021991 64
8442558078606043496365651061611950698850209470723969107010677258 57
9939671898255838158118692119936182456565667586871062247543369363 77
2966359854595050685197293418309486766550081228338945788419597844 45
6593209705024335836921264422160235373089586093076548571070992300 09
3431204662884913174143731628604568295289788829856331667873814442 46
9186244785510524497953557195650462046443534713501998363167173395 86
5951171061997575566708654652420434338645361698913819599905050796 135
0465570207276130944294186667230482838811731292603831121674378557 87
7045643257442888492108052524228161603304083418266278130747106059 95
1857579582046551922591962124897614716209286677043061789080892959 49
5920337197542975624178546401274979775018171802360434920825658855 76
8227245300450919409726812594249613437989962607518841597872747354 75
2462924318167740214692611665730548964936296028422783056237021315 90
9217316688556629983810200365462550731376754430002014691681551421 49
4158400057968997724027986881115100032716436042047723049275997419 36
4961436809257846758882440919135533254867186607342593522828834841 3
8605493673665077719397225322944238718288176240607148274299861585 10
8493520134477242916220777236638203794131977226353053358923911629 998
0385419071245897442964382849427767372377697463304374597462430638 08
6823995850937717038152546405631866825859625212077851210409766920 2
1840748832966543991295978307560741723127764813594235336387307387 75
0750468404267726122029285571570125545565039082701858571184717093 23
0355974794186562617978799161985358000964707178449532957550680442 50
2983540503455852914406414661253839469941510749529421477565607764 90
9394028170426194118620017638317371067855421709658533827108816661 28
0182422561379902840262042169941652463355746284266252203710090131 8
1937732993017108792563767336552996737338395872823770310299415435 34
9347852026194004232830188168907693976661613514161955821268731557 08
2985898118326571222998927035631811314690622159529041642513044295 69
6680567934356267962227387000650579451675169826154719763414... 
```

```
8969921730970789567736974639904515130020299283621115416870506282 75
3454027161185906662633228447614780591794210264317526185721084702 27
7014000828184755011475795668789267411090782755751526047478564726 26
0082135508810103714902614973170473765707628188925237368519420403 63
6496306930546523550062054395816432236612490592948946314061455545 22
4704729975764688257159390988601575979240373668794730530219206765 81
9311089961971064051960551893186554029194045862995577829091453119 82
7481689651501182053076927704010345432290970970087414368590489575 81
2075640394128218160371266278888243693317732335971730726622724837 4
1358098799910215553373274037047531489686865400430322432557316380 28
0018237108755609037533617138220877700580585463213772964762793303 28
2849264504903062801498276927325948286453030855105736642134962086 25
5095474523648617229099931314449570685236410739978504675342384678 01
0644003366388293021334039019095929525099211953987540668615659994 10
4631996428331123244650807024212049479572121422570677799403794790 19
8168765591965628550311613626877188910357583392402523435045720916 69
0343368301404917466707757145741811533781016322883168657674634884 40
7537273340144903508729797607268833064939369109413949972505146633 25
1181649274327447889302298416996262692565190271209360895219839141 48
1638024519355628081066460994270889519376871593057413078577550803 48
9226539821555821614097192184890697934995250296167343012959433848 83
3833644469099046860450015548163457983556995725828319396095469285 89
9479965104095093803985317817876028599009950270258583229089570804 07
1566945460184937455434363606288361773168500775443055432918503787 83
7516173062090644361202519446026384608243065395678407389634375666 3
9916455814556307803183574769741140177286499269195392344829586395 51
3795036813410254178176352300269610550779108802339545287089812373 28
7068564093189764435127498650858118692307593158503308469048622117 91
8726590936148709125638484225341087726962519470551275789850452947 52
1554310251432376317451579093805157019125473588003346894737454491 8
1840285340137217800776862599908737905214570441338021976874426577 64
6223240237766723769858750598482209286693375405204323440592606162 58
1380973535290278726166664258630022136420875386458215326769283991 85
7220083329283637511623003412955688714339241142305525825649671606 36
1322517316340725983875499923972772596447226573595916633399407008 29
4884332378858010136369717134369772809340109689965811688500404153 46
3914791107313266588424163553984821646331597075530948124676429882 69
4989702195015249196673127092161403572958431471452741650004636316 59
8551161587104467184047068918090049670237614457087269928100869952 65
2789487069896390909696572225824520233811514116412863946949573807 63
5512714556293752383493811472575272323739677495420160640148239388 83
3345733666043688334869502615607861140991211117132749365990691774 66
3968459879433762487879212090461214241124058467240091092937225020 76
1556576354448499027811925181307395329719679982472278408645963126 7
5532462425645771645577934909507159446637688202373573694027410193 81
2410925631747019252881457996303460376685426419506076682836990036 94
8141355848686704179649151327264540140614173817619517118525494797 81
0022012817437835814788641734422792986081658268633143146878604244 09
9166876817349795457764436126018148913458601777126906894745899405 50
5826423655664580356458267557029358126957113541451492795555504233 45
4079418738364232638871896749775445136550660124780250620002694686 51
0360390779520068350932494526456587562688452994686184581527597541 74
0643196391046656052105972011862680009441318327639609622566095909 84
6090639835126327776890824633306455495962231768687117033866188947 70
1455358405792499041802118109768138853311438425212508002352886256 49
0280429752734767619267982765862269649547745161226214815379656416 48
1026625502074822348461255542684029580969701146037661246138329284 58
3717266469005891886448447614816925448792643909552683627265467106 47
5253023660787526927603897043873412270262998756728044929562066152 09
8959622958945634453124545509876916108002875315882587972297985583 455
```

```
86328376179214878506257519191825967943368893028790993005718175509
64983509456707117964499397525033658484901552758759334433624864238 6
52556138772249341698437208274446968948286855820108033641744202747 9
17306820647124104366924186774168321979474515565819004368754585831 6
84924381015342480565654677531515222155033992668526381221091335912 6
91238862225565929828911888955916187919618485104988141472548993642
92153901525766709447697958315103165772334669190323332613067351097 0
67934858602395975150744975686284550048060377305000677496127700157 2
47517430752509922304889676074999632513391957629461922663706201643 45
24932758077196788858055277297081246480989917418839567274546699826 8
86832538047620451363033587538487172991219805335311950950536210322 3
17381055652010563612603903635445311557034336952905146796094082957 9
42746454094934521813661409051553360937460713166248035717987396599 7
80220856561819031486662848777356561762775228702465881246888885335 7
27394710937919543048334541382094847506309114434621154993436592221 8
19792201445617382989140654932725429675671598155301231270399460867 4
66314064012870519594668875471549914024718642854989782266446479580 0
28909716127892669585127564830857057557659700585990424511919264132 7
07799082682361426915678851265138081282550748340067669408240728554
50340274395365168273390756162908543530899526613008019218824750782 9
90330017978932026704826231148871444223776888274166915606628928424 3
35778662438620341119156390939255033954165153520949807540331381949 4
84286978636923827806380225856880677653521006227993488474003596493 2
69110412673284595003376024881220770227728840864964294366876610986 6
04062204097940678653282917265142641371669882379663966695924931526 0
82540806953065700542367309837458936000561072697616785656297270276 9
22295744763065716574577927530705323528423476470952151653008282984 4
84135784680899066115865137306473884601556383759570566328515775503 7
07843599097485929098133245001440805452203495308654171308772598807 3
46897368806541663001313719949171179085426925030177343250336720736 0
51471096152591338365680003886718032726632890566253712586928342620 2
32752295509815729695477057977671873915595912817445298601142957394 5
79708000884835734758895750926281745155133378292938414513708009134 4
51098431489867098000217605311238603678474559629979534942428552364 7
65398260839737254170168651173108360137006902405071993978851575876 3
00836045692772094411910064461751595506043985073331329623637979885 1
69008092493682383371665407475996374386845037838040802175647675638 1
31350300297856951441654949554027200006708078907815035951104857979 3
13052645732688814561463754191582266031901178665904622157183601933 1
43967006527851287649339208983084174085846444343778149447465262523 8
01597709418582255418934484548034961172983290183791987199746878281
68640066663215675169523717326766281391721359304759685782286908689 7
93790382867130400418203992536740648323991924814845710564340688454 8
35723947653741570853554271095701471010348208494650719329949331186 5
82776419818238253420120714530744384690733818593585245693254699099
20971513747487462996072651227211416500145011619522438255216581880 8
70271680838639965245029915316234570303088226804832482059442854071 2
84772141616627322064250490231893068841793102770590141036901043144 8
08160855148091238714859653038799880971304592515987273892692940757 0
71091516366684667968750232086405411651262060019217918489415269405 7
40466829482517131327903562618942083224491026369643111069012057929 0
30096531288605050100132228393891492853008063650084903487523216826 6
20061973987793316709676509110321427790530725941023718848328477393 7
97396758401740678430017244966978328228243353488454477785314651450 6
77629355387437008431710809899953927369572969733764383588187573111 1
11232253728049514339405260783243649208323274442958700407229473628 0
37489298249360660979667026665812003023688257317488447915631042781 3
42989240818935759876805753481393498758424356703403464540159892259 1
26603888021591129857789085259404658880971848921445600847511141893 3
32996828254457390248524301492500550739290836747083390004306426340 3
```

683193834849588606483214369360524285729114372573707686210130101518
781077267922775805179191196411274845017676867112631901492060135914
089329830294152212033875823134338476536464301742631745434711690321
500754723281121425982146596732444242757154296043663736685916052925
087583853667843849363352242792198567915805739894644982476651824764
953668690524179074343006166223201356131047440390133106755988806298
818361294975322444251389340785470268022577045299524241123728508050
025904509460067689134481310662793079124923341763911644738442693010
503475791805009267511444016611451415952498286537442712132581123377
994862975271469879858071619946900299095406517010163146359955927101
901452536929254266571889042229418712795479845849153856437837156288
092930332828168377195543829275410670399589836820925541301937017 30
684202292551977368284176405053617957577382711816758447793638770 12
443086994630144410400233603193262562168738442956304478155058678790
818272531598171598095101747861444812196875156883236213400627461126
924019521172893488327052956476416466563180011832327466601038020583
363209864491703426097626189962362806368928420387259971156912394347
944973764572973611396172132908433478958082485672479643864751481153
945399493898684723190778222980305982851051399539884737152197820044
622108583309540520602502052287984635446920563850818226306194 82393
878749500756013320880678309430323336481293836761896063684971246687
630132720612109976515416803117269417225539177194229306077457056791
698002659513872576519834467504931479514651508868860973516281381097
473718870204287182061837349324833866578518934329639600642337056824
274174009139381246095981484506665996289436854363858635882200684815
408975870381916546477690783442584376542509032221427367181825316482
966057195816588913392546026614204955761840655945094268630387 46259
378026076704429856167186059411922449546916415509728602885972551288
433076862556403883195694263722403041822713957474113416684641757463
790860062649491986768755071215473640580685871848035943004995660420
870606744897612387489247156909748109822837746238382275716976054278
383475461354665576721654929291877287766079192308481873009550591352
654666299844363710373401893567110644680361069796969786714950707044
159076019741235192552927935848660485608957486149428352453372450289
310428110007039863256713318526780261896056678259862766309083101050
715124962068330814463707231297692692588354514503500793233828118543
693737262281563111189970061746998803327352855979378087951258012777
861007248481272360573196267001910997134100943409214076603220381132
478041326089480194403488588440836270098490573644404997223208165352
664823324385231620849021259773776794691374674516705642940967609377
060633450897977010132545433604350550349338531212518189402981735056
518484910818120527345029594047822237474726976564140154076634054968
154758665217346135457573640733199169429161225891516213594596192562
184133604180354261498822630165813139476895597205657665419097275402
112945519603334757279134575343162516742434704156801384818082126198
581652673163161471813236289239297247412389896503794178684460220938
074430000462800189971980813018873793798117108767099648633955512033
075413065681713884379213907304100766601517212797185568540196728263
358971057221443492666815193756465759789093946789468503521617703245
031959643690512223166649841665928528563876110223901698542164128067
447584162597794417835824260624174123316840449958883685107797691583
464654616157109589388383093291584667742548212911515977749904126062
509237845529623931764774796232550215776076622715957466273668209335
197949842328125063462580472359195516802795130432967353267575462469
348505037354365942229006331260631393393514699119380167391232114865
451256716711848905591194491877968797930237610501773537891797539705
493871533787655957821653599867308341605208269342580982571664278139
211928187714335986973706905479761227668076809108681968150079588389
267473329486629285953859195987204169354075048542707451968430862684
639020274081947153873776236873209940397337911665123213445942542325

```
4639539093946301756337000727312412341977952378119648331503469285 78
1237548328588952230089523785769916926969395702002460071030846840 54
0786523268102071833211435842102674952179730152448119699808375116 35
1351948483804752682291380520265600039157808440902029571404934495 88
3664884219203748980865649954206108232916947411287904038235427177 26
9882330694258169931560889365290035470057064339139315905695775578 71
4766426296552628439276692014160409763796683016799224192656271028 59
8149102369528404375225278138413334073241471616040660672940744159 86
8793564088933339044248710553645460393377957064934633595778669004 29
4053154750285782577029166652205543962387036502887084883222429584 20
6312497323448043686446980140572131210124723401985197242784817097 71
4459701276587444318537212412250840909913393374495535852528164719 79
0651593717762931696671156804001930290232135909749123553934560253 83
1906442829396270496491347748402685071922638057209132622135482780 11
9793760697264716244038587585234890728751804382018817508249147732 95
1753898983007781892892674352757489761616604903747873546297143128 70
1014399228706501440759609699032917243929728812079883874222681457 93
7228049446491021891612951272544735983305525810195358422076217966 52
0414181843681660244518333691283635946728081825950600516603868033 62
2724325159621215343796124671569565640498354301193198629002403458 41
8518384392257989368724652764091014903116231646696751942148025211 05
5937498347200694623127384343472253860713828671424422400225658461 802
2177261305484493396004047783655005640181708755577841359278201812 22
0670003806609103567360408174861976084928301194933631941637421817 88
1291547538583769275717764118364725934309026307546518085182409059 85
4047454822911987922572128140457249050285691948246665097790415996 69
8847137143012716673179206568517920655557364212320422303478492915 16
9599438105973430714834282803792157175500670264318443470036421375 148
7898334446379393152028715680014892615756665973505879766596700874 85
4251679582632582439346609872036300476421700287013810785438852199 17
8394632058822155374425325487374708922809957410724239634080154889 03
1049056721367774502280951489139599825985287649226708817481259044 33
8044507406387874472403020122942509279917410996269640655656359424 25
8284614530604025515819062079326015106431748242224084806939203115 49
1947634845816903302351956458041549948641226303409353407160171506 33
7260555340895162428103808994690695961849309852995012274374496423 27
7511873856293199024145843353413687136106150330339850350392793287 35
1181769811481587804941137085036811430443693739906138514748789753 36
5585610490822998256893107042692380423543650430711459654287702603 15
0597871471429171692029445298526968880920525788165257234072179663 12
4925710554714486720848068563545697175965982030649380338897576464 22
8628626308741661929787557396848123901985644052610113566040901781 24
2667244191977131532922463612316602940317844056018301633149363554 42
4322772477365268836965937555036939278042638456089255665510134512 205
5234908550755240664357568751738490029860121079635993116845769809 75
3710792978794616248609816314544001899106095272580122491594408608 60
6493791813224709859392681429915479375906513799731260879632927822 8
4643858268298295439457797286787475858157269556756077944971005840 83
1467222306457327526027537049501267510052345286259392701977265865 26
6749991501014685484642777289003938267809757930552072483313983541 00
3674408357410128586451056457206319336287393190944582495915552931 74
2257528868489767397513904914424402462034091981169197287463231502 37
1649972309572347477623259117006809870592852023626524502322800478 07
1051581325560016512400430510910973828343321680305011966298090039 65
8459142238226029330337358986612596019301135522692401881331653228 60
1477338286194890020879585874775155335862501757614016705128166568 18
1224266561198111246745985292595711065120649272560074825833802411 22
0362285873580281797990730786186766807892811186190399023620895605 27
9335269809344790294139689213578169304485523704318302862322634598 19
2539770909442778299402809767173807896282160369311903844231127873 74
```

636490254240793531425322196346989523433928816631555474988634454837
719837559447869559560177187759090880055632436906694909022682939284
206088286093586000318603298124612429675358684277662118468996409076
188893912917771313887715308976278334355639359387511016041134409872
290132611441729497789796896581550626048823161772404776075518114324
659458423284348580976006917620684031831789646307767314470892908467
746240942896933833733902210720589246882672017576083234168320781862
846702896712189362835027743808999410780344011360160391581386512205
597877189496037481567291636728983511512123465872415845916263840092
992449819325370795776625380560419342599713903739258326852104897992
475575247703508430304793415195775325429215222723723267982944143850
611283446649243075848602978087846522872801219129348938537722791596
855795400293956935230592632115923438041769286999425024783310627075
945198121135449558985963165647278224344946677621826946548938975012
741969253230045551560649276120366986707177629842090599382428025133
126254779716710063875263158946176577256310292902669275806472942551
272749265766471925483149153207170006791937941676951649438908464717
333890116637155412050416084545719264834888785608424839951665773227
479213592119700228890220076597682361569263570876072021417580059188
884215782680657684970604918616792587029128476751340406954049741303
316204269388651668130715030775577729508591472795888568594989083999
900520311999134159990760516727963047443265457583837995379855375771
933019893206974497329829383755521123577466635879163895125261700365
502448697369211897300469755261538836263173810798557601549271345511
727897482104364192060738576493230342916200071414616181694735799988
844165033012110264556322266376882127386253526824194646484846625219
522831997925746250740978213637929340316845148076176400123184301969
763775705824370924660987627833473794630225923628403481924095614901
081154162857467481678746514239563905290987577784186555810936744848
469016630387888282877799974790103254083279869099703832460475922754 4
096854395691547526343778662202222960886147062778284691159579061245
809116745389441690367953338702506471319273878485328388360052218418
124926645520989426138052865218992736528468986639542300831713970507
715736869140762806671489728578879549962312693976155347040912604292
861151974047272139628256851422009044901093029409167190893259060206
159503165894334095601280507642964093666679764580230396640452753600
529403620806446984985462208792634729248089518759906947398789032992
771047038272471648517242246008872549644875295977652586751914045817
921127544905032180453213190790793431039101582804910890838649859682
233706260752979353548978663455446141513267832044984890265778346060
797061684264136041971495676729394796439173144293052133525577990645
534740470289841278194537280749135704415976266167417220044002136777
443438113641223274705885200554518711226439337940461647941229677113
932146858716337791916514420410253686795770824423160040719192442266
488351036073466738577257958580065996016761135204698014401408608689
075823941768090997382493530466935134831609197507646682455288652916
824293870672174892053663116026136976038943608853622553228124054261
653915005728772743220414397818929828282500970719770811175846446331
328979836679381244790001510510210855347659897657604589124106687155
708958832057272464398353187148960289781432655889675095040792134211
370832670920660115452027946973384889807590927952957600580947761681
301420595582744529629859794505639770523254065818897674999391452106
779958911813888363559938954077405328013102866254584210666098540487
384106644710072993893586221463465983827616252428042397701184191937
795156822845677229688487624124200600782503370054861728295206372105
898418479020925510172026999042434993532106535481678523078250104721
312064892237608900787780826035459105716407343784620558314322475419
106854756315537143763533200802176952266829035855200124218342853257
365220527577320598331882029848111058786330810133992773701107080706
670191055576090303127885905593197519473628255860957781347923550 7887

```
1639974468080770299026784482814344302687300422796566791707575692837
9562077547081379111211761904730234613470848929090479611885562506955
7726206079721730412678537405889874606542323317270342310405066109305
5132694547146576137195755023028395874692998743213832526555796732035
5174100943713423499519842020802932688119206390178751063691003437725
3792372996108649565368536555365489723890973076431863072453316727835
1731091919821709574372265073448414275719286183942140005081616012635
2544774033539050485893193597429844835331706443624214576849624941315
4367233535163191656643946699252873856881464075314678813603712452355
0880317968864901283131987210011622587637631233579238229155964028395
2155202029187347224958276126067468736911456337325118412620584344965
5986524391331292965545933454281985011727138857702095798721309509315
4362018271291555368352757717182614263707156010449245866129322322595
9209821397632679114559866231493546667485716233103996675307167268025
2739685162962913012499873712391398748403950122467672396616753850945
6035834022693516227297818106764867921016675659355473506578333896015
4225006145469807330934781420457579589580694718428738256078552921005
5711349757079430114496392000355649415656609404536039645766292513145
0051064488223604318486022437096279600494457323823510233259025109985
4684353993638965520426609765253522315635674766395362187001680168905
4056437966476545569679920242518794369389892701300460631182095884485
6205671093399633169226221986551857822311869440892694736082403473225
1416381856907024559601206560850221988199005343100487112335762680445
9382466961272794503562065123366849876204875717857992690868430425045
1487475836417956140494650321595638391507902895519530844573855892055
8379625287213159095277595935317017394922512528413072894575610341135
8085172692731241373398534165858192470013937501629150382651874598715
9689996286828401870010832536612273928252915096259840851559674646515
5997048175588486570647470786052034391627741648793856751738964855781
8220442849377180061760850525978991277053995374077642403033729121575
1826163024932942446984349490402776858690700687257758284678920957605
6721657773361810697065074379320911993402225841330303436759936085992
7553004968012886120263010641843616009634331173054228879959331830495
8425030846516465599792836864656919047898896765298219320486754468295
6671615962016874375798527152125818518461211285779921681533609978055
8207535545070588119274518661796543886481649224779636286856480524245
2204320582626352891747471300908249482062636746988241137260270085845
1177212709041152082600959411994278496556521344561840334342480735985
2876585864481390549551195453265599740508695987379528548631275773945
4924443551839491399936136929692685286851508384040638327193999178395
4718448734007646742873798393172181652717653223134210384201928499015
3991280476518999395764021776703534923445232321594435600142024262235
2789426071440786386482306734659404603312783005133941686969862642425
5524597350573065997701997412208844195682445311227413572049446915215
7615318819431071558232072367646577185758508769052309124484510538545
3438295619838111978169129220185889730749068009255056073726120434745
7683943460910923024183866168056091277201691696709557582736268001575
0553795921164355578269426259032244378372466219863980958563982600975
5517176176420400949720960180972509084786051998936292984814831058675
7308552926267850508231848190272093838450192426524599149403991030295
4927792032662507230855679590881651956458685417503041341115641322975
7752479635866869028660786131164406492660612090490501285185082753585
6842528335000915873632653389302327303671667672247637762271847566905
5503289163020791937882717759396899590479136918898647873899535893955
3746220840529407068591654765536490500622119314232743868990565906885
9568690467635218935985446295147402198372727820145871520102282891155
7602521782146376406150038237093229829927812940632916968279735414845
0263313429860504674540511358843741236253181141725183822774455085265
3119460834725932090109198479396916980203821568702691630255266438385
2793131812211333525622015850775646110455655773591932092130688245215
```

```
7956289162140638756061120548514223576598522118008769883911641510705
5675877718751924040650063964904059124290005531036820751862716389795
5568102582760948588267811126251479230337206335225454020479927295671
6372408943654180644776327434507744998683087458404106226567321038454
7841079939613953708559476790310882625624450280137134466092190161299
2283239207713344765206010522379971985269567548378171166426242475231
1951115860998579420309359563250659668613461920669593594262447535045
0690912041844563503992709605946875681767305832080781454726009756772
7779783215511714666297656140999049898303237179632599200624896552109
6958931408080823924569325726261445588846927112774717791254271736037
9318853374126322754806321897423944190197557505154422203435253800129
7506708223024216933888802341938889395554966949854144521172363606308
7494851348193592883389384584295557706143200293963986536262754821660
9254373860682736525587313438821539414686681774145670557377465178894
8158211891701941818561136346201003867176864969521481862539060400863
2187299090476056628269058837938015807515044038790073167208039963216
2751496654610401479200404047842003823597616353086379301876663675985
7320551484634747560708998811491036875154989768158441583172696596302
9404718801881704654064265482321187809896005193222096473050684739682
7237354098881921551086274432645727780550506420551638730342951805759
2047000561847222383143640686957384953974853219254831992952029774908
2442249062223670878715692653364290848548569342680596880708628329083
2280506171208084751226320524615007286828428994914745287754857087760
0313754983504633711763019958072434252994761738607467282020171043017
6816168177372816515429848884606499627138785677833098737055775869202
2974105225152421779253656718598865843234817698036561676306671823196
2383309380240278325412829845666756449514537551016721882426235745279
0036170580518456412510160829077956514146511697515124634395196002740
4862411117982371021458663854697900349075896937534225132401283955539
1063929504719894727519438315849950088706540481262337569936340713833
6295967537653483396498694569801737770331943373122844891997038739504
7304013444018247496875358506006613692120003627940195347192264238868
1726858910203942896800160627386338123439711141445823448187999084210
7502119197308367829468051535418819592165939869069117823225080147427
6095536259766577896342047881385716559012960961505867146723406707910
2749273998349937385527568800176807783497956012837011685663259208382
8322340583050893809726318813770394246750491517036946407383029504800
4605759269764597933077607654906475533167842812250481142507183080907
3650890209599672671155793624461303190544168800838030471813216949550
3818907240800630837919127195643660068666531424126730274461009333108
7013620316400362741507995361124424691294928647973016962472817999973
4009063705515967356891982501461129854220945851285119928986000359474
0089861221081913103246427485245554627991162780732020756483613831711
1972563815791182912310552051064206750020531839362464898034158647950
2773489226633806016698330896479809601026201620998967031723002554306
6608807012254738214669097515780232076731900428693325419894271172142
9011111626611291884971034677961335815648327578995212490734802434064
9941942371924440806583103340508013480110830365447897057521666846404
6254441182280671538876757155351181409919859581040311176613827124710
0585036383588195977633044922443965910173762215578069635028170482226
8233708117863351877440747722878780286865977312149907502758718219550
0604991982207180664100334894326390827201729518715975176809735076965
1194023573614883257870519635450053110247355044079177844062869077131
9877812103124773047989321360482281096806060717411817829957875892717
8464857448602318846322504854312368108691405698687835650507870408067
7235960795745788317524045599780469077625565490544289733559947638590
6596231179064574304373902753790091537853737317022784056908378812085
3097383387550562757509204831819587693391144019832382940903821297269
4458679650258556746723378629762386879838475443376765516110597507238
```

```
524827249024444177496566299041395735936435853754980455100335274067
918350569481039133701521874574260684150713230009520034850226384511
212113586952408556022391549536351162230085789053624569004207722178
675422827364447220601405528866588456295760334401038155998612472 81
961550022981647600334970167657611682539508835525792634203486127111
838929071566190396466848729075564120816018804904699126720058994193
809756522965110536605746550458397815442490898657700719760977915878
096158789978518522069645550325669985256415467455502148315939826176
094860589481173272993740044523564213354459834368877182228361042584
876798582230727546612830208670924237158053067859684003420589967719
154011621749872264042056460920937575762807237463604733761333217490
036589392770705970992359892353743745011569717461188627375450048007
409264087561618295958078037681369139655426528358825271450046538221
767040904799329046924979455000281846898354737008921999341789695665
225003336997499799362347727099349209913657823237112733070037805695
937564446310515632769571972613273245240454878364608072611070809691
494846136820290509537355088987554857580167684522922403951505676179
600447100300534362937124945852438385187268656514648239149028844458
860045354407746807053948031533236021283192314786427432029702227529
042550106159462106457129277558010987495349313176263790124594804396
734215504890117059626333648098719377448519937687965322968959692049
501277273359484611456291429753378114870694459312987793913092 67594
912252521801219478534608709736248365468959473693055913371487 69573
726198883497876894168191384023429811250608071105175424537103965868
246693265769006694722400601172748654381709441015355969792805702440
504972188408721057148912549308735882875283637579474091379196712170
607584463939714397321097436190855714832015224021306315322819502364
143285081526152582710575402428573431822356733456084486686436024027
993415898817855020873443920540744181280291090027241266731022443313
296443043123222475812010656733008041840441289073773042251373554035
154182360471315013311491924710538351488409576591144392451053639245
210253377557147807031153225325624825434082559714221130451039230969
224403100195348358544175711788253848947655583562585363524855437406
845632986116536796131227409522533109852409354003190408956754480 43
744573898073259535606572066722460973500318193539950473247791192938
769577029441504526152886279198000323061528665670695758626118605 7860
896026400889698180342162471292649026863721394945203879363810029042
810361064359919391708460729529566213819356252032866769191526983091
878993302223312271794620784899173866186935651807973322800328606338
425114281968700865088906775099371484474000921968041826808 4685072596
782966672751421067698288804765330317300128181628826183692601427685
093980586495717025886788815330503825236459493300275439184972859731
781638517718170572825431143240271899193257786139370534811616206789
137825070794809483398213041271055914948123145545252986172966 44020
136189306858815592257386028996258992389426938151139142409180887263
863159804700599316984026198939726396768977903659678545879527471414
741639699849664078970187176400123058245903792557037510931706689544
139424351148768360271414100384060967555869073307507955793521000610
947782587714935671505407105742874676006380126048503184897388789 1490
624908823709452673763255459610572034474274380429064983662508463340
998214282806941873747951485960412777899542263150646150487865780669
866106836717500589053105633506282001822650959689756791041983763313
886792700619598199369144131344515012779471675462955700895487826993
281392381991613268742035718110496692036767460189469647284832840667
316026822089787053744425979457933754997792942618633673792154247202
989348617079020180981729907994537957091752142188705423671303291723
708509485099720365941366131238577742117494590410137513317404875420
416331425295559583132184930050009538455494883737767372829474328 8527
799573611042688267489044855350588839288925405539964683116530093108
867340706146748770013430453873334602883354562849087935447904592502
```

```
92350441443184966215506944169585452922579197350829929484516 8292890
84763315168321680217448155980598508703168562542702360960673 9890610
31128765587876072307302304551969684322272291066239507631971 1841392
49636514682168770956827714891908853808021446386970665401439 7606462
65802038590954770266728524852757684408325682068135075623259 1144731
53560824711351680133724697231545022026441983339002051945014 8565325
66281129736284784405066758370635840084984973541388267022343 6745525
19539218672635813986695872119928419929924453414212058452605 6148482
37613776911861023358548247458818642556969133231514988084827 4678246
86614220317762799067243579604183429721703229060556865710600 503745
57523769914206633024657903936236108241478026185225380090121 7785237
95947516472272450235174753962744207321830752163305622319719 2378732
45427816155919002729390873630722049287588479788620490291014 2651011
60736170401428884889865377458708520257336726960271972072973 0135602
21525694432986407589069506738526128410835565355211061135187 4464711
16442688874939164902071754206483341319235278977550993017340 0078132
01386545142193253860176014955445077709926202030635823086469 7832553
72052055953140889389132336510246091438298895512497819087296 8189432
36983821812571928175741746792609799253524614591635895885705 6487727
24995324490559343621256545550778924781024428672251719882440 6708079
84412790543092995992672706317492739884128211342899139986635 98199
58850320026418302057645697253815189062927743425846752444092 60536560
58148153163326486961319423182457209195399566449546064727493 3501838
29018799508589873528799808279019925777181098105681395172507 8226312
79608000233326582464058782130279697801806296640451892253407 3068412
81591612619280447215170885475957574452923365065002436822131 6389765
37580619836960517138509741859771620941368340606737672097469 9035163
58992628803937160269903489033363743026884082259017842819889 8995944
28211725561411449342478984953961614434241255979668659946547 8613233
97035341976738545601825868981949833135526786671663527662546 4693496
93652583114685653928524569301320405161192303534081859121834 0680689
55807593875960187740760947331728763612538310381030371331420 9549548
82252351223344909863450745192728521368884960890245051192909 4868293
34099435461995685228567366405329598487227931477378389031832 3068322
04132559987251670750825060596194265702912134340059974661757 0068820
38366777540329399256012591121062433271549749493924244563620 2722829
03776880578003777563854945593133050458194166420726013705977 8167835
69841628022434125148970576999091957703732775255869412002356 6227903
65464586759413550205597183015669372152728827187900509139503 1029656
48591156673629029786617140880568319148320307154492974391671 7769954
46227081902705979608958588596918570958752893783251606936794 8336534
62856293716685091253193807161341685276756116434973645582994 8714433
53747297901705938354381115318027096418280143892240542568373 2526360
11966445773876893724735543198318726321733073004663573183393 4425355
57126235226185591488445443555477808716464512464306348353232 5166922
52794749335139143713921965301053207524844079339749865262164 4098097
08063149875822755489892584894704780717618876527987180079478 5291198
77192695294800208807593803772893184767069329916385630676209 4708309
50294299084458480565750892724325555835685170520887941496095 1356858
27589560155975325669015685783588909679388087010500031244600 3484985
43927412465506813643666246254877720450103305817109966187457 5993543
95295148823011790263268240569574978327192851386394282145397 5404456
00912356706495796210333076767626796391290210624628474902007 3685044
08074543532955415175647344314652350072050118119043475143232 4642605
11134537995142748708378200075527311419986078212731377766124 5654137
53106692278103244422602176696969831605235318604838679193922 0147759
80994690326788202532215624808357499375925942957501307175805 7139849
56643639109228457133277304672755758133757203835525725738224 2558537
48873807202929394806221839975114442473349493161699753673088 05311648
07005391710368464446901561600737402738728073725675072196082 49534823
```

```
41104904776391547513607991270718212037207265755354668956105297 9435
93621079118854586609600247000622126570102831909608912419323273 6493
51426378009720161602081001953785557879196726684278299048014787 5469
44978906137985849794742461211694573465137548006094664594907207 3727
08243008135841427442593031461698684270807558134402091005797075 3010
96061242196522140632657091408112100130422743002994668073578370 2921
52741849760478246635434148040705257923374683237833132375877531 9479
80225050093320042293823338541130098028522480352144883537152617 3384
79535074758574780024908950852729610429401714735489799440908313 1252
46540341814154269261704148864460025629189149230302758233733011 9128
45250251931449341919459289324738262127231061363344305690801340 8732
53919390015414395893716123348057924680493701516410698629728989 9069
05576365714809388348195406372743815341893744550806773920978237 2489
19573966920689567805558897400640729592176891131771171066073167 4700
60157044955215169626816801322559052391293983515051874988166037 8239
50919051235944956724953388642175018715295466424711689197263688 8136
94889502075576967680295767724567661146986350651442778398003626 9867
72901570178121874385265666187111014027055067251317640091284629 6071
74987887152721963111192310127260690680662756747962960312318496 6105
14946767424900005336833767889936069198884417580767597470796238 0805
46859024237442397743343104387224947802355040596527020192744709 187
83958384407317094497098178603092587752503271421967439670247138 3862
90866314093957720698242515866821475720972747756830284016999988 0730
17384289335648682700453419302825446974361159520771607309767711 3778
33175980180441830980401430929807840736341949271852936827372273 0492
93741718520072380661239481856769083060531177145777124000893213 8021
06086202776646692690098279399503011084406460670194038048422611 5618
75516394456476573128925714048200080403229112737914106419930079 1092
32240154552044271221678696969115844732106631848257221146592276 2874
76260885743305025280778565136843207522836322450368225536361512 1225
34145515332128026505047965649146246484947149055581933610557445 7995
17946496381953959523337687685360520252083825984739166085045426 2951
93719477741755670697688080176317601455996760432222620102449420 505
71872767223623814202538500831544111065833378574552341515475420 734
59738583019347859930067886346639546127333562970219795825804914 0592
40536951939727057771584705357989791548026031935651596026291556 8029
32175118371645643868703689721138660464980961301456374395044905 6238
22452080861784757624067162979890833429667274999369707511306709 3948
32772432685125889819605655129511087111609151363023640731493480 7847
62035431007310154118074322446930161388688705371133639757171749 3472
91056275163615396573976830983712837629222995167981304364181978 9279
29916063533459093988024959362219510287808391971650999756852014 5088
14961892834505995094434784132037970270190728956768983003016909 3677
30502359833001514122827640253168488984144786761798171890381598 573
24900407724669527630485512622319483513691610060135964021014119 9430
52565201286424465092216457510585059381549789445572811545344831 4415
00975238617175049471883546650863851132879315708910585136444299 0928
88345331244767342000354338862994470392870924112218359901625634 1144
01120545391409455346681400250417328716309262333455754884714698 0356
45514648184799429412127407996604132134429022831360033831375377 3133
15073203725348629442047316070569792956506417887488816887652863 5903
59738755761807281323933768636172665552888195099735937546106220 0539
15174238477644995733569583293192040018209378493489979432060241 2786
83850224449464388430486380734525280192733377453778776497322023 5413
24801683536914512198596072507080071239843507001736805065860465 9413
03630285256736203214099962800613738815756117216087453139304637 8420
59629689295666355346630647539662207912700767475806001946850639 8068
31093980579508984085097903885381106063191983113223376927653471 2690
75617024455469866887782531812160609672964046503826797585484049 2587
60384182127324745063064184775679168390008354726516037403370439 58052
```

4205913348891278506004412932227491926750364708353965297789649131O4
20551769884095513308150814022155153197348165169897045424847485 6897
22107544214946313206643252006450442889879599042150922461601 2331466
515012114122658867055315951717690244293188455848083881388424494879
578680423642585404996461880137484531025851986160021242097 97885361
906554552722096788230686238999043429961320010628776791449116218789
178246358789652700595822052521174956710285514161907887684587236173
723171827374881510450984705627060378001635944216961342586051631092
251470705662826726083619540347938270056450599432347589076494573220
685166615914054230037384189744365534389495186971631617180720454167
543581001145943175342568965194986643064562712541228387496723469335
631947053928991902863585881563630498798538184076601591427827693193
350283561828113273132450479855177688872797241480555565729058049182
931300884884259650950370298096746176549289900417197673032818192366
526415861705880834308839322490069861703073663516387365192726085086
365488868811102173368430952799182487498854900132149612639139960907
977894856718192669885866972240522896782490012484874455386598823302
170800488282953466152947209947485466728497638434486607498963490574
435597795850451725116160518098296628910445054770059345936307042680
645749301267987649000752537726185963635650733813081500126949597537
505768889482605603560954395176860973667526590156805920648893206423
903865126040789351871590998379442031217110850981630909953700475828
924801280772495220897003728737103269509188386542935134418931908773
088599672724154248903501408310875221946290133034700435110341867312
185247325024035133254799613920299700551645554704919663310314633309
077239079544820542875692858157305886906391267273132368472101389255
324729489494083607075238799487824458112943791504658220234 7592585003
287512641819102463426391594679443876282831654861021409282951819176
116643297562878066065328322593490695129304473449686308413376193828
386995778259653261774138122889841655646886819922734680490659179871
879805515678826371348673236917947421803493924274544866313312238124
226715520629406899535052983115187026489617361924961400414668757542
944950331436137481334040148741154333627418973398741813714409137211
585849223846927197344197959536793811539123469864160781236272602 3201
729613583026193491421686797935944168077158706030290478871078872914
294132279486298994659581453491689878123922299667022487917269912286
301595676045115120172595172559899679595600129402209262099289712958
463802022888202509337717997762294109060220070102320848298730213 86521
527011720160824690916856123594617819445088921931873050067400253729
995257944568195895295675641477185341841852922043516007161877079729
590637529172380604467884803793254909327870452563058774792757039504
447714012779413673341819141015326413829761208845707501535158533936
575247485099968730477240769292926933048587607318135538858498624647
248456494806976848850676038470087132623608902040703468665113277291
917279230671782770898716162401884517364688072733784845847898865390
556090819981007319832558843262262256638118858218043481022258732038
130502263769398723898792939123426646221226036495163713544114802170
540499207368195416487614791021630590648494340869865061518818405403
236758128350794337757664358638567718335131925387159887412650949256
263019435863357744992244197270976996652743089688075924141086244909
722864926744903457747810777969445280066670516332484638232444803596
189596012311729480057445480399217265828383405718264464692254797595
772591020794605966202879464826557009605684764176976767875984424 11
191975016072076434924739911894782179423470803883227447578850484598
414746767798676618485667156065093841283671038354253409830407068871
970901518917882423690454156317651177452261265661496964612615888444
604583655822775782040435267028612665728391197370137339677392862444
031140941533813001465210583118473961741521015782469076176671743268
856015606498523540732168120062497410969258560623076665414064738136
843612867089535778576422090474079469160634371036831276652306212850

```
9299661532565110197091585335288759740969447477698227438817703024031
2966173346775881029818431293947238900152905277290948913290288138182
5214399492803672067154452056485233421988755757820243069199369369117
3611868084423782175198341984391720774708074310301349550381693042654
2943156976283335218166281430943334537934147928443145326284868261748
4085917725672208767916291536238482329518581890823601993315149917231
7948655076811759379578689968071126411369581941829159528262963489134
2856404145287318218459447364814547646842988205686850540005519402445
1480269063153619618393121039222641217411100029230511042789600511659
9491126897824671423912872037923393123541338807708201488290605987704
0826330916729625243216680486948147095944185795661575901580087408668
3202145801267978135685953551844655682865201490648306234662417226699
8536224828391246692318277684431057763006613122617966799989884305306
1759227383310948981563294798873411438175311616117217345391534822351
8780073202728808525082158476969253625787121350813618768178474854282
9720346934847138545696074272343355977297976552639616220607763105370
5327229840605109727360336781952566358892629092755402084366052544857
1697937494506037830734892560249678496193903548360950519679362388238
4421860980286201940652116724191148973626415637666893788643405108094
8321428519900911014751679582972686456444721269574377694427768152467
6544415250661799245808703334311660376472908801941457538524604110213
4409438956560942156767481186076874186163910939286148043159595881722
2654363731967835501289791420526179758453404243452970475520850193306
8516720459305713853673080796023618743322033535399895171487903153889
9726532034756943882068685954038927296863186078115770889740758956298
8064729614755698717189506628787548351776300054717220525555884229739
4159153018904518428071500944556633667635016415056500419868780225482
7447937264999600528728908078508708851603290611826720736052105957644
6959689294172268986661830779235266411091684425713260145583462243720
6855005256141499859384847685097967796943659998339395588586887137854
9234436557009147689295026742744840342716856609714629703037320982047
2572097767411253024053812261249201375573871306716083858077693870193
5503294108963124263774750046437600581723195945054018512965163430134
8555303891406902722141406416288021866616135805457997220824039533830
1255451556074542823664148870650020695161445528903878752682074520478
9503435309411339518841692055989143122224033200366136250873814650441
2073261875895889696598400699224325114561747760081565783618440764755
3319107268324242438683827410325376443643832028168906058224299762342
6368146012958702749782699439598497202285787146550756939811578939030
6108030822074109782592686664365355571284085386035082248311977007390
7621059544941404658936293358450305968964200105817413450459192586921
3265255069274497217387563430353142259456885827635021677639599028968
0913281213648138000328724611452087524689696478388720914843707165509
6383244154870612565484980392236722138121170888034169871327767353164
8349810870216453651321074889784325130769654478412471905639320191095
2390678209595629318237601212753917340035721301969604154583755393316
5466988742044017436720824736459429000794317842940128734020262567174
9922360694899975907481914589521326528543495060837553239283315873683
9804859913625715943459760624708636924937058166353060311552567345953
9610819579768494907751850489800237451655232467906681829032829045994
2287223626272734843190814264734336799156867466283765350885964835343
9556058172878532711502591850325115270185922481845169931668240248137
1993276822585683911550318192670926164438422355955989506955921473054
5806565278080964684969252714443578597248428903484097405420149036998
7144866699664888979651062031296623097275669461209275558053048245440
1042033485197111313066489418291606366319850509053710749218583651148
1642492586513928206850819961948623945134155675631221025181199535428
5494412707521699030987550046244698249546142788402725045249657986993
9778677789086063365503484652899486514823226270664312084607191854217
96384904
```

```
728220018931882424001380496459535464769879061388460769928166173489
660569794285218507209597265714905109745220831654816977895354644722
356814482942209342161589077356226401332499873619920583307976429791
413247840293351512033091266725632218998637352336375438531808758196
445089950410197026979524809161318299354925990789762272840383336047
599615043864476807691293664997471113963052085438677080203272450621
789906352489155999395987445647218261390612148917319641856157409102
602486940621838160018814813289500648833189330226410654455489160645
562766911573591987099132250350833606342350139311924882489839051742
284554535348600897736837977242247860250377444100187617248714675030
942433518065343819819200703921949071160273190653232945568673962960
290799032948040647012810773338196135641624116183665740332063968818
689377096529421334521713728094180803240437416712735909256927449833
150707003509300889589424048396872898580140960145145468933764739586
937645736009207282222997520512227276122617412100382867314641696497
289892035337627004837498006481604788133967963810120882501451174372
126425831690684034270998089046152418834947390285322986033308765946
084516740727875114782835734380237932592385704332986336603161331540
604459152215131859619734046010662912135834364122745530323188486648
789102009603952541695567741800011446196697052372799068338666490399
962805109899043382570559927890299613515195051369301133279868054148
610498178785421136189492218605707992983384372251363130505502529497
206164803406163409383081702964367498313713318859773071660043949424
073354723274292640421218943074245105937561647923959680851594834713
287807330630214763969161160115214972092942936078237724628364892341
045427275597743045552196835491985831493325340985071218015444419106
493277294438033495192106188105735717183542203254571718334763090399
517780287578193124003816919401588513506349226209264681700986862258
369514693682594864746399989784438313893496827396330389299877526393
112178321346752047754055118851268322035082011353992750871842028629
054006130876336745740191057855825780311222782899184072375162441769
636787312830933073906029193207920098956871023123455389071816901726
376867870129996667677436861874700292517170158815385898377994817186
391517663336920902932228856077456052883005255252156982486466547128
417470429022638281813331528553436534228049728263850524445934300535
433089197030063568772961280441733617924517745920484139266985133190
691667754934022133717510186537951103220424162498891901925298965435
850612540406780591996564661501017241097274853237434749235159008432
792365774057631072565896365660916605557817572070460316804486945521
595591830421793926856549251651773902368071095218175387705062927654
577472715151184652204152770201233150571364838906041606331494578773
975739543862929310371041014184418959679703235210896087286743793030
718493858334858337822861743615668847152890918039448057647136579817
243729024889356387180171286938516880527310987616933056964979885948
523881263663522809627625011982289654666122345102174031388492859508
223594500382661718023879607936566348424731472229431518823165786479
936023003131385961207082569497375182226182494710026090752974069430
748769902100405378227731687136403358022926751371622821717645589288
807186075313360871120149871781882567495443875206945830494493605928
625487223669518194364634253423732960246168147070893220203520195386
651844543809052579953659814828225285502848624730798919419958088128
730936904863145482613535838731757753258148575025471886346098640593
643589571784278531532722040410652642484943659567547603149176990750
574182746569489037345386658696152026841293046659704233530737390138
156747682661307073129200501974028964914447682792766481831725886620
006128683350249379057540099565266081305377598023516250829161467 0145
395939353328063420484779281191905492681876519905511263150112974 72
182048752290188891480034450868433471865912512202191729830662425786
475411735426298126753770144308964122083930453725165023096746473393
722998623711951780437679172113989146822380794881699913261137519 01
```

```
35670006802240974489602865082318586906144937479796866717957765 6100
11192302698135427149084816692978956222351410069408071383113831 0245
11116405480078909084253490837870047684591609824304810335204754 5161
81639597772862841830006792001750693438515274656179739429323131 0957
68949505841221760486555052819972517356654387984657111355031465 8481
43367985646620190496771330565327509868706559776026760974108666 0042
53918500298581415414832724182296235217059564836084602854745511 2638
81535758031199718274137037858038895760593217216469132625101266 8842
60875013453223893440239377567221171108251732262030224457836728 7870
98480785706283470184226674208041649447312431168329085602371605 1444
65313915557981089671317776682111256717928869939264407380105329 0378
99565377648417735349465980532110854401788498211447960605854472 9000
13444808199376113094518952967637470219843983013031155093364902 6924
67223038622403823099085356220778607955200027596640052677273385 2524
73359539925797855312139903725010698032564185421440961077231249 3687
31658036406146393896731090480127092467047692746686084798873849 131
57055831963019811378448093431912829383825537207399607804932399 7072
90118243764401781434750916435434869927892834205097586188622320 8724
89831954521822047008065618872723883830813791825144538131145485 0348
04252499457420554010989401080144963771016892311115516607472092 2974
39514202939259877068322397416230021526300174196417408708203854 9914
40146632566061968008343479271012089489613895699277878271100922 7898
93612427503490099533115443473905370753730763556734025596895414 2915
60409259676179778137535070070543388300455546729425674392560971 4870
71465191073431255395551150438851133127520679490064776664197455 5906
23427525680852629918853333663037382337869769502248672019985194 733
22315556483364958621045042441999008589365222378320011955726813 8518
76775923957969466135228726451958892941951558913460409917288406 8795
88412800718210507565162440217783027555195368063047240481175543 2724
14175783015478057228533790674789952819388397231784097198534570 856
81718101164597333087505092903073533050122442143444616158099546 3065
56686768897642925094710619042325618273626779966033493863686088 7159
55181278597946437198825399255682037538988182278610976810540969 9829
23875462207787790881701811828842623532886139542239439507923954 3355
48503636348818300723749632124931549200662453446793317264389906 3667
06579259581492651822620049989171841812327496537365772961891355 9567
19953044230779289586227326242554865958169256071156914701457316 1348
43919864879680956320333745748201822761954041331941418187009644 98325
10672261450349794502688989880780740709166650524582656525289181 30596
70131719453749985709891871380535354403087543744666142156326419 8757
03800024054536409248562604100846733014157387967904188345164896 9638
61309473269354705424334626298396978556656439331992910820266356 7320
80240055668536948943765566879505398835800533899472695602249916 0719
92380371169429400106723284476266290299496788002865303975658660 2677
48487249508634218252184028968803846932166491561628046518309490 8073
24445552511271436014599882575974489989859985626878619506549168 2462
88524115566074343962911855049845098596718429102846408243805363 2879
29077009953893610892077562374673817559926233815598036778537132 0760
53382599365094122584016590435917445672910174933318229389524851 2253
99811602906172407269815570016362927697520013038106075754383806 3760
48487277215728301788675483503363676006144618076915192883405321 3016
36052922007283099433187131670385128444423146041000812782609241 2456
19957045281236426042198428534791020944661606836946162250346353 4932
56117879455710350982314533322771704099071576158354439600515854 7561
68987480059384383495476458948911741000861227735454968659890454 3903
08498037357774349087365533542336027353690146871592464999796686 1343
98489225884986384784122873716469608250376014536803842340543709 1959
22711186683449364425224496019603552380126959344801645519462630 9320
50100959082203774682951083592051149125699793024201285883860293 6694
39642608580178827705104580208966228681248693423321614713222992 5870
```

584931077744941898114234275475990438218463088731156977041314925189
610756665967997555036730710012290063888032440871546746968118231821
356014623089237967514944689268234257391101004504820965193374642197
410580575677213790709700318223830201499776113873358237603559137136
631600180039270986225718050590752999102440274481182049321303625212
791465802024571522101998135612852921241345423176007302669204874140
847876434862792396889144928751603494609640396263822353589295831453
484740642542158011167639493856343054142643318947755097427608728927
813759992116021491346641955428348033088105726531141987378625628351
973124471032083564356106529736710642890821811468084753367845658460
237964029540856926310948205860669975983252522484674088789242409045
273703590758481240828924334394755371556634692297326465274789169792
674927125930878450542356300298772054994648857563082418482173382810
161869848421670658140607263170575634932157535728693137523201842943
653429566328366896522414619681626670353447475345238514564251435532
062350217785060612491811915818621214213599671265149387449813487622
559695185191720034126150204676187641039359288669954512765066501965
680195988026602338117800967296819798072273522478036370023172358442
450604716015143993211607957491943529081315773564771593151768934933
444382388462495376353514650959332284372249042234185071259237725796
634235945622059172033204945485977572408338159716007309690411046453
219334666606334666958078940148813563568283844497023492072829372198
077416803870776220475145531699018263217926759901959410773090425131
492648632094118379192160518451512504377028957397008273136418628316
334649961693337344868558205759248821722851785109991023312716215169
89431231126899791988469541514221601165745397910414890189063653559
306367357791891209535380467387629740448612138802932000744361644850
463815878408029048173872891330641075354088267585598866058753119035
3100703113657468331661824425660884505556596931891473473987417185210
713455192681277248574583651734103924896814123764644576527552297878
965696502977445419439790207328745602240410028255729130427473527801
485982599143181842170048306700383660407883106446599623123729808163
777980055899774016470265242847151445919515325037368661587706367090
48468280950659560970842125312262979679267660679684370762560788440
61830769061015293056003152324509144141017268969408653874945871448
385774556835483357849457196874270613191496708146984821953797094571
959563633479327395941622531202734072946031718486302109978380889844
13656604920313231925361825118222038579705374640223930360961888401
658749431616702919831178288344495607425740521491777349464859564639
132693239110004882805966166739177971700949025951229233165029343168
319809936386333053603112064265515940214489937849196647390367308703
10898747693925568362859777450140537529567717241775748087037409804
398687087202621468694326971383921728911594230798126859538137574300
992815443649244790111754684276878254397110348578455859670907539167
765812480908777079507473339315509428364719869240231124686904137124
000042056618855017682231764698943613600877179762077197732179959907
887186524353282882420304024286026790096476851135749719444686806954
886069962276588511084681605062027666597805766196644893777299680718
247662469192195624147286152083965338672316628767377633211076341737
939028313221055093965850614239860644241880925118049201060701057761
793425371946888090491446505504462670471156078492337087375671581335
961747578483245010398140188768140356943183596240183916596322612756
763528132921591038732531688565752695110318435361190878419652411565
7878890122199101006102483942290361679741331292527273388969357952645
474856477063880988275658996525803639527927525237286546062229466816
794289868477415288251134137404854728271513645045965246606762717648
750172309457768375524614588421846900620211495205785187326599572885
487398725889555715978911415779085605982909421416219923915397826418
638281338138185156286760244470046738975433831578812175609925857533
5519746476091598361353290827037085939572182976167992225132816652426

```
855252345061975709128952088522229403836314932585753988934165954603
409213085655489072501652477692294806262425783895158199535952528448
235683504127368461172474210856666882371408010856693763931453990104
582896168147222812965287792427817118075868604318895374161969097747
579419209377022167237840802785002208962684009258765688468051496381
250824911112933364583615534796714385563649390284953528915053831259
010495166405752864600634643421734980924994781933004047776150215638
232444277470663222110931423547763561073334560235450464178998761556
261124548511753468259269863136145186338456787206716698010518235 76
895602367801228437438569318700895347858855528700826212047757175436
758493596075050893792083619892593266225235340779353705541580828688
334104719373576884355438869549219853810679889359101023821962191001
616482365272254794657551734616740291274387546109581917375385052727
219971224929586979235472777506357496207396126256718671142375627311
209906081744520531414122107058379853385955336969645325359981131703
820288455339660319639617745732940069215612414777258684325192550462
908455880926100671260119167137716407537818787829553781702621684416
100585925149456184441693275050451329738408833972558425344453349023
836047403432126921395371046170482130722741568760353437428795469337
886958730705832601061041838637792778967245070089174210773082194506
861080734917346493064530748844350532178764549579220401873871011704
384072875881032706620746648915347772150327535224478236925787441488
462207470795216859122498005198640826514425297187291991143863313072
896454481889398218055602822982507133113733247497117482807948017558
865145538937389472156840212471611528839050229523960995886060843303
836087250633526274140623333828170431299321840838981415107985423569 0
556578275090589708025600786452317285488166908550680102966309022655
796817214097968317822385419547385585544338110306822730726143162723
651731991991986564432994134401191014577039936578774453221976441048
513392218603221598931974955151150987189166365846484978642862211323
601671054678703100229367562320244202401102006220349041470701224825
782056934090492708190922167101685903308624854525030754270269391478
282097201223453118103939531129367544517424931724231918016667960604
445072641754488483807854470133441246035256990804018001834439084383
734703514005145611448279980997774473226966096751169706945169249642
292955529660851124794790728535773709141746622249830142978657243320
734264606440818702786962924592832461418452964872085175740913923203
160303023273571462363083318199798454553472661869096349740384201676
010562598713319027919411086321301800427069918803550647121149593864
984187596530227030772746775484078754396946386933272563109020950973
452471149562186028598340350346431548913739484811520534991828787065
108073074253727125042364779473389851664615335648951881899876830330
295348785467737663964033983964646919682275980778601133321295837533
732055599892381670555315514243456307479647201615043248731043211692
974603261342984427435399182461135219554505653089984153928630062817
133784795160707037136126716797611343862173111291664796615805165088
611567141553513528574636302617022570136320541011079790604013683006
797939426615758113683557243339524940212239783711547863132387490747
663727654410558448594644499695220200517662968167443668786168935898
669285447747878847235360102421902805945608724299996904378466122 8755
302004040069609232385507416448904489603464740925329187693939496401
833112402728773205128061663445384217380492976098003528340125328728
150628961950682751762636954104778593008596987345635548773436005714
836362187508303417209379308357651120580390208086035859211532897889
694072657480267343074461108041798235435690769206096304626085362087
755944413269035487769548753782090101890152167425321503042062448624
502661188961893125128939543843794468581916368306897684969604404314
405371865867894399038938495601834658219976893311998555251460203028
148034158075721623328712860009309778665169456612237312449010683 12795
204423199993829314084816098145561472128310285168751180757702738091
```

516248277292608356983831226254998599461658016754468103763300697710
445998985915613467890792604125288237714200585524307022670366379298
354810927436272703415234154695864530450158248490558891960015740807
662851776564551263065476383651778755590558760877025381712502706532
130164430055388916696394586940787159590249783843542522448778482812
131987003879186819509148655488882615328436257829006763956093362205
762146635207596470420472220839216912509786862781654607416241281054
775556170433273494636592991208511473295185277168233138727986353990
730438558720938065227923716016178157338683710420709214863192801845
105609409105458701847908309479878835264607516552790464070236562826
668464934960991104403701315909809749091239539697545205715195308386
998270722345208715272666243193619585538106628882805195095618350229
659878306488785725961500139877689157066437370295558139300832297260
796561980549744003832386454626431673189326950131846319782591714166
092381197307357345878538284121760390449252145713652289332985882328
423561672957602045152330973251968722053080814362593689627759903982
894672868410968840012379452007298048106326850738134758174190777859
228302744924515717584677602208580053438966117670344296373501625409
266866374307285731001428325909247034880993875943275397997349299482
028064073984763007361915961634608165801860599708405826896779495533
393681869978992983599598638694708705568416285899666598840585 6373
449842896575132610328733729407782050188658934457533981666553 86384
558623802876230445038702972286547636716291714349349737825829985902
903890891498612947465532434160939605396900954644477618399136534938
190184664992655090077322151361475898489150857157052693025252 10935
015699280893659764240639412765997103541866428234384728797125754955
566109731069855401681241622338036817834786229004456053204896013889
907292835049388834855478392237106592799975392101480224408367899126
068769876960509116227784227831573093401197504312558795772000604448
758269145498186189261252927349901134971332976673666247113527144040
394514387909228811489406710288509679280342383394140480197753950573
714257849237419912933606789611843170844650639986642092895352973578
144640657328383666840064588684621715774190431804174359388423143689
640386059675245077098509548291580674688854722056204233101219021150
413296847584188136537557700680202077182955419874823302223557992664
939668897606905387743540693366185412871603235844095693558094652460
059896614266459412819186951402362448005360747917739144749871260860
915992197594561976149885384268116495834603210065845734647743297931
871748914844499695994177380347566068281106151848379211963897500103
061910903485662241778666590756958354701484072669812111440431756498
943311118274034898762837254574991213174257618942998040651215259517
225186749918055083858411342633465174628967589815544398154843336624
252253037162926932085051005929710475691384754860851118034887786981
617653346944983548639701462055949768406388238124034991437090054916
583961285780544696698567544736863905686263121366071349457910370051
535176428057260666341707335368138341683582006409175507347135610335
423096634789541009765749684651853309502613946285309617784629509120
266382203953926536106186462154062373530264369575125106451093890379
713922526036293634572906714848186685614524908674265217052977470841
658464055338879222060706891321832543549917505741912760524353388044
925491764660769179296719042395525800219264173767487003365512714 45
190574165566340712906124284971908928369919756497228595329755034971
401454444939711338565583248522467278382004480269868145186209807096
002835864440298465946566816773882537934606871774630792087101256272
349480366172046376744444753422984503064957166804070789957231850787
982044268293870687748651401843201622457134414231895351158879036412
424458855097616948252657322471422554303024057005394593491852283493
878456434717279024623886908605214295865543661675758103887853802886
459683306186010659866898494860779400331438024440900397990860880147
732052483306358954070538146956348786130029687797259228091191573099

390556598903779654437472436293427871391354278861697623989202061285
446070760009641787575310011107270430278202361522255317905146401703 9
586876251369357694854979938469505032126357486607045502852528421557
136887899857362288334148070957420243876260389214693021340812106536
812387298313680401483817466007376348419036862638466654080888038702
632301373199591209343058335692623855515812726768942745269621135636
275514545655935894710497317222497059448404832714509192071196151292
177268582046129728425853433433282519059722800545166752750001579141
164357323408656892956796620524765844913682386003098753319622899773
913155190122684302023302556141912248105326480320733874262889733285
836076924893007006189191935593100102229936913546825819489708957331
029912788568982274098448652133894138213822978811981777646579131139
608372143086899630656494685853631665084896588339172659474699591702
895037196005377226238179212457166260867507230587005137390860080670
014194003874192675540688741090822950462408888075029338579838870095
340597147974259688676852069537964789211358437916187441369753360251
134759300518000033501495424718126620547475798874034427225938459662
564178635011366900101526091066079150211454847187249462310809205994
854784781848900411720691575797717664639803260016321383633386655978
324895380416720031309475575855702854655098792151688846982074 3374
726604623928611113551495961664178392210713501426064670358316761526
059217218325897833880214074701388213101935894717563465029438480312
693454264838649848930228043768410054171712698701961799867703749295
533217615575243932822688644682638868553574552712868143092902855903
450075478266419305939575489859718470663623593043462524496917131240
807158714265154827994714869659932706386470112644159978950291026411
634185318728131040084892164764213837348534278958180220020402487498
537638726293786567769051074100463941882283207785290115744113261166
996070603959549637082068598615028161709223846505390944497265584101
477134821398683505651898664976879964736814748112632864752869525718
897172010087490803890657618291621210629190831679328986790392649194
507376720451249672828738084337584741229144526468976746253797814999
273948640958019392686513422843863597088780296934133116840292361424
415740295454320443876384706743624125021509329238442061974647807051
394755748539913109773284701713386222719311985505687389161096367219
882696404444362847877291604351993074137026272294124589315133919524
994532343464975521904164444838192817221422897936697060374093250547
354680806747033360962266442322440220154891676124159484911143693863
730359943990654722702043230259116373908514429982629493859838003142
573820218244686857539785612558432332712487785093872667704738822244
927923405533052409758429688164725453190915023149916674400369208145
729297004970025040256526405339524784524153479564210882979912334297
611996383236675051970993167107599720353762523692502519382771779532
950871982185833269594152286245650495296492783497768514919950128322
716913438664505515692954196992545604504497582963079385242837662221
183230982332937911497429618504536395406530078250511484927898146207
166392943211835756128413611930733207291752955820392328150987232298
510901480713729965189857239744829253390055203042907303453716450538
232288292695645294820748386174021851918798854352199646281846233887
972820115051987585069917040689125399485667965407671373430092950131
240706496354234349587339973690732162841624906036472788645012308579
509671999379345557232499996727283140582979828297166358674574934221
164818959970523173630842694565078833340034407353016984163288388650
916633167580138628729464511439952393181582607112755421201648580 9955
026328146693079363345697138104991038675139575757537317351279904 4821
738428466914091328978643655001545243061193884063160312325551068217
446729813620593689563149594034665082740478088395854347560655276984
378899697756330162803958189943265486199702319950449497726562873780
170464985247521462604670019208604764528792188841194394464204369877
791753468323895233441123668091755705749252307450645623123252215418

```
066859343373499875969194321466575566597108063503518907271771 2757221
683327991770936217337035558075273157338019335815517447965949190464
311871618990321029438830035541039012362901803491596712199588843629
930884130315977503173958731979815451378504271734844346797930147395
994680057231744682706385624236987996753939883360425007899520860326
479074657668806310007233906201370853071484677451544144777531205748
362353569895004590018568754345847189049290732647513559297699545526
686771551721271230382513141315543248046290063011070794249575766368
382712733188596822291579286878786947058828527004678795536293232566
884981166150798463067786261513132112218304402824611069345348278767
511622438721488129667787077698989057651705078276904666215008736609
105227287998037324904868906576548188174222349171907485030704179222
924164227149876533930740482227511576779348775640705207617167664131
943956719356069699459036219890065530131417811624594244841497978601
429173653994985842280437796804483029061534335136393097339777935636
557651186259983593177228096128714715317891646535044499949631579112
515975867441975918437335602013487866342166138492981130336243782182
043410844152319365481974987405075208288368468690673619716000188724
172646287175575359582797091749490353647047487882183909326142561836
254925986632059274716025469940585952699242461166045790192490063429
531296084292810624670072751657008977848066950337621120715959000905
700846942632164621589321119150833292213624776160223765674547585312
703001341206756571080484705485516910852032451543134899892558830375
661922394308939017569583620858847420816618721564012079368906952165
888755384453546220114011927354990860274636915019776531884962695164
429366936017260342724754027281336102613032491809314707300163329193
404191324779636210867575758463919264253279615046427572156630679459
018988236198187080091560385304231990958848474578625128544391078212
519962993888607445680287692606281599550228522043736493517621652208
127692782514137947113304775212156916821827629731023847622649534010
946666557267635273214020429844425692700109172945027863366180683804
729339607941242381195297157395226157926464629405333310397402992 79583
514794348583601147339355864763600601572965203388432662864061640504
329768392257663041508483015173116545863680188591891892088210398726
748093663790230443283972349883246653839602650155081952393357422665
539894238236412452946315639754598767631866736583353452819060803560
562142371640359659370993522807335486500821621451926855273824497251
846054469928132582041944923875728170740273726750648094568371685701
293589674260171798395337114342724243589908740368970323487101214575
608026244996530540818021442686608543732122032134034313536024476262
145083308940700728458145551227201778475724601187829309387418614499
037770862315122884027073107758309422538024254225977686929003437831
873028958446859811035076226471464835706665915637436675168134094282
362929930062341220441450206442465553293321245901323641640429676355
307398980111493754168786256831497937649386243087904859284298393148
196710437820011003559968132193978312353573786479444045893650109910
885960691414606707491146322196157735259453524280997975179579133050
605072181269106773611888282818854264624943143039410117453149360525
464505344537808930660708592512717512294103783355448907079330397276
216552676005791117739796912006980627485888075414341143279292074349
412203198687250332333458359408693838427966786170221469258156451662
521663279229917376968204717558071292713684373903550598840290555056
198665877441126341066963397644945704169594697387442516156047580659
480486268446376389546965202142590535998313721065919989750262635100
681670836274046559145404594697193513854005845888750988649544542560
433522760987233910198940115653846567965737746015677721183804205611
167170272123352395287198747762296551314314637416435557018213769598
477636319906327519639113532219170174221741692030796969394445170703
860316362264792941934529344080295828956475297269557303344945511805
908353053775599609898528938484628149909037978082706324008439017192 9
```

```
2229275679954759820875262447015885268471434889512263030773193050 29
0875692296107935564096580851856426713744157104611896331000047218 04
1831405243399793760888611718895433470715840243161875709482911951 07
0195920095960331800678218616908355168673069284014075605648553811 09
2227103296655812882087856460848060176317468312120880544754788183 47
6225213788294458760639947424825122756800623937936834232963652154 17
2785640434793793375289766123815157145534709540737015531153735623 436
0071243585495937390720186621811364768365206954495681677757178758 77
2971703005656320946719312368647152355228468949897883000904738430 41
9866755459436243683550340149878662448148707499730059129054380557 68
1767685875665534170422731497903682381331581974729327211743005439 19
9144971330213975606649611136925125402963077476761430750709400022 064
0242969638567233024444246184380987196191324130310276978429321912 41
4323879103777244854228494149384070395151517443282593813111610345 10
1038993341520994154535542152023053086302790409407457279344011173 76
0306599517139677606283288218488913495028143598689060692530042187 79
1362025502880759640051762613908444766291413577541559270396378340 08
2989808322139632728814610185198200311620563766612389091409151025 90
7596174539092484357659966866714950684318786048778361586751722869 13
9206036353007511132780551166568763166409784820857767845658718040 05
5128524885585571069490908618907368453435332818332512960073022336 363
2375074731356994615796227139539574120354812732666904489129317377 00
1630641020006787200081033827632628247422189043700098518742834724 39
5241107012895460827323461125683769495130947712570998211902205675 49
6281346613664839557494543018135771717559686239614457130793259354 395
4356370500843910690103736100554345558535028985757623846534334574 31
0473852406327338171592474733291232838109042681508315276107861716 29
3736307042565434812107669713270704534818755844924767222594952318 02
2398553770498594120310426214371804350091560093797118297404449236 56
5310323153189924737112497393193673545129844220944306503687770485 27
9044267763927301803207523160392336560698865627376647572801958827 51
6092060832944517234250021002421498734923161711862770769161569592 85
5510860185840819397611773534171247535865271982046158822305202907 00
7729198672594624791829354756270505983893883914314986069254105507 76
1633191359742768724343700050855299733732888806502842536603488131 22
6625149314444830894885296392167782788369376909054640152157920461 87
8204313178403966630284719365503677857548528233558618847399540078 25
9536787936511220746389425699673888156294505067078393872200253359 67
6586052299630228555106265084899537891651672935255175480368496587 85
6219868979264381570860924959005886346663562763226350759274805797 22
9799890433243219925736146104011464504048785892983884140359403075 00
7436628991783137953183954370534751223417454220818619134790199514 16
4858509474543904235214770473020275127390123003302153662239182900 58
4270582740606154099361253912813330249986686205463598326769215297 37
4409733852396661224042521562533723549805440952249838752643250442 49
2774783147123404330358826469672126823373016067009309427925693181 66
2686840414916756850180631799264684768308557421552026374984874366 85
2966460279257275511674303654711733999224046465624637338579774408 49
6274123595958531420309180804465890171210016774716493785389040256 48
6326851026536670454107946253478335782420434551184079660068374840 9
7851793048667768258039328331869957215801776549245573087128482913 19
2859382520817451995500238497028541598757630787736805377170275160 48
2379821173343971036277099360494166410646461854995230696252156715 465
8598313375107796173285048738937209390819353513271579563134326679 94
8835731405964911983630400702517971706172194392428954817859371649 94
5741988795813394493316592790425462349506970546473710619830427552 96
7101477107082109700349376121631236920704554555517041832955799667 44
0333664525896884549750136907025492209999542937043139248058869508 97
0122266237666935946709949221175917411105217719818573539507732234 95
2224525469204117157311958247342612067849698811956381969778624989 29
```

```
6681826807146582020130664780519631824072559718184504022722433019 36
5695834736185932277829960728584913671493612944897460810270821 58102
2462576945082709104431458400731201425274058357589835870732290 62332
8462617653989925999887305846787479095083024810280589632343782 00684
8270032212634479865172132722908876170771700163941786585235411 55033
1464776965703901128207378762440643653846922703373592081959895 66437
0649462638323997579875130864793314448903768641623408792805390 29674
2852193608630233453885397668233261169744681080689275900454599 16024
7912497809160120022371129729939595343390666988778544863607117 775032
3715213175380193944326207873958400513964844119861008278725635 17874
9302066706194898005576149571225583141655330836190291801923098 62243
9257746252364915062867277334830386269267243057313954044953886 28540
9394796557490707193509241556046709506448290225710896029818298 26884
7945309054524589562790086602414960068609321191006193487714137 21912
0768833834186744184319898129592636595766448795770510960158230 41446
6330283237203229648981427584049471116792567659123918544330202 4772
9477094608194467283186964953244477983617344168656744500125335 98065
2579424202470082390065678734058813894290823554328765452230857 86996
9076067182402261045827308492997589551243480304822755128339380 37989
4927795469774368438101782398531595258213806762204025354986852 80364
4404688646519856845912556448212382402365699996497553465466735 37697
0492876772509188015822826211041574723728522152475203531206323 19710
2982372189535848035588541145627157738297635151214866627323519 91316
8193756641288259648959609671024051377518472265799569374383329 0188
4003003652254515943515977616265235657100896720328077276096109 36120
9048381544254905291782484944933036540574185180295349802700074 71960
6016791213116271236268325332109484737303927734701764511468579 9475
3665536347676123250559515539899642846278896701459605178924760 00182
1822423155641533149856102900019683975139515657709390829171678 16180
6821531750397057754878125929638295165540113719277650116873477 50299
1742378364715677397558786791973169949963284561183495821535646 2349
6097272957456249408665501941812848420836900802622126699551236 04425
0102875367488012271909243562017501198118107660707952080921394 58187
9790736197094037612103037989791835963290873437504043030633056 92423
7400479746945637233899694937176585824161527054833814106480503 22425
9056776036381131036605910766922637744444165188710985725722772 63147
2425740547630367727427046861236859227217902275927769120748296 94166
2407358823359839115435287520937651003400960648222130817566376 78142
6080573010075451062871813106898890937829776575403190207325352 31002
5296242461762425588992275757342516095813863665277955127448834 36297
3357364923822591300699065107120672810114681963623520650264204 39795
8961478604021962173589562302345808305485917127686739731442589 04788
8295430165983148041257553825659140699338672046635481329213601 81625
1783070910329204087708434587299440877164998524433015324554799 04202
3085497628444570048084253504880447948101250893248686468900780 19080
1385788200807012021227553620997245847071347477175788785153525 24280
4298125656334853498343452211353335598243026786251059458053367 49798
6797791764191635535605099745334444155942363356542955069018261 29568
4501923137641824367199704946853288128802873277300922930430241 99767
8521193026208734478482545216695301263570431688695326109511495 06513
9999475732605745994381658597637862627946549441911523064554412 78042
0212555584852319822703069402532061563927130533457872267220769 41231
2645170934854315774658794738686758770509784349662074001632281 29512
2390509461683913714018409951968833539080151440194906691222995 69313
5649963152425120247882387300721096967753699313976688861216153 75723
5641927806628125510064516641988122337033178278646807743629006 68671
7114797777308202368141996056966151220417216276420688811022432 13647
0678484998019163358189922684131434301966270405603558974810311 11965
1448529153573977861967659596179204467978374603524487875213386 57576
2138335450442551602939235093645632301733908541244609080605336 4838
```

```
100886666508928925771034698948090435288289047623997204146715786161
286743437747505780777276958506887142380341495686002180875182253 67
269915712343160494269284432634384365278037270577774863158824705198
633851823586981545987165440709120339806391947009022569745580796688
325549928844995547059747174312795946900465262096701954641111451631
718631132424155156325673449468552527755669438589181724762150313162
752379414000321546259276098329784018963983630064000898741689219358
524528502279912725978852919296611647962557209144099066795320906435
919691029714184958431139639610661119047324084831696204890107434213
083271777777671266988965828747002800354850842819582324414452 3300370
338342553837210332149281188176305176045313356500728149650488445633
867245296385350228720002906155819124579895067955687282447108676408
290221737193086942522185635410220213743727246315159725083105423327
870041813095887689782852834085480623384841147052041156163579055768
229624523729995087313750594413825910372128489638596054900992888451
379705144535142769665965146415561778437000960970867167897129859828
735961824973694843348608424233036731821104744605256320768433201052
261807372761853884930271134544212649326901488588726963804850133418
628404614216704032825559760841672780982744568953464460188320504154
931121145804157312660057349498607185538920972153890740401680407868
856514857013503358704391540591935747975988500961838650314418636344
838663180892425078401927807163664212183291846434202729012490488025
443721442419302749566123603059120488818520931431309655578249124960
197587736462873131308595947191283170769849294279711376602230284951
641789970967489017971913162283796915662173776040284399172060011285
550737504440739796257349076656734857180005674266779141433742668470
770820907049279770536131546064391087857116684950223311574145797352
717467378742206404959131374590801670983487460444168746899772693091
897422650758858393502025797765985766225954905745873802122024893525
229808025367961249297177644816534374667054595055960978376569549989
645209234094919497890332496834123968446555330005856877004277103959
255221285556914491688372456215053542221624413977299156488007807 8695
574642692824962609809674483177081659220832150827572122333921822073
063080890699965964488709986203056316264151434494521960523116874088
247549066210819292129919181489176257896409043000265148547323270205
529240432771801963351438198499668278903940562907585756246151928743
123691539935817901426132047107093889170431622124933315135208009 8008
548124398358166694814106464362441612342080511250019448583333428594
620758846845900688729785918013419848930428975878376580135133360938
914144434675292680165292227878777461993273680125199286275810339164
035763668664995009327626332574085968343166181092336395445912247979
886700934873155136198973321274590613569151928117413012605296346991
021156507651449550201116842103631396455768944282485250373907540 75
492060070597411853573209095111777476514142942044227539673857387940
614212788080242948366789612398506098750222341757393900684396415301
905824313252097282619538048219997591005728206269299992054826417234
114486896113050140225398454531631345933289324403428285540366465193
029291566650871763648052382534785182422817246170395420907171056601
019217891200400002277280842428126140919280167640979699978147184212
642966658004707832904426889574239409698672490525190394912978704646
967420761169174447719685264533658251572998412354537084889088379512
276980246828152436205941022109971524098181543491746451966648056371
563804536914653443705484213885992825447728591883619063288869797183
675838328369798988174214376873617983630396240059167379709032487735
257690260628755927586064438912628023317906906913827043966272836787
280161032644939566305173490146117794447518874347507674852948972660
145411052910067661657845631309885171220331026765753311606965972789
367202644194951639297490252837711556766019400043398247620792493171
053805347212452278036947998443240299344192076709270045967034085602
914651448116401236099911478673673700952481404214410954342693379894
```

```
2104457672396378219504473328552859178698120214623221161088357367 5
8341890972068532805690514355148201525708226820926036537597364517 29
4115750501064651096282107686952984339939190652429475780561948207 61
0195250363048854013730600064682240385770819231395537181950485612 12
8555265623750441700536728504149988378145376042374655411800867603 83
4442248586760497211043599125354687661272844641216530483000215081 90
2406241651164355987027100114684135737413675817078410123330072918 30
6811379637218660293841942648753184758725998579781731831792819267 77
0119419898583566171762963164625934287383552481616987823687457749 29
1282730801013109981579539185901835507234416931080416952579699739 66
5916432364546751948279075384879391364191677254776975716582185353 98
3646077377300537746014872273128061410769820946545232606778157545 80
2097248312303243631467873269197638291433131978633522077710145300 91
5446195307538248362467769552079557471880240122021090175102500339 92
1392433825091904275795031745226416207035001924652528122237045770 76
8654878286278760325072361191904921408408000946144463889850644596 20
4175976049941876449788461713925428875968537576195361002225061268 79
7338139220458822041623546758921548969001326060352951208076339185 61
4125389579387253027191349238313158385703647873010132395362684313 96
4516564709646144500529720845194320439256151530366714747752215956 12
5852990159434270814509570821069007681125432757172198685609786598 30
2451871361702423399520972885649272594237669497179421500544317558 76
6023434997764862172107335843131573211639471887088517401717372729 63
8552279997468704838666889586563616444690547202274389288867636090 17
9544231480752562653472784545846244426008393940256739861393989074 49
3895279774782235681638900694957542299176585515287857394569506542 30
9356655368567762790760168404043946031118910744150965421682884007 05
7148086764249760506542843055543997084922814888586472622023145495 727
0656790775159294259661298683905388813096079246835639448295093422 23
5486085924188493682142896971058429476833786344676180659570071930 73
3109300357300426726809046976216414561022911231171250642720530873 71
5274698395451581575727364738703434910422297110377988861170487221 15
1555324014983973907617823468768959433305825640626889212645978262 279
0373428947812505100030315038390338219658936077871789054659293606 95
0861736801237151402364529590281545937772366173204376455738772041 84
7647332930070240150396006898258699091433050023898372623200942992 74
7892114874123394993268760520099960747774769312411785643754684135 9
3003874960799196148829336866234433986178229325295014508362539094 94
0736016945964293777098744667223386552906493017186152102475628936 89
5959065384058790975921404048078286807149823778777543938934875829 55
5718238114985891369554616149842445473843261940111768384400797330 58
4292718445090372920131397729573811022740346966555872138186779401 53
6960728087343763325876478459645617135122734773850603915939218002 12
4102598741860175380318237923304590632664927901972873919411203839 06
1085360339056254545779941863034518148118972731871278607986194199 82
2220272534417640929177904994880536946904802061748154151235724005 51
2586243037772975130097129753817132795488810989900550564885956149 73
2614470467847613419993953952378274624393961527741402303057698030 36
0358019287404966081776625341741784632198313538206804482134573435 24
2110929399021679639545902364658831503814634204240512362100301352 12
4586476238585407499745833817337208693217309015732437058474208743 10
4068514323098721718223941406815559383956228010685905093290977066 87
1446485527971858610334721117006622414047051377040215450192493942 50
4876195389509795782872782175917907117998344913777697957939325249 0605
5606105840516856921381763681664393846009943833929623758954510476 23
8741876836129148019260186092806730653246407941502667479289564940 19
0537144735388604462177297744083537357769843388334557140856887387 495
0308633023530715018550269020224188751555810797602891074852266545 488
5868144473387319529841588238484768964838971831862638338814886550 54
4396991440097122069839869567176271598334764688883227416225666909 817
```

```
98913877351696741791483830833826882261949639895717851100125501827
74004095678207843724870086146030820772132728368131873576160451729
32814887278235919201266749409223876677984119535362900328933194279
87638695178685099781269640639425317643538219897951827032566504970
33357673666930269758858357978073243501268630322186784350212379569
95058542804412698024181360749298904347864064053487957769254230558
98457345033177490847852303523136831550084605374115272608707269217
18884012241591241433402690433022893147093790065007446198180659643
85642625487772605007290676798570620549256180635726524972215063123
89089384221876564195641307318076547219228610945903007666537449336
01000019578782724928375427310482870152701129292716197841050268223
62861073641281701917630657455288749955287795050224943091667786306
48554152976153915012379173224274678262340302994985205192015698223
54642415805154796537784587981379848019252845141154975451987656240
98277942795410665211436673346659654055135663674532473519393158057
87154988528958406220992922167465791799224980988279271632505584226
71170712017268410514697986865495280374025072336190127932518258269
06606971725333060248044106354065927144192595455810282583589895247
31396498779708894344410964722681496734234066775402693053992547782
01367595739617450575057903032679233715128449500812059897680594500
40744272763822182760451670020088525069866375360187514049851598875
69290232824836549883885940690976653901685993770217895575866492930
52570779180557597579824464455915128403880615136061513413219928019
72437549868319025475158313310390844839382952778768142710311862983
69678367794257283981098212814495832063704513798430876078960149195
66322036817678367315920646766470036581947795524936038332356510836
06368692857956414999456004584964978823261020077318801153696195535
72187425602429322877650496655153578386594306079345966982934876643
63005369143632892741409663843064339643796872468495132770931460956
07633165657072917303691852219475182611533322685084834310507156911
19562377160573369894975068919914301622735013186661872369858426468
75234832543118866065972888690367134527367139725053476012201104777
81707532013023243419240305680561195318425504513499340893114163608
00996103997589429734241520321731411616539687176234315810289170863
93515456870025679558793750547873859832754380482038145030386526586
54174705770361895083878446389506239568802996624278204317767156011
86300903081387670430705907811944059486685023216385492636296047947
06070291471443738104986155989572602101543400168981803912633838635
69168754328070074313991021299339718366911984027984953994979274226
50349897547660607598165145234962981571473256859084678806804675691
95116303924912276930654524538435840910629407583367633541439578727
51274576406968359668680173029229080359232338310210163889188658847
93092448960412741608405320279128180434468237487701473938215423373
73309153574707638873326667387749830340377282929421507255634060764
41021182944042433892468160041662637703748393771439000616256634610
68362642456701522932535601407420361040386909110440820608849230960
40180958887157253879845462961959698562358875533087477877826932904
66418246926122748517834304078018232024473719996281041190681622666
63175039525742338944488559046672052891360204715386145146363814596
44475989196720985457119301850721728235836328518628055227201290133
25974211835108611063744982596123643643519493473967658503733662442
46926337755193587485991457587436320036295507308988263864589937076
59271932349518092301457576940050945936301151205943079484937632015
80122904849560757998998429085662176182939696278652462946305996240
29673627391665538462048803735648155813460870020562699886686077064
46671537541537679350933922421215289175870558580013049636114240945
69314936041489588977384973594416986767118286549696217440692014999
22291885761832085311872416936576901336842123537566611682146519090
79556958810607758884598603928141551803060489123234805880105594264
90311372293932698939119500192574402384762386567466542372980374119
```

791292079661513740775569681860006036968255554914459921660520527905
432609633982268529908599899052245596677024891664061100106401263543
520413152920291666326929146564541790600022897813534433356194918573
693010458860085216462200050007613100481161792275789987134938728800
390212240505622267132137871434340034587035325561794712471059508375
327985787745082234704073955859458547702999710900559468267169646288
542792069944703182736685123499795194008132551045688859165339084431
041022049021040202448661973280850743516007349702203209129215106671
458425767481613136593466782806943117024519860313145879235395104352
918213938943612076530657471362413092874211911512381432835194921159
835354216332986133810977937756269599713704953152829983461105193117
885571593450686864995981973856875270690016788082811377040812021316
130694878312177410769333413682005426400131598785740236093629663615
478576871492806006833451194591347470303385745229027995419695205288
657722561420053718067848984638353522223579812458537980852576727295
415990641808359980482599213065962246615383042978958209467131680070
509310042415822279331044083058485693686184251384788786868646276585
827727387430307212579918850387420533240897275696731618650770362211
718556519016156172627640821537161828665192865961188109681401396445
666570524919815812962315913638208104666181836014411681708261895692
122026315973614889460095598811485147747572515201141585698743202163
307481359009284892078995406611035962421079762105783718935803663018
776740784389013310524114831734074326821053500485751253221366388426
988036718707681217508372681035570026584903795349647304227380135 00
534625300365192950293073233013696673734901632987256727824187158536
887045662375435712971990292114201797296357904512452162098388812011
210730243774751211916407192142824538291215575103807708469567317903
741862830005845407741008707473645924996983464139001351166936721588
165269286449261671202510994047548619427093525606909377526368303471
987892294431946507407641829731137706558796417997594205691710233106
059385443613742325862789163934803068189059695405730569934322917008
137744073312923507708766174122051604075922440031952882198018893189
633892552089297772126629629013014328996357368835644647985230069635
352280406450876166781156871202937222526032399193532510736984833072
984507253325187806827440346453510907186499693387149120496080607587
491475799025184475475801864022744699680122706192042295811662422991
874809250039051877018762039910264411665847698872566842018767507006
334032907318338794188863123160635389903566857511271527984194418531
574958290829324055054968558276040336280922049509474481193903400213
555175282391282675785409772678592332133787102989828926385451698542
084361231069380682675153039204189214200694128883345218657715100019
936371456258100469600848176632689847225994443137183018422144374632
242279166316526152853257705843635185464007637119371613279170009984
764429333086415767890700567625575200938387765737691816891333109927
808510254910086221261211119557122492734690740157728976376591519323
603887743228139676479763980961216462370739021471314791577724274042
106540295473307932511887584859000599287746891581536974576345906540
473036862757465613209893995165209792553551095144211926515159110686
820034080371867367239384758518805621076730536667596468289892080344
487613458282423746234374475760938536615083529724960444657596301100
150493733035562017690569260389181626840079368395329188848066289113
005555978166507130122801888945229066791100925110026525519437093168
294811856000920609217300267885782685613029492980166389704513029790
392290555436844545439493972148328654975244795964925204404073139888
656919446750663526433598405627984895673217673760899740739627931811
694879801788574523643580541340992286714288717245325652502806477686
118219223812280920259258473175937266740733642381339259392607888748
602944605050744292841611318803424588388920043592008641817724991561
880289243907044516958013577994384736356384530662316952438742529381
035727572576430048411241269331118494843162322460464915871165525825

310508415886841208612037368302402495416986136724929754790997629624
677795826224892681847480371440970385420879921935807589816468764167
443281461875880101789296559796793546166001269443942839444074768025
438759099136128556825665468785590623778338312063982795628539059019
033369496117946592946261659035973974893450011620325918333163620517
277217521813154072539719942860083270245092222586656016179101497321
269560734597969379117422356840818122198072497599829653997520781339
264839659820639845695522570811352074811623631034251592536743594113
918137153985080087702940798785838278062960543236760469981324702227
840955503792837006708721750847953308728940305418000038825023984385
215150545229796348738921572268488095635193152334800719564110181178
383199653950827942868371092625383195756259680660384272041879988694
447260336658953552025439719016894363348925593346356466000813936158
495469209573729506826798577614591371346547462409752603781823750271
130376320978297497739534743816014809922468176615614023504155567746
697752193575958822697021180709145805440041389770903057594436602370
568628660603532119436820060666991074327700532837843354725610 6666549
086921561587173729827624874788462925840717870473504501164506900511
581571120518658278020951288539737123299384347248586756400890318181
417880206511205335654974281079003031972954515576722946070488 46619
690534199809984760954727260538978856306326586477487324165 67687014
489129388235736477109244629958454845619288751573457110208648658944
527863443777711460823680259699357889589737287522042352068673779345
659217036054393881636245025978869003310820000674863237425 3975989700
692387641890681401987294655208038056133839709533040755455 26867838
526682543391244814245737325953495497461419192785586972413792093718
187887892976044765733793279684526989043437646902363801220 13446088
300116669725118088621300158874486481950160071441414853564384033592
762380222125998785023745771399151762907547669200505634430650425742
304771082569182297111123609211618819215004782503641072183560713375
113671020427895594452374902658476322043965810158066553061826747703
288162930922448562954861692967737670641034350998749628880110273679
725115703770397937665511178974830970269115374589426777444295844590
749958542507846041464879109598763733885943476936233295538691620976
320135040736135123567156255284266280405058709710684469001742104605
664848331300997431113988386769225459997391114705983616475099881371
932546038816894981355537445421413167242712220290777355829983864096
035139308958356646396317059061636520177042652747231940072823371379
449296741585899302463689566005180256118256062053698285695110774131
468252585286851956794429353603279956075612683619187434405297166832
503095784499860615025126071165825097778436754926450696639837127426
437437071224873716127870759120070045624871368357332331403886791493
287368587123881276307099358949982637867800585477609494255271862736
888556614729962870898516981845497136879127127951368033490679119501
430841090857932580252918603289862802766997130036459890106624196309
967118050122140994534092670735144589809338559568597382065573846815
329704610275407246775512553809718171280938303404877235270690472513
968193057637547377706177021651566826621765281029132139706164856532
965982799565622663652490793964466900520037654547924920439499687930
535570015625200828287380919696947778602579589559898350795286966342
645731289614353359283050200998638601582129197500900581707139 70219
629977521322145434921592778359737951979333774499635347655284975171
970233825903208574669231119546311032670590986024221174394073318258
157739882447679758439110638443578620846877070578121257073392505182
024668968732418219295836286165253127747653825529989668128997612409
351805227255048743855305616524673796059994362346804981771356920042
885416984678207839224969183751524310223193856251977734350314517554
033532927270231194399079288483060110664191119865642564942983625497
337422878035359162393592072646109893333932861954573175029986746129
840961205789082693070483292543414197315571140428306814152147551730 2

```
07284724351223742399917298269091164539820961865527886072722 8029348
73470899775445185646427586152076644824716814996491572834633 4907757
57601578215397931364811615622666736978114936243811240265361 4602646
82296645537652390098757385759426545063913419702721187592457 1097309
20378902190470000941062344992948623823005344319649272207817 68961419
41104402295531988670321389741396322880251435667815273310766 1514204
23777844604793268807603836239012889078724515286593443621052 2230231
40089946733352069174527409799531917523141816415124862698727 3328105
55534305045763151705310645598597201444297755484266228607478 1680123
46613930465921678658279051896878671304624995644906173686094 7615827
18040307231434345783506710451794731531498764064855497188407 0686139
68965245851035196502007575742765478000843215999274680094585 881912
52944238682229199955322671019784959981474423919221510440534 5506430
00234221375787828668907742819953286336828220606681325769575 54967629
25158327470540152329147349700871254638401442941175434903042 6038052
47298212212507933446859352264346759404739232886188553607149 9127045
03602050251510523750170600369589626675170050147552910597091 5058633
54320463119972197613513629130451664926601312258241935520202 1940413
56008892861441933547515237781342070665921316998516957376581 9946638
13734983264425691983968282436885426033715532094691232136394 0501288
81574835688501713591322816093650456661592336649739518103892 6551384
45793404356046731163302600485517637073936228516063756599974 4752764
48521646341737732921722645478857377718356040297732365574095 2362517
98091208343404979662053784816243517747964455283899004138672 2816623
68308787947624481391708673882643426208888072617549180955714 761184
31083795532409774110398919010191610846570442340197199875699 4497209
71527278483290452356371924005524355489249589380450600235916 4053150
69860972452307742002646717152266190315087570199906568261223 7475075
58392211135706975568072734818761514195101343927103997664131 5094589
72097522250731792943575319725501641154910316749272657512824 8485711
64826190321243972309729177514765662763610560009408826685309 79102038
42552875877605928783292483643235530537574044226659502893356 9482036
08326561296476563122833377091298516846420998116439844909012 1982130
45336097868527165445356484401970644384245178761951591821796 1253187
01096804854261378899323787846288378435262323907638035722735 83926185
30931552677718403395113165426095156359804913550062747117460 3371231
11847580753657049382518203438846688118047220083691086196590 2617384
71607169479380810824046497937121714908037388086066552719401 9446399
68943652924188512950969041625727595586857300318711221090225 9407862
18312223646357656918925865312746216994865152170031432666617 2644511
72867309435004483234556539252275553922402904582456053025244 5097056
40751156555203355087626924447811269531088185422225255067870 4607936
99607885921352966670091806612327468138256368632330156439927 0096663
99738607598861411930861638068461604283265492643691014707041 9919029
89094651992763389314101989557129021023049877267000387479936 5706030
04882982068059309673852748847412493816188371653939192095897 2029873
38272649827957099934673104807586476217753125859996242519321 8839603
69862265650669782443985069814152400843067977418216088363827 7631379
49706480099106102848065698852614081001266226989273967174425 4542693
32832717803003870905844944111018586845795379532110777800933 7613398
23935052314233226786150279385218468494883136330557283171695 9746572
51200494491612896266225593161079045148758908690131664984273 4692231
68682058402146606064383408318480332851253191217206656902831 0475215
08275218728758998592901758428131059263239089628840358636467 1646816
27666942164335718545231683544824818169244691845660920938358 5455535
88034456637562016680238668803775024506772680416334105789723 1823622
16230784242624874156620443867443519258258672917798164298538 2425859
03937247473886580764788348421749763038809240818752398112126 8016542
10922655086464063198102659510404666644988282522729620731741 0631248
38375782211476709991418072862076168339147629703848661705336 5727891
```

```
7371821115234965122404867712079389727843945829138236413125263650 86
354385895896086070195719937940777128516477315613697054881080 61103
907438024851813542424834686733506439385414847172862660259646129258
4842028642684106698460667369123260344427765033688706639398434 30532
8059988350947355273206866548093344640881148402646902039446994 04696
778884796994896816225260617599365577948375143944991300248891 624433
7549254048753518533047276404302963680932701636714690008666137 74733
861789622857392422185849312066102095110719225945845900627057058 354
5428527429394077928948368788705373428296899927017258215807643 32998
4187871176349627405045032389809215233398826370361175090034823 2563
4716760502295522582907750128130281692917908039035481194904119 62364
5414552110842522811173320916588214030367585376991334466373313 02177
9077243422188113728484978199802513095247259913790101923065730 42132
2092718329400922219248851929732651711116595570008092991011082 53051
4181670380373943053916672649408336474410496921352358467299551 6555
8671240986021414860096983287269782496611508430070542674851718 98740
3663018492334671518342902774725066816020775254372882978663576 71129
1170964574394596522695017273048764078250253442397506872720363 34331
1747432354673480052835820641406678783341307438762270596948810 48373
6660230841576205745552719047608684397809741444435238919432554 69585
5484247911825928070524850625437844666588259446123336266428129 2458
1137047799826980797782518908935346632129881961766187868300423 81760 7
0449508057835589316070504372295216327093663631818807476549147 86106
5639686459134245195001663300781608754313945030086543961908264 64961
6077559098724108121896614140867538066417856184469942889731396 3599
7428521526523657068582859864726892013405622460303896034691961 74106
6799701762461493108951164082689869640534939262098031288035878 23048
5167117164414428319484877107060643313089894826591618498257592 51931
9159090021786767376098089683278011263566285826947182805521160 18169
4051853823607237138364236668994910831496145641468891161463312 34638
3691643785822500369619739022889659695192651107979992022933069 76015
6977256057890093360809074836488587391943907946789828084896820 59559
7328670744142464709179987845204652072704707777169858611169275 14105
1497264342743333332620965768903751904707285212276173902714077 06217
3356322451757683683826963873825603346634574750154887378403135 6209
5024210361051158938813432507538321839841080256174024704976662 35165
0694436554891803962059035735229935981665381179843883831835758 14912
6868516510841571808784270591320356462731201701342776834309234 15308
5171087295059460334678351146414135225339464125383701783108252 43453
3372823894179270112373905933065078718211706821967863991326721 58433
8740557342820929096571195134232835413710472262458334321918403 05641
1745230284954727675976267944071113379043059399529629707068877 008039
4786914972032086667968744010591741977405359746401134954391796 81088
9207061066907073300557493754577591720911467425634052435975905 19824
4147416530178263834335919917169711858570173706871568447892927 57052
0096784080442241598254566036172543886688118672587218614758706 60744
7929545666852648078730528524335462285053526766494900730009308 69656
5811585274169961312465712951758348074973744002370233283440011 150912
6407807825986602980412822085315741457343609054612827926813062 22994
5615342533763384061425354404383865216688068406558790090495036 60719
6525071898533854101267182471134809867779393425285864299061090 71700
9617794708272113336673909967392831435058617120392026595558872 42923
0571432952775075031182320803798572413314016735304865946515578 43767
6716140260638466729463457916582119039771344789842715362827871 10492
8167467018936151088533354959478784739577375014675073385389774 81574
1369856256667670409258887212070233980451214119704076710284141 907153
9162700769929154139033706557185811382581580804765877513500177 23095
3341856540995559046108162973319957758399635065554975940965952 51492
6064438945834344676335145045741009442384261697515827105518667 96055
5744020982216602635985321467508685688014104037702019797912103 88699
```

```
9368178114726773288626999205133078979248489350072116960681691936 06
4290353117922305178736698714217608218767155637323587786877352798 32
3223594682007954138403045576174881845405676239163882398422175962 98
5114066554999510560499595639102438029353874016383455704057695566 70
1358410380861950471065738445783302564535184508252424474043807853 1
3849690794792660198422683526899607082200153347537515607523280705 51
1497431105890780211440069318000349937094512206586038642388656485 7
7806844235459965778586131149373166113290057128744641223827940879 35
9943659096794345095792655026734560500995483703519968302034076962 96
0890166203033294605781734521962054309778415075971981401246606221 72
3445435408259230425572178695511829962250166322460692769666243570 29
4898425742375900256475266778713849386085423117422302215863558446 14
9999086249875943733814481800432261913133817555592945390083309913 63
4810149162797199830259730814148490000860022108674807603460105182 315
8156776618225879893162821901036928122639962003686833030583443363 14
2827104566148833448751605985673274151957965885639365825277284339 77
0568215852504093995941298376728135962487653698402662837031667365 95
8112014191606098479019058684876971623002008018618935537500159984 131
7771305866811235408341765353494145366509953360724735824756022200 62
7800639370265982870603780156309118168796505779164410664658634806 70
8243501122537658670245346732250619580155621819936160796305522715 64
9496955632180720171330110063478032805788268303808358276145084112 77
5949289767339652772762432386894680826557008163670921597026348247 95
9208497570753573752740689868299758297182507352977062875758753338 6
2145282105021190864691020103406499472703411026320496671168182032 60
7044045246376884234205042915945815400086447975389171390643660676 4
3312992815064967380453847615413046638636029514707624386031740162 61
5460127235106453909777091886240155855870267734919403695790951138 49
5991435992739228041543186294301242498578771702468611394199291998 24
6121453999931903143387775845492652352835153497790727889525063406 66
8811166638862160853589305330877204021756649323488119233149556135 26
4921372384635841884474376849088433946503431468230786958900497240 10
4841948629171620361173287976404319980384112665925933188996032631 97
2707548362116780123891039035406890165351020389843919789511023774 65
2307917429420865168773306057297410775491424139287520913668622171 78
9663684822058348443524082924732664634558815239461074212989839533 27
9809704638303883609981110608089509716460394928308046669450500650 40
3944995203731369376828386188989296195282104389613795065194277304 1
7292598850724362785967222012920211173458210483676696313177096228 74
3221451658118810793762070809402547330611520154267603239259518136 14
5027501232188962988417831851356036419868226041615115790122416889 96
1519088822600529154454149176500857314562777631123998788240272600 71
4411776230307487292585489345827704021902197507647450338097880923 80
6342031458659830252614793168572310609292141770066770999203012519 90
2216123517138484974744116744853069968644181283647203696595619561 74
3005761834232444179878864552744110639042308406528863311555023777 76
2536311584945348894173610626298048963744218963445301947950287325 90
7226856547340547500756782124995243804648538782789642968653279845 87
2450579724026814245579853483792974248251709702131855747264731905 58
9540574276010650270236341614890368092994571740608924394066668978 60
3881539245456330574998285258048772332836408198619051358881013719 05
4488073336119773233604153932251276550775978637818261892042567235 21
7798464598207695796171249486924419779409429644361773562846894565 22
4832489231298873772652957073016830687998802000230075480076922177 10
4502867242437299037794304505141062195837971661658631938305492119 37
4306845880709025651660846892500154865565958714415707318792614609 92
1518073679300630342811906399574040538838869242165942272491046151 17
6609645281147648864211714127233625524896280757024617306546887476 87
5609635122037932858781123568767682540976966764973037673724107302 0
8350437835326929801988557038726110109706541564974863206442103714 94
```

```
23368326576683548067589727647052028617777585285206453841301398952
97486634495622217994747637328482342143857233675718410543026921775
80029663121599484955862177286888229804784285591265833288067735758
30031444701769864563870959934459467292274966202717718034180674494
10825840094007114289963732250879261210348374295525047672020812070
61487793049257934048533312645005757461674933470168741468361328290
83985612456530966558263550999007959716304897096691929082824640417
45451646106456970721510664721864849304884059509713918302246490963
72803073835724941206271444988821044653752471297471327657681546809
96910757651606971750098612592262550119465471685831967131649845318
64877188404145989991075639620483197118235530194369918788893885013
91663676961850757864551395000416995647489336024494709551207019825
51530826759041072598503913170226597501088815920543860051505993981
08485637382557223992672053711148437183579625735056027482210980274
18220792961465454024006866359601793223813846614336075875739281334
23066827047745190931627629261321420286849315473446420001719956588
63207110432291148058689626915128751724065156014633096801716030235
97940468732708195068006551236183347169127522061638150170816784069
73652419713146456709133685104702413536639533454531796046524622171
88247632133077761003249512280019438797673040055996940698902470031
60587865615315307947831170589932309909295971333
18858503452753516979381170539388605840035046836335285346296242436
94839369535759056589990361553442833663492334732163883222736613626
58028487009946339053806601490250856376133223918104049889346179814
58298357211495864557938259421430348236220156455242842262302169966
90902457482797996284170634749673479176441191399946893415943761661
18996820140481020112120886006173659278544279519015092900501365318
21896677871361056571548698719222074011365067153376715651084600000
88355568602278127523196611735707564497906356052141917568480475445
49386369660737525701242359108811659934505040202970106816480513252
22234894798901829465694428861470219499827912939216421875065937475
72579481022216371475923244373242713472190500765327069743437038776
84787037575990248098767015933951117469673743210215852738854286019
21127830121756268513273918680276372160765720517711332496924504774
03387572735184536421254084783431880641626377014082082513452642441
36690799456844040420443032332107079025521396610619019873078271586
19545870944913305332067519676310664648059142701337254219100513512
42704140621450718992405034885045331404377764330633365726780940260
26207767674434418119580258700965887557469442299762546549711793310
24990908892608819367550581207153160022085637917628939918607774255
29730213268412276765991962106617364535272508583891758602360833318
07921361866869745723183073131651017200990850710999667706175465063
40233517519666402193759050142459316993571789052329792568625225282
34336308791917010473017442557296784761993245452756277036314921446
53530225394529167340185142475704183985734277203684310353482401802
56215366402833620725277268604531669690485815350686696194526931992
89256652824504064816149472392717422134278330654324644774310894076
71836872578061704962795850575713187902720675846674786519146088424
73847261445417963681958226466857891446853837857172035232405230925
51649758996177578239046508704588246897536048736820062807942668874
11881647642281317209537271745324991800504604341920726570335086576
77485868597147804331259737002689629532968101921190421165732448429
29162391460511188094934665367750535218156723600017417243192078350
39905732742158761072663583450284683286953884651976209147089077113
13085252050028413196010217666886437639631651401907991845379868968
51504163580597794990929260504895892581693162205248211876112521175
82170247677771025277542978431216265330802950881730692967349874808
31036386474173750724820187249428906487364784812824003290429888475
62671071275406476802405693849627916960000990969206142804358549990
69473030335628482370168468811588719497342578060268178411433992103
```

6269337982040445198089201112778626066923605564041786498803422086627757143132016616016457809153994323172590550819066029221589620429336676390504831513419312629946630801356058019194412811102944074570131391540653637442323061761417539515595575250541679308945950333084142497072943606084730035406825686969893038735832307891618937877154002547922819111364087428452971855293043822189148480363133395967676630810488942798193977541231096063638561927041826360597318999421676277017630683864157746749006174452324401127237237009681444279501624037529612343197470198621237335000637610983525543960186755823807817579377778894824396048672195218908988060361397222667038450375132053504382341861187181622514235214941155543103061981756140482908799023495719222010120845510711346540626646029674918682050025065026191592278829030146284850836732352562782558654092593939796405504819519921476860592173609399718264563829747872364544904972855777329197166680970124201802735712334171696364116062258274280657716307479121442651956327980799103693492571291485032121680674930104363544428841398167836487891762673603603015171114649575004833595238049128453948333193579501502324550736573945441885173101849024010870147017296315744406597240466636891667049289768105358570726692201588232140015294835305709471027583611189213844073199039661244220011911163844167084709174288356728247821068755939281804381314355962139014401825656595497527621678270207147170955014166693006270483314466405292042777444799612148927343973032338581042089817348342402048668784712228137630624595711738887660900030458535116416185304937118935221507271908990252232214165310163395317312309991527343845044859642724706782299937873544067735352579014264098017490449388335009195914325725626255142056921198763201918899939446327447205187914036304668681813517864957684835698713306381543601726908659721450193814653032820556636849567489840744987589601709197660635630290658779638373854046342529514700058080495065823212711257021918953876951862706825880891594719607852126095359754474818451181138214438071189115162641386204780980176417930774615604427169117605825055916211333090525821308992150505904338129027044098523409464691321188895829037279768287495402074376206282636751095540417603497651600920559988957323041384892252743476108676565469996710454431967045557609667812448341307249424578230661027921577310983579953578039565299892574833258692947622431613332435274575904566816266143512143732120741434226977287207982591008361503558580296671444965237287080981571142723112547669629145574853276288861248999865503699648626185569765223100624906438243969665719764870919833865261838224036291767000634657838347143498275167664450821448793286590346129357114731799670307036092015526642380233328399866558695994638602488400319572488236385283958397585300503626171696961956502894653506264036126212053987307670592168975610466320925529812833813150974237860951282717906316471394644464593868073078814261661434891122652686165957338956025126076824145905448354359637773328950943933638801382883891639720798430246923251966208773060379183140485427611831581540447449895092414756623704846280855893843256884400466578672910292827380785420443528427100258220325089396223396015458391524826109565117332296387256511877553078152820098077270341744855373433008708445777570202438149562473725535587452767740584699449926118902782810522004718089447006884655999628152568577544639176669233805027558727618052598678442987313633152927114617580609558619586250083679223795005601698774899580022049826926125123915416703226041995002933800204714018036033344926341595300925581189387410821506332116030529045191773516952247773621283370358275375326294965372732145335345548058988326664371736825531254229237094598390200450628641449889073290545322362267981705957101175714091756289666442596468215947708757931950765193001641522861875878265285928728790509394534288780322007321255467983248846031671573393508569049729029393928638292959106190507894911936046610759368585662641071308989408385372 77

```
3266496895226981292832421667395153796256198388725438883996658571 88
9273662053870161013546418915429929570541763100510680673963285459 46
7988746752777804666947715324522688445118437275203451931935639914 34
0260561484782997090595248021192159080877390141784961496428931202 82
2925027270386994157666780899761576274294047301624165848128927873 56
3638428616474319843297131374420003152585351231219430953953726426 52
7882356485148713412600089896350744056115237023403061628303717576 97
4862176087677877002505767489499162920264934466174175208491927608 07
8779807870256033715214037127037272830798131220431794811083189844 47
0541776867649831805647326061452407564690319820736114058287645705 45
5419400602577432696444622483902019034102603561899400619553213527 82
7653970469519763703899858451875863596340659420516530144894040442 37
3385301366615312152766515068199004841241754963204666926118007580 8
1507688616378516491943851150926045233120997739545467719052975368 02
9560910201993546598430430035119434335114094898444160698732279428 19
5370242876508063219572870826587904250796319581042240802001654969 50
1266166942468197572297410325727514783599243898244381494714081519 24
7394074882890806773827903413656853335730070899137294416069474957 27
9810447462917787369998266496189020206694854820275302908447235579 65
7611906170698551563456567552120802007691228697799693389486131386 25
3329206077957727398859236435601189309824253557254420324262338204 15
1093148515954529935917378549085267898076980392927828018487960899 97
5611889235390848509277533324827157144662197001403475330579547790 214
4289067064167961381738653666419851280330096296345479580207959808 91
0118503020615988084343013991403588211011804406749236592767012051 32
2708220191744987167262127467398921893258359764004484358551351850 49
8745555873428583160636946881747575748644547847382131483501994685 04
9684522643260787738402489268987176366245546093623004611058338287 69
9976906999711424833720815913907326594333069366113072587425893388 79
3634721690105053838431930340153540733259054191051939202131327886 99
7788046868382270971862025296635334976008327139364874984890726962 66
1986001177842523007531287556302595390222592383772952909241995316 1
7527232435207730162796444868686738312331106792805619190597488816 421
2754736094738763761549877609459933018665413972222805991310487079 15
5555981076616722516765995969077208398047753757975267608607102789 023
9123197072470224287209171063025057762592226563869181576893542503 762
7646493813976538928017029211114655769231601688067980741486798267 56
5876610047863205617258577509843891092558539805647391296019287943 16
8978330511429810401960498856475210491169057467913576428079112832 11
2854640686401597634023632206710099000584535295749366378072577817 75
2779572081667403452978391302926848521816430181188670112594776061 05
6632405345137055048475343121817379239794125558891103206925626791 91
8949043703120431016873243939855086660279543711885367220426149060 71
5274551892491458855839521468475267894589323520744517566606350056 75
1094721737082291389888964787178477168982005062731317551320403502 13
1737921893397587782343959253683178416035150586685160209653050646 215
7098455757697463135908540160036503895606480846780335121842329436 56
9392567443523173493648607212006720217852089831835144029346958401 35
8365835027094088246585432153234992053664510237929167811174122439 46
3991900980506970457999705240735108650940136186232651877319159885 61
0324087947122314776307211523455352696904981878717144333741391724 4
6219642131835839378153090596517931921346491232183556185755496069 92
0994084328751636306076436428139041067375621367880242121643241462 68
4673749903004552968344975125369812092025155413182713471323568157 73
5922796111386202576603579139077613260699381731925066573332539475 64
6736142755050243830297016029218696745408303336454937712140941162 41
8832852555434614092793903013586132198210236002037906243572074658 765
2115546728475487237864906935914196912820335331331195407615489394 37
4650486586348561386910331872665561572147541822067993465679249209 38
3811927930011757343643685816454154264685344043413482657382635577 43
```

```
4532847335074072394615768391448318351792542096484995010092245044 51
5433232564179548540924447431094864301799884583222532279607233155 86
7710110359778503806767185416819795976782384587777539951913478074 43
4303270880768918303300162280037801806321919762777978765106911106 889
1681497573669988301537289463993987948118482876519679117937386415 68
9734083207181171239054374399982867049794468936557908502087951648 23
4638789822160931020441545188973426273870749409468849973350631139 58
8288025458253351358167620229310596946954350908300146950352456115 70
9964708124062008022429733238871314352098122297090768432903928651 94
3696255901256435352831139962331815791330124444656520891459188408 53
9423784181460747639818990276274054275855005457142859611758837849 78
8755933272247024434036429910348230185174816602063060045167611368 34
7200678679283359157758692337689589205585184484710770850159497238 19
5938048686912995766025390123418890213233191934073760365773636023 17
7792061670763678037167693483336469879521035272742912258925966976 938
3088925436004704935184678385660560470090985125046197698767270533 83
5289799830004524296793801329695752945331613828472337916258138373 74
6978379630291231347257839979226046208548537835571620009958234665 25
6923979378349286689742900613542690686099307963635494499157210361 80
7263976496840594677362645219510513052035897428434197854475401179 36
0945917303751370962271973736432763987922019977292247124650140058 74
6465457365003769261800977286185852203843550007070992122294959787 94
5651292706587521089625700504500635571456207671435825051243370623 7
6540753076885163929084172626947682441066803727721322861275286298 64
4536385946366800356588684178117952021933043289177874332416626119 02
8802492502328470288489024511404394002439021290036341966375740525 91
4197618802069174565017123145060356250068338398643834812734576049 81
3642704025247741496296440636015956057513810753333640103939104322 02
8044508674850863540682478026453588668312602239581937584976414233 46
7451613854785314131191401362067504375685317647797857213488585632 91
5965659530374472949617624860523018320703420145057084310324620758 65
0363411934576049941359321896186237217872523749114762373667079932 46
8184332942984000457047116393207240896935146095396386859395660869 50
2087709126392608675584472547391527294243880047323637131555254074 06
2377695604764256138738203049467571282465818240607713749739439204 68
8593081375714581035622305797897189126329819093921151288716141043 69
5336829998623608422182364279767934466541141537448246367999502434 12
2315258499950875088898521665568847466344939676317890230802278149 06
4118039749914516527834661945183314755245631421732888402878542529 33
5812862926982742792957854269106543459203792060789181395691108837 68
0989077824395394966899965409725260340519830938567465160311948828 33
5458524041299202786544322100347667274882902719198657779572013404 92
9320940443075024878971564943028689907731610406800670609875363493 91
9434522716066993735966561804379778390253080908499616999103756202 36
4920786560448542077604594558114174019007236777023093946757935448 81
3323526298244468386604987329294605484985269070382465787111659123 32
2487777070860245089967669790165312431705495660672788203713397352 57
5361440790637839313227069810941814034079972839951167998098707227 22
2481320937831755011737887158437821155800107121834710190945396477 70
2883378282778579271572142855288612754841544958273824626571247090 96
1477510515372621622508861772108682383571851431491479261821472178 56
2758045660514940912601754051053431712224750025051264812534867009 09
6865383799044687832764110233610043264474345479422725897524819131 37
5138213009025432964163774298295997625333304501752664535647809536 71
4639099172684654473794069311553464320113090797258717856049058603 64
6331858968193406459359565796266030537157349317087190208530235069 43
2927124492484634798754818838716750441944824768607749935011971883 31
2406840954560837553361327986863855138490935048548856685973271036 40
8530785317016729153480121272387843993085642582839235520861204063 80
7886313766647666748901127268668124769081013308742688202607440480 70
```

```
2012985458392847807580545959565652287530120180720046607207702 81516
9695890402119696218521962666438867875258203774454065469211895 28040
4905804074753958283994890607770955811130919033950275154802357 01477
8684849392057503863101931513367650194150042854357007664597636 52257
3134298314796005933975296052923511789144198796043467451307504 81735
6440142588552118151336365698610939285522388093753284369066914 63223
2724917609095213563467268500301702972059631676024592156521610 23789
1395988329404368863326246062688060041792228282089364164713604 71365
5059884673441672522668295384910843510209690845289521265699614 32272
9293631201506812487564399110898134701278085161114208634654660 20530
5292949679585278126649020011597061232187652555161573137208887 71893
9628244918164446951626568051793234050543674831757991109423614 66610
5336808008829714786307831561097902407630482231876573368504155 10049
6479736509864973645307299616334393535494899644272662982655606 906282
4233467947911852552943436536247368928961677039568704420344591 5600
1505529848937545880602314633052680689278164939886807934318018 65671
6428317944250337221324529488642536836133397236603217151309078 80741
8880229123956141534719700230657537215243705526543118000452967 25549
2570148348786773300340445565204837540378735971179160861928711 01723
9962448311419344030872624960523313184682526709828556664693261 50064
8484294827411191469381519899298477169623780283165690895115227 6224
3066375118087024030296002703983832784641172549449610110764671 5797
3573229578826434875364603160834297198446558569727361013436142 08182
2763818309883310333795484311239325309560062951645381939277039 52028
2550823828963567512349347804788533231142661747442793997566400 35448
1277322937665220503137132528669502744548182985159115851650093 35142
0982077932914601102379421083899526600461944880546633944713189 24288
3977496600263870216869504395616552281059644923463424285321788 83867
5812048305672615567767006813659982712271968346782433104135515 84029
6992545540991917060186792265896532888078659219170244910102184 27122
9788399954298170895895554083648374086733052783093398740594239 71270
6987667010476907698565566587855520732304233572031076868640128 65448
0063175235300231210384911504653557959464800523718947106279076 31472
9432510200034681033566198021576420397705140340845500712762009 39513
5919751825872783562298427975142405820134958367456298500711232 31810
1504997427673348945497469199291296797948905351891258439903782 04617
5885904900747966149896628744292225484816692018955456590566308 04143
2023909590984126260869473448287312645131683314605527613945222 09679
8867846246386007525533283131222641375263552564734464888375709 91509
1312358916402054878907807379336762856780612586423443185731768 29823
8513143057430593853508382604206789348794822891994121134457428 23452
9060296788172965899286667900357395093497271433445244791341337 29706
8768737191118469420173635901668659310576941044542761529916361 93461
1611546705240008716975067040449097611878439938267854683862925 55649
7548777773857131148533482102624265275345190199525836085671960 65957
1817757936992614885618453977970847993910917756265877032325214 78759
4889888355481894005812820721843035825940517148391167345996681 33008
2965780613110448718893118146617532370408917207485890258210135 94770
9054176536441131962544973782440088681429419036041596095340011 79206
5459845583692477764021760366071236873406013153580581994763826 9814
3043358341506925462032515588907291758211146826477894738029733 88942
1591188293093734803198271452094131189967300568899578828174715 16064
8260666803109911381989653229828772873199254402981917131436797 92458
4577401017077966038231499864966320123984259320315394825913674 29272
6025444206122280057290168970782394278658321365543074047098550 54610
4349681775417294654885914809577315131283330560904335221208348 23849
5481147214833414637977027720192614270241694489628273149921622 32210
9980594121362792440370993118269514901400150212556706463369436 03815
8282100716121306202791864909929327362204083083111864301442759 30232
5547600424247332130706797303256426933307426331277446081150656 42651
```

```
03089988917209882189682696451430030023681424424075296003216806187262008075636770847041204912774758007743784078753012360942269739348677930828118479949153716439719788775985896016953877893449940762556681798596481827218914448002971226029079180345568762034903393541399391045528106334628385629149975723830675627362550950407623671671214810712824019804554454678654408248615045399632832834409540095543259895220779414172819376650378594059270096084845969115878432099789537159480534730583598226558408445170479654184673186396442096296637654181254507418033222739148180332009594181371000089489096326204498766349791043338466193462453091696757833099681801705044533763669583679801653898884876899402298373424091742680021532185518694764133712770396299330274149263616270456717183644656793428012962070606858052084640045921464678329418784466215514776069751201938224276163988675941520057309432282716894380971048947714461330980816540346575212055799657778163656676333468282174381558351889159026031765193393714901514929850822059979645742797326160609454212441859999217346056774460033827069316370467345948175197178401724101521447634637296654345138145438862982526828905113186911935598278482730987550004311617192866762009492306021849026166314745381527686978775467873495480329921986701924170198812842823694769468442773070147996417521724884052991437595920129510667739533399853068290389795279289545691018948285802770889018877404634900936661051389777494542491944012205676950243260450598021315989515629427160704365346603310533674663491726829224746595009793040298671156367034400844936731707688305912066550461939371915558062938639407884457056429286649866441300432907321650997838646947472571247137341169629567660669944601143797558742722444496583860219008048638076577777858121182612721704734933764838254433109108290281092976361977859992509293708796972408615177335915757625339022492412548124295287564746245059985081428434598618564193167623301426709823894682667968526002596103741809316214348488630315457244897748103123138790475123528252661020414777834861228125697054848111143317208415084361228556816802452776898673154093735674703671882110032490603474956899773822369378015494368073309547265824869246899264443141070564285962178973631070856030139462203788318173285182643491542381475651682804880400247683821304849075144302322599242159066161806554977671389053253812388782491915017966615001807443154382029672778288555590446082900645668645746001641143129337755059083378484590053041010134580583708770623051305510767757405554022008223889139661738743852347229736391629808806123502329107227891840273725220539579478703007189547090406019734336014564714548988331702506148056091577148601491475767033498188692011577686839257975579043440154021056920541737492690939085432465497693674113184540945686511494874234405229905914062572451195184539644389098901436310175867988278331946888818624137078799183451128814400172637678354709749631321456734365199799626082547310487669438265273230835220164229987109110755664350156135606585295632097428454246003883054972372531891626238330784364343186479892344264320078706290911374349399021195159326355252822201431760445014788678731547309126588974133987553001385760566107026014551699537478517835617267099623491432622488413113407625070061586816266121109164640223154057599759122063022710308445893972193195568424244483455007814907772930562806128251429807369790036245253363036009601674090049832188282317510439320312807621725058041069061720413735524029672475278754884575142449523378710287470659116598780019264674424396020277797728831164723217173202259621372736943434501299335338754291118953157560097429787434626520672959346333673393500462717419575356133821972980759751202140376591362664068491735641385928892480059417820718327242456852271243880343848166481112588332557361351643796006561036993589355106266651752257019126703160858227262436265629972376212764375867115690250959993240773897380315800648089005088739679102002575298203312123345751500863043390765701463184501391238682832596400681762636127
```

```
118071412991668214064993488794398196159152595286484991062040630011
617069958912558666769526967189498065094855000854727297731017409503
335517450247209687316913521800245615702226123766794269346958308755
497478160500730193537586687848882733316337692814180901491975858104 3
182031432646189479453435650840413999605194089518964079986209283203
391837883205938737258065784231474624362509112417558389233445184355
396967952454397144558686462338924680126103612929423518320022897687
847556412051247610262086143737828377892285553271231903687743441635
901826451174820720924477840997342675052698404119988776635953457 42
702735214672114806026553731578343949095111385845304057956162793563
191132018035608059290814797729852533227643098921150646463781903862
875321031123794674956199291436228466344165812282814047226183076 97
364594917159091081498620213070692327624143642470885242245567870299
330972719329597626194446426893226024500992879218647913074376732672
707978058688221297591058667823457101624315826141327323707333099998
898371950065400292225992679446990728528056617834198966256281070152
550729744200900737788782363533695301215521800606949403283161225598
920770310781759911169617198605604253331205705333092653612757398655
859630038486754922266048807998165383061342576281067275731859049 91
759990772868668880813071095226140065229673983912917793723769303102
235354914161269570244538500008524083373196809547282152754985770312
581954873596106210731948206673984238393665937183942196535039901046
604614362004631137780678461451322761221738666684390997219115915278
208469164372225609305365797352365029332204053727641053784247599347
202250750277114754096372938888675566023418734426074210168701098610
008661322719003405725988006822053082971169537393254333616826226200
415849567587787687396320466727767702310585457770884288975379385284
134430075537253998821988438776233022935894387543509851928800994326
388859188195344241911694541715476175940458793473805199413412546141
240217289988094856635738516831632576272639005555239024980284628731
846529011547308063980231152227275469165075924872443668274881475446
054053547072008959583576673203003859484478911430704588605605691908 02
367815454427983598517523180352589730929426124253398803752094017 08
258590640825586335357415745782275500296123420884181538024530303 47
822532301123531800958836160306752272229782042671692499599357133633
587209255958228670174180742665654334685615441771384625498620418030
112419781446923516869247047162101672392697983449624950738555069905
093271339203621842532237096198767401098791914981683374797153495654
619447833134004808183750966625473655882130819225827659145317026973
485925525610309886593165769641403511697438496779354255212464473675
284454008781291315812112119653842288569449397816002658558989822510
976889750383776399012502339548245271506474730713887956431333895624
711872403059199439537736792461906193484724035933301177625241439458
340269060784065363718329365773730244189513727617241135926596317214
392929708572341757294029160480416865077065363227941496564187382426
688157415390747398861120200545088959193286614360809350483240359578
043438455783408203883278172165318680548756072423232787872083850811
223778573483937462725804103702777296181923806368048732291646665759
224000435211274293331644425810109561046338346123970241630530736007
195561614768221867316024524950298552819196249645431360454951299629
120138782512527506636107470545108156174498800507649304843087688157
233096221159123675556569979628645804155181893311938459320054392513
409069407576322413999348616924269965630167135831908669957851982 79
125300909170562097583605265105176539538230943434609123090710532 04
430543334892000829053547671195669356024042153733544241091985825147
796974005293821074507538830213954101192023743434596880569731524363
788273839233788407335678819087495287282333233681382044614692579 07
792859831193215133000401750105220542030394804615028813286834409880
896870055721219293906539808685542359893960161977104472429553718332
651765552381966663333933070303773894563345383843467864938436020049
```

472793306548870490681904360676936455143389861219439103392686779447
843843780388071692478541440872855549668297746658300965184122172852
115736310118625447027207626374929658379201462748949906323885823350
140081917931937784584203825696034706944766569532599597415952683206
737456567997532200757859243512264941954364826555304627658022030115
110130510384896526671533083759663767910754585954087674271905315848
129934502097090816735458753196764701327055166065781466216195798548
846150474738026504257219000536127914688281182537680659658975697183
538184718606495312356589376946599973712281293917392535719241964425
867048828543241652731137206816736703570583730497026353035021528182
369767267953461690640750591634626760304805112101658283107769849463
095938379428244827568691843027635256272621533637314890322209245182
918258478415637457416878463898321436936632543741300804055614687097
757425366723047454682368775506685019359590977545537689901383031112
058238455965918120561586685036095204851818221576794711897658869371
992237211658100759737344172239616773772709794791547334204322903244
425389634582076766767196560922413665086850044280197108644379874126
659418079583518409680596703961874915034665584320587976838094230748
345386664526087880538704979674742592562218840821926013623951205435
296521299146258552924980035624149853533721170571703750390367858186
184907181442585297770728769183200284433608727922398676649373261176
329280908755273765011531756090031459929793364618227423811619152950
047353619719525397417023732747985843262414410337887932830067
427341230164327060145079647536233138549050904587367866404507462785
819625818947666745705829584329486502941993037540602301233194439408
183365831204601857880579917484298277948176006334804849924087834383
650137525191327765708600096913968844579211034502372729595394447800
627842437702815794091556765601915131479930009625652204760387686699
582839586298784459304701402369593940875762652839035938488581394344
616178580667161204600141088350841466845732061829621301363574471069
802409202354014921052467944656094453958967877208294918090430090452
978171425633830357302250487310770493378415368506469831234038187649
266812349214041914526662863946919404467571213705021918983543732005
365175744269421634442836225853497350522800650513364202449631355771
728033542301003988618983828470852815108952304866986096686567148271
305823544948777588490152569331148972752793189449014247746410132346
097794040067586012399996831496655943717743096112277760250432328962
259198684599020336791582353887598108795814444488200879242423042190
299696336525795875980034878647271242365175018661980551334285103636
213477379410701739195520753540306687125146659368452309423054523271
954175176600157962342674057586077291648593263618027739050636398315
949264471326925724345210986288355953031143138471578084168593134434
366222074444721102453013737557404128541669152813245405632115879496
076073618946995742276501027304322658728412587410621634615438217996
572102185144259338694510725939644295444667813602272242384147673621
323815616957064790706926704921190001807923647133652294081491169440
146512629074149735243518739878508198559976717150368718246080748365
285829847538975347472088453108596135864905406481196418607950101732
299730780038242966880541550774545328512593137289396369202934073267
550052529230502835283550456977790472971705113529819122283181048349
333892047553219911420131453203154849037569229787391703208830935550
035165278495437564520581763900043048629493902307971403039200503953
439667270661064211619780316817216738996309439672896622458660131649
481643911317380543642116013523608193902897215299131811748944333428
323796416075154958398439717531260969286560263451447903779467717463
151511756829678304516622133531593181530479979309027971593842705728
879892791276997461371024279326834739249863235480265747468101891894
860690727509485032413684630341830714217308096148892443970285472395
969352937726191666851403842181586184898941917170413706827026244732
141606382857948147506788893527165552671898058329251710860890262777

```
06632613714812002332208908867255265398362652361564616017952010721
46051463850539636666058424304770159702700529638223859329438586033
48576903306780302355222878992088767979704107361164266601207022535
42474839234309318149172085894123638239482205401306543100211939272
80927293239090662017775023986008981824886320065291698638810483079
03363213502126657245671212443408019269296442661064754237970700627
87844631268265280135636652998599334583070911465125244445129645980
98325834598808103552774176693028471035515200120505149480676023391
12567708918632662852689477208034561854162998750976835867487831753
27106731523338980628718354314570128556922652622454672853050025717
70493026824067896213424404469437343905893003320477897457651445153
33114206706834778509777706975369712852380523673853043722137253320
42709713914808802861291321794559942658531082808832360197770654656
82474112482745410833392855051995219140114899367306342212128989997
26407200208255296219852667343932953524310767672494842538881626070
06553057628065894660202543598159425713430369527164450071434863875
97120002577694943811133009847803211010112946740804446799066599384
62734573375824805363439881404572540381295925820852424453048015954
40848190736793068639711107737431751491496384156695201989388042902
11684453825613747993905103373776794494494726422798658365829677309
03095424590795607486211216235246136227389650016917858296794789126
16255649932061134798626459125290831043060051827882743336153617482
63772389150192518049605445353320956301981745865864960686190570801
39283634968284096999476734117653686392835058972143465207349763505
60540047475681063941924779830266479911137381591552749424232779733
51649374760030277150998014828890284476233297533314314212579868956286
68686032113331423235799269223922188867956243403066594768486265390
57790692274669598842041061094135971607588159695984945635935573718
16476638209111555128487730968918695182644610488358885226226100301
43160147743156350812357265545719204018286017536117949256412612660
18987427448111744532573254817639192883820204228127318713388353367
98523236793781929288682700344463827679107652281317917108206280407
85659992434121899727171017003261924288600565558146435130547976379
49973992014581013372539244970475350369076926006705070789509390651
72958168902551045092345326258993048873471166932888283719100564813
06083403444098143040402584566930556436403216440194356041346738525
87148655520362001081854478857396548947436797521773340722044577916
17397065530917014720501744971658641444995665261098410439525042129
68258090829863824014279799516701081624280759171910962574708680147
31192110290031735069725403080129141896602088649494903970487760101
89562227140199822170383386212847158602456840519634270398250757953
83376410160790256789841939507306187648062529057642910809578933458
76914859654348726595657667005264607729517897277555497708728907500
04248353256387146220209994164397105721994560111135277993083527720
76892274665037992133728763470376377968894729680598095745114390957
37594944436705124380854861769345612812029380708089867720453671697
63611444683583796599786924503969580100742829194443594992952400438
37751274789681866860842365933432003998703429627944820318454545079
63962619432625995623778090979608103121163452900523678917826897455
86806877464049615182823225919738140750814392028096544424730242217
57386623037813210669128915504316144988444774037391673804979003456
61237046378422362316079655507077661875288275030991842022336503253
27140300959794223421173381758166278429917941110619213762094403864
25136051819587012426255066645204227387518035031753445656816930429
38840272709880304257986176064270476969299029087042034428521968329
29917766911861301175400962698214557445392491200074898737059534300
01591262262619283169647868711944334681177453296587568777386908375
16376526808680256501952009517450303094156902794081986107690031801
55125152774786419917019372999776022801746413159033745404822693315
39114961980270228259837442741028378020124519804924878164700358208
```

18287524177156594682890570965388845038450029560719617263515512607
51666612207896168952213462659880140065637447888383836647546922565
04259371271335918006087225358460863137419793213448508088066465201 0
92000102067284519046912794314155338579478800309959217974857822961 6
57479926106363449011546147742162703854325978769820758271944221596 6
21318623501209026537932562380614656816627054800114059357297227017 6
76741410145324323059138257556253739729699535232112029468952047413 7
74413831190954295924525037351583765817995316798170468947985626817 3
93182519914782601026727735878860919069000499240628313939212283533 9
18234407629439188384332315248406916047343301084355166318175365069 6
11054181565372846087871562430178870275112050612911247716253134286 2
72055737386877870715274985091724461069922343333681862396422760600 8
57864297909168890283912543859252151045995363287728232099168750957 1
71020188975738660335010361998675299393288758722406884362280168537 9
98463422929555424779532714327766611509065294466682848554487189263 9
84106988285519449639097128520737389499630428324859834528827027005
77266026245164627730561269316974704009809384155303917203969140759 4
33338459973782923816468155965181268326364991190311161105092289992 2
76907244921246119909241007360183932926665711582235247666855720205
69424409727133965292240464292202796678397596603296452496500875548
10986062803000894423764711680405630914843226027891115616750595340 2
32113917524593163653202865565788574154401225931464802354533215838 1
86619343272374879053917130570832999533570232611072518687395125033 2
27963955687873596766438808282523202556166469207328790407268399078 4
45497676077099837152154892743488566842212805246691659428745886644 3
07969436901661432441863644848520405444571033178713575858322065272 0
71230319972520572383715693241299010143970375374363976239009143103 7
51382942223182221453452717244718183839540665913733695895945490922 9
44150833111391641115363073828034737963062187452892704182831490167 9
63447427772184570188984055414420195521982132492526180821224300942 1
57848641032462874839685047807780154872595720284470824203286079793
29515725950132269161956074492367767891306180564982135493140868282 8
76070306664742763085661483475880850468035951617696962521491413652 0
60318970221368392658911269073111732624589507208681682309007277680 4
02991592738011531075688589893577447244700068567354929064677389979 7
24060030425907893649822992030792899935849094239340577725720917416 7
10243108663722198297672238087371186707607182128007889211491602815 6
03253124940555153659160474996080742356016473417815013206197544569 1
95577376269993643939393543527573911515038327594895717433688475355 7
74747109068105501505318269153845359450668931243929926424160812659 8
28955827119726285629056922748490815672945243369676693445243715718 5
12585148772520077919164006264937721940022056000130030659100675461 3
13372957691609046381452804907693911073555251055167356006276441334 6
08086143893774626897346412974517969490982802790447423777397970060 0
08191134742728823392827045455225160099019251410925560857348847833 4
97397643768933421515440760172543557436755420075239949280322112076 7
50287872980304660339271584927980896665776332196145753134971651251 3
56038878625518641585030528077292398338942556349265624101289799231 8
40734893644791590714245417969614041210283417450514654259529800905 7
35716289966839739729374623292703407134932639361737606368874621146 1
47703322243387307806697265398325970090741833743717138803161312809 3
76672375666826258999459780900895446787263872734037954228484815010 2
06986770966769668039561527214358658722066413052362681532211976324 2
79326083255643214380483359918081141607805648953672954159470511798 3
24107692500875038466535956476416404921481924138009876947943995588 8
96839349956688781008859566844980464836416042396982372176172977997 6
46171276691273169765126415383581144023768254482070551189900872764 2
25631075793948250785771836970174066313727765177355788154557923448 6
42271513361259184158937112308523636873900883814169786573112525679 3
18689499904751510178139206542640443597576654728293587431065907806 3

```
60962012038841615015275395365865437968808171111519616648267214 6477
54886303353095175581023689850716930069223019374053545690903106 1839
69918569138248359953855554959073641071938327353376040340918242 0806
25721359086529906799856049092868834539850754736770880788425062 4167
45808453349038605133438162000476400376183293421114487510336775 7696
16366872627544009743080491034745388043049857908012384385116432 4774
33345274079080360009231224260222950948426184270942076100178576 10341
60127572178718327331683420077783538581677057467672418521673773 7695
14187272275276541201887001575331936698407749211027392715343391 3418
66051988591723638118091980530761284754099544511656790671180229 4158
47524496008377724381912176881931842419387657604704563675001782 5276
83839056155634823274297777335099544352130573606823372556934004 8300
44508461250949932943450491839242624760395360495372847939479247 8566
75554702070080292609184310500103527495486577555211517441565608 1505
16892882387766585662856106191849456228383022790452263080802443 7000
11894595372114620990738287400497728649828827614409795036617614 5499
25691483466858365934041728616053690662989816493928521537428559 7129
63889362332583933009636543414179142758064122733856276441488390 9068
98088262752182544218955651922578833382148459675133787928510905 1589
39228271581670333810299493803153239515067251941742683865144140 5761
40188357958217499113545790171488500354927197692953976973631573 5460
60892533084866950396358355697588715235737995621894370521141177 5453
13637035399820044668863761194293257246183862160302753338089294 54338
64039269900601416620715319842874374199501983585958610692585449 281
27967426141379595579015556579699841108749818510544685114080020 9528
47097403255362229578733188167181985343616212206715859791082012 8150
24302573630542713645480484047677747680523204425674527945923568 7425
06232433349988711978921529807182254159471411814054831641021018 0258
50478755482474902891800561240286927094638401609705368029047477 9010
97415234870107497487157000510451330374888719306055176040693644 6698
77633884685334179994728966138322938000688723580118527522918479 0295
93511588671080232909549696282815390408153689094298181499240274 7732
34696878229251924099381733881979999454178919652186386982011488 1746
43627951798631501750679312158502592350509233589768370386969447 9707
00738229012235760671088713766817100169500729605626493086858164 9982
48908447374183630040216121132688449257698315488101449121854386 8852
95032727597295712871602897278642141496231201575111121381832454 9492
25099603489315704175494057816646513473683992052842857345817787 7120
45580431420154517161187297581903301827086427288862731829489188 5842
91850675898890717196834169367938621397706654407526873777434417 6146
80277866480968453476548699725211843754887968124478762560934848 0781
70168974333617770969302555665691644378422615214271976330993872 76712
39088122794785845643354023806489833634664579821132027488086374 5236
65861514910350080305645170038098985648678874836509261634566166 15482
31363438575137302553113791847786049191090766637758940836460511 9218
17509993397203244232129749059433924217258575306727488241279551 8327
35349421155611672056707316130639429531303467085757519015916966 6667
99678388439045692572132962575668208621122772560964686138766291 4353
28936732258211611175075414821532521097681130388072165351202862 740
21808289279349839545269732306990856143270496372291749340178559 0954
60388283897492440599140230548981465735472493877078758496665956 9255
74290839147855845985325306186520880040735463498018382937399588 2370
34975971827627159247054948057096293587751497697718756102894601 4424
49041982190858550014598126290164303185186602675287647780911207 899
86197456044376668422649556644981880577427659694935767874394736 3524
37947494193125887812321097765945411613433243898457184926457592 3225
95552996612402876669125033059213072654165659743565851724779662 5915
27836818574127144719265552001710717333249627178730102089174469 2744
35297112449301209455517027325716048205598270587984362438199877 1160
74940492077738342251655753394050047187706191643252859527064271 0342
```

```
7920417280924264726827650598115785761736685907670889621376064618 06
0302474157528604720125520710440747293466576604253971100743608906 73
4014494692106569293069147085176331513779197367455150511905643968 50
9877770084405503501410784969369565023830808473782156832444833713 51
3445712056012283435474522448172734310527169795097153942770388609 83
0795755832584653120522249731040859631155412237056561974043107510 72
4799767781399317712957313634678946433868238389487844144972833339 32
1414109117002337423715797021010718886292271110252614051384075087 50
3960832633837800692357926695052210234157791988574840926255622190 12
3023744093914599967195875559390361221940757988887304689639949224 84
2548798177338753330365424175662531734794764898798291337578507411 2
6691777591751814577181657201298773352027561996581965295270449558 91
3092892677840957785375060716158583038907090837285574837824130364 28
2659503272805764109993314100019832197860195435912418046590193643 30
2207914675707806565707070530523904278842058406612328349791397509 72
5346233365853841292202254256707748951215299096281943810299123686 87
0879277115246897761642493983524886209265650386378735754635224677 28
0576229991561001096091873295464569674413409692097819054089202362 26
7277155738485875863072944395339774123263721233197023379510584979 99
6843535583789425092021038481277203527200639481426012884543869643 95
0452138242400353997872189182940801046861079694387748982253134578 3
4856227818276543051404673520740517764718253898360090898254805703 31
4166393455874112861774465922369276716460476748861071609657261837 03
2635714152798173691797419907686519805204533050453383525359422299 08
8964546653993641205632314624428167690506761744129172367062339151 37
1937253010053037656662310252231169559619645589402988855946706959 01
0528415243922253946525819863227542882515386954607135060239089866 47
7787486908402265804623134622970288066128999368172751743426206607 23
1629032149174769457403471940233940189195437974952471416669241560 64
1881236699425382648469632965230764497785637415655253811258558282 8
1658325489104473815024506295803225303488254291052138325574825829 78
0293642627103397795275252742521946985868886434441144817927865691 65
8065859707723581057941603089909551547874463177887209315259476290 48
0423170029056165336818382758672155810057720171715532442276125876 909
8277076116584181048702608790648814931602028935602181476813083304 60
4746462827318586808555823064952816387251619649066098633711864237 621
7788340644098780972277392684793302251906788038790690055036766492 51
9494962285137302502418530195808618442388819247828799406099200321 29
6249236669176784338275465421455556860038366530624201384647849779 79
8320896734034874824806136561943668855548711678844788329005922203 65
1112893328675696885637680800201277763136397922527054785795031860 60
4532581739672063073891186499061863989422800817305551815542231259 48
8050663139267106686471195606313231430999331123179922390788057692 38
3069993572504469527352814236442227650335374467893826945839603233 56
9570410847455860262706426106386963394749527039296987431831668063 44
6073550010136019568828180703285136397477550631501283927518760844 26
8409891483891109583582547628176289707891846641566864001700607218 22
1265745266317094806349200360911775779661803407176922524770609021 9
7367799322463306132030025953905785289043250584407584681618654108 33
6407800120740956041379456193905545971344593686896733860801813883 73
7277578344810630255068475089360676701918333055858527933381468024 93
6138763407591868271076819198772534728607574876215710017765788527 81
7566638553177054812621212475582664508666248264115581738607278036 39
2921008141388674907360116509090461417828549404109950439754633975 08
7539505652488445366047153296235652626990633949500660807063203190 30
7916865845496514997555247773049525856236149292879363785781953433 9
6620512768790589129193510055275745724909562129338278442302210789 92
5058479712991156687996681466195060170097869802116864323329286703 07
4001397629827809751927670514733576584471291954856629381857223050 84
9272230897956541964659856271869850588847510715837786692100486873 8
```

```
67228629756223689468633861222332640813852452945648933208953889013
54208196128731654012109386023881091933488247172075299647301627507
53285846603540540913763145815056153992122254523936709038263168718
54129134835096627046122370512203894715171857226707406961776056040
23693120370189803401977904162427599325082503223237413157500182246
01925810022672275080415172887522490596324982502759047405574911339
44770425122608373558301173125279195497252279410329951350004446937
26136559807928391976430786685154321579236077268528723507290407292
00241594769610868488015536265558037088909794257029568134239967562
13559614330545363818564583073431833479282362168172873212816645312
72284418341684033891490134125462797890597700933733887626947606717
26342798094180201953939766184790823280087020177216633756689624996
15388435053259510705027615132287179291915781664264555747884742860
78746366509253299570216094619536757097996154972497266250495813856
23687174939123261377715617827974451437015059235166596508039341350
80774049593830636346967063036031042835569050755558006105077786567
34964265259998259971123803733142471557559054288565734382299271549
41289124595698474393339784372384086761170646032679239918010593347
71055346054448440684115309451465605450314857074228691013668217898
54088354585684729453374543141225809201555599533266304061326458005
00924473945467135806262528552986460282477134167274648927316452422
88153759131653969710972089709283076067173353906412687788860452947
89131703664571699354191523535634353529229271682074640805376754
95991782012140276202661808258148969675416681232699776624855074432
74305569753849281046552505541758924984811566028046901539663315274
72597067037195598112851326062438439611938858671393205079284801004
56758232859585926231913658429896528903607394151359844894085856233
02281483753539950010240114764074885140694689048495009694624010524
64767327615642446651627347074703951274390859864584706906638555556
60482496382004527254963131318899271644682067916146623441901689852
51450706990441511389495295117100043007063963035691197468979039275
61117677235340548560996849450457584649565004786611144563653291163
96063880142825355815681758618511578791367344134067333542836690031
24854142610647401718186665721176291153860315098143426137071546629
72529161130058939148836497406903170163788542805518599636858415132
21858567532099857766723219691846224568828995123362978744358460473
44691567835752632868833434081982434253280162099264909000158986326
00920023413426696583238647889375259161021928753910172460096401765
88631037031300723187124741385334555578210877708369771576068570810
60616291076439092189165185745285949138277725176782128572142504945
42153239602128809095806714845654417674630420222353303742610262376
17651966743589733747840817272201777145350700133359778456216134851
19201572068846303542807392221688181133483538168982506459557530279
34930431045190881894198005252169581741134298961410156441127936742
97467479517159881791727176167226060922790523189469353371551690150
19908337820260222687878137062742456308065618207158630271486147259
17568846547398183267708682927148390221597346262613840626260844304
73834724974713723599890395777388586577353266753434244436690611983
48875470912655860072447666835605864733024368504615401554892320532
60422353666654667455350092919547139214367099091023702786202272116
58158988179050119579257303794248556839994081940441597742822851649
18922524631821811629248015517992262732462890170780222348886171006
09731546150045459796989790558349441427626436099187919623585911822
52761349579412400467398923526719622929546640674823891345524362554
17671452569320856611889620650110509412379935468467989809696901668
80843952453662970482527316247357300292460831706621462574838141147
61674729435974961076660975202018119144863160831343056841730672468
49486324442135876460903277766297812680837019813313833821472729930195
64221273377159415201221898391747014628467729736168148410423928200
00899418570178159797195875532086188229529127378823962645411684564
```

```
4196822215470173255351029859873200666436946162513669672111374 10725
9233871289399517817271324594344470358391718954106617961969164 94991
7718743695391797316675077127054289003841324738174197370155402 74483
5894900026943105860017852377261818967064250080000725209328666 24701
2560151362144945286252008443862696754256051302822803586746610 26082
5148753999560261964372367692692664220145021250837078886601609 26950
9232514498397593166530635557944713639811755985467408170857140 98765
2521121885739785501588335078116332837661864879869890495706039 33378
7416353651665851001558013180692641942632611828348301127800810 31849
8206601254231452474202599666013166179545052687500253085363398 21588
7598655787075326127095802617192660914906586131786302477507892 28631
2064108315943312003968443645913246282970899100580828921125962 58249
2994251126369052156458700144013718177456303197754707988563164 41797
6585738593965922917959829989610531530771174800860457189869728 05951
7674776119773215800450015851656819395788240436943593627977264 83214
3400157428823450923854402661395963771249100592822935544246124 34018
6690236872543563739654936819965611953988175417472058180586517 76430
7344357635993219869075801906251670892591733904861076640125256 81206
1198950710265101240129991791603201214187321983898566840693439 28939
7231729936790159665053467707931039820066977856830844997649522 76686
7407299929837637385438617144369730260193632454317890366679488 19410
1183075521352965468108435672070697127636377153418283035893678 14195
1912032575147639902304368397029313905066723971948369204889971 23039
2141776165837034460034085400570638123216088975077351163237027 97466
4517314421950478674536774092607809236458906888362963209234272 965459
6623333842598448681160730811200153940509108933556472603899223 82106
8441036858633628517237147621873741647175011939792330920457827 90703
7360921692468026520998397826538831503839561894570695217943632 7504
7384326419974857326706367029429897338270349682369318424983518 41263
3169833836205165709581828201223456886495595182755397779870311 66691
2163748216053464866633748776649206366762244649286528837600042 00258
2962770006109944226706051836093617546192139053882148296750581 55845
2792209544940039998401379985895270766817443716751108461083583 15199
2409785424204248397191974982117651475997512952930238109572486 32734
6349070146013951571251388002335308636136320159567923191735499 10870
4633534858591910379326254059806556644757011146662032627185659 664720
1643214494399500460697606189304742502582762033992375282747784 40251
8491909992915675855969921289199314335617564922456672329562376 89496
4711988526492693977804094813722918491870703375084163879290444 58506
3897873356014625582822808399053514275671692252238546138804669 12971
1680645946514012224058743785570392137189842692112777532463297 35662
8708465442307238941331711426219140089279664951242378912794742 02698
2822174117826984389590240307718204268226765842980487621206556 23793
1476946333668759015362209739287191704003623633658711409375755 13740
6092965982516793470741123359535868760772537633145073205326992 49199
9155941864800577478424565402405155991132004660404245007812078 94789
3268196809441212280826365172562970099764501442639746566888743 73445
6136072563086482343728091221536766183687572967744778475069508 67484
4133415424214312017217494517637589402014503043555010896967799 93061
8642227941010144945128868302026512623357959285230879884664398 85585
8919097125050787011094645235665804105608537183027645361998284 56055
7602323724977090704294924901513379188916077518872779382151477 19986
3304985163311225099552202138278046061483471957143658983331379 19633
2110852293317414696353664290697739473140329228669274024424781 40352
6528712645735617506620985332783444538645241121696776186919426 58423
2538035510349667008582544240990402852421689571295543258573250 86070
1776251417638211878436803386199366933977378588719017350571510 19654
9111224619519673697904511314383818366788581859047343970513940 53598
0816094899115283724422005428837406147713090664439648600570322 64102
5458753754787017021005970294264124385049552321132441528792599 82051
```

```
83188012502601558897110994733301048171238618047011275360553596 0199
77047235769000744540132959432847854804016377788354935396881818 3502
89552732450961581553489488426617225405143773217917103750684211 175
49573655665763613571740371980817240294958519187294159474475693 6348
94246483422192057262540512230756985887126418476857789611076159 5262
13614440088689148639682445082242130694928163747977494739363404 5430
94971715848314595166890019604769551588672860685367440537234263 6971
04761847564268364160422782861202051036578601493092920602446582 9846
20504461688500264576256524324964489345645828818718014797484283 8528
31875471321157253823209628118362090798126639431617330976665596 7114
32577358624014511377059268548732883405303478137986544151034025 6953
06674576224966360556766203314533006019174532010245978936141009 9985
35920037299542503002407058272876513677775509903583024502065688 7366
35144668154502156361675954648300801958268129988256046050232336 7749
98141838229437131875808976511260352957811338978836952543412388 7284
05184746456296903483253197063562645145738100700719988130539821 4825
78960505251386453577577057222603656382844333572063128221324097 064
97352726085708005706732060827094894607345339996549260159199599 4204
43862639446212481926856276165252752223265793769131398350930285 76838
26494990053269162622564954820653309888370449708469300670364123 7223
35615624644681829630894675164694004590209115767520773468775559 4240
66074737564324239351470501684985509854159376161805493713189047 8167
85816486918289366393070130298377270414904790306264125923486838 1687
47249703024760033225825323454110780258074864425079801508029179 4258
94824003335116394897301425840905742876359824494045701461599914 2123
67028271167156564450888420576534675447420568079303305209831672 1514
15548479304959144628937205992520970375275811718301300517222198 018
06509842358120698054993059070032685607084708658129901031301488 9415
92816402909591411673263574465185872169838421703038172549326017 7470
06569947581133248319153908595600050816820765245447893138641491 6336
84727224528047165769255889940975428052711106952249271743708678 8098
58244055060900033855348944629798784910885850453630618765754157 2691
29417727632613794179357281131264845122523510732076439093971092 0486
03137468286368084582964345854798027956507538250551933887352324 9751
54096206531310875105212306422615986598016701949982109931428782 0155
69968505852731282699535249377490312265675535857721293439799567 4709
09014642821024587068309859509532970517661817932934530417911243 4462
14883335533223003038938851949806675671236597380188410156517685 9645
85290237042394300863534216067818198827861966463280066378980536 1273
74050360387810707052929603074063673190211573179456834743924836 1207
86882565551375119960349844834866384984881503183522307055579951 3488
78982976899159408552038943646287934291477318895687656388147690 2411
21940452523306887582336074731067064030708814474169892688332109 6817
32972997536709608988853118873787862986256404734582048810433249 1878
67310909594472174236918530976839971134374658439690473767431746 0708
09436769914537187074282375434061754956069180819390145313401259 1020
55105089643259694012237058039176316522638869681071598077734499 554
97135411682989863078050569258966423675811777569333754572784631 0161
15772836116259934345516217143123665278292850967145521971145756 0949
18460864349570488863375057522930224504373374825321960215355017 6753
89630007929250617326503294968303677778804599913147898577904517 4511
30723178360077427769775093577780396533217997842945194957736463 6332
57329073725929135626975370563610818716496830178732617813222527 0591
81567585859438410773198076966587963223245783515636291753277140 2357
11755272053698422752562925796538726866327730823121994345234880 6242
40807076012841887695639291029418123354058913639524572626368854 28193
02759750983179870190690008745043189602039209402021579394150735 91925
06842610366271703764160919299158424249393687388401786722477161 1016
42355346866310943019859014790050823536232729310314512110031780 3553
68344747282772605668843209465311037392948032715711395920881851 5108
```

```
6594422099846069951203081621180136508832077111687105145877369916484
8552621489839047925851600891842718168967409457042242032737852896951
2937118749970151124971510863058691462599571781375076928723889765483
7882282693636241042073365595176242765404699094387761335124669691511
2814080517996390094661690327471226095113315219153984851762534387113
9885747562270140694845994542451655310266134982498354623020632289874
8060638302339113652715982437025978246319764625215086170016951140270
3953059813750851067594167952757674652790840120779214974277609314389
6385505316086166453458903047124391324726672036759684846160362042542
5418001774661817818664920879452822002644802781772467212895344885149
6926933693065295417931168293087579460559197460003671360808609007844
6779618827693047562956409037227650586960838652144963476708480148193
2146139194058943292842097894405697384450581846837458439434496621898
0126574052306588268511071788883265633514594159110618005103740501326
5564627270591570977419963426144420665743353026408500673515620025229
8417673304397282065006117588389085379899181782855124572800692721409
6576533291135336683619561000118028422614280313436575163016145684704
7965640028417199916135653150776551046994122852681452311606147657866
4927828761482941983370515513973237150299505923813445284370726051019
5251475656596493709717447978623254433282733695991674530567583081753
4188275502292182980413979760544210250973817161757897208851411554186
4648746893036645202642737566295547744303718361360118736576536009666
1557065991037892041287980347993806326688322020980264391783841292591
0681588225546376433698085908366162096923477308375999069439020676223
3013489409474075536831579193458195827271027837161635684249575227584
2810879696216062401196475066737023369087612627287863951893431030807
6781009692376867786457484074768688049462558044101018539289107020579
4241170587192070532888904733233781663076191795020128966939319030476
7435981948011933426065482201663246886932427265004004429575975291340
6187975652016724508215871645890100064405274378939677097174329909486
5642314055238369398299165913286343247308310285008160819706948187055
0503813440781792559970917119775586029139480166538909382179442884052
8935489836556012094515146020217006102528644474276854382945623866097
9806400589036229811346094404180547802639600256892635445485281398032
2768352495342678253023742432766114518390252734838803243746364464638
5469256790365009121267427483891147729574005999526288493463286990865
9117644799926329586871026762673118123618998079497966226841594726880
2135651304603582169814684676419217968138611128490802084634146019181
6226607207840284853665568019824968836101239879426604956200003570721
5216492836624872858644689788060837522412335353705731318980100822054
0757522459324373304919945190417484659417578698370791653300312592332
9827511833538771597316219655099072950362559861122855102964209692423
4011262084130893636647116222258436381049979322497777337626672095253
3359115319593913266556841020526565406120283601570660185614755763930
4876064972507200678445878138191384962302186423414360703057565645524
8394168542373837996422031179145552659275945440126686415677100588762
7271453681119139731106111673341706384902545333843608138886827530304
5721626859115693466563048846977560481297372718081166419635393944763
0809306011858866522027490429069557133147523941945661177831742082339
4510511610940632055138770390567042196091157445798779011562338961283
4515377787287944818998590340328978533941023147766469282662083365162
4957586745183591060698327018865656278298091634981845064405555291406
8712669022579563529498767377915041639980842957184839139886802330094
9968225770313086198851914075052971116599980860638051514496952534209
1017066653890271676076575197103228110406839038102903518948451475732
4707086407236631012650426709239232540165031559749210487441315193391
2477659132735791091361045005871793415815800307133380971849672696760
3307285402913656232036989999302145680641343591495897098805660187513
6761150206304520909921712252843900668632768790831746208711056652889
99998537625
```

```
9453247861249632534667098367978115571025971459257817367977559632 05
2845300663979403424889627332670267531561895280926571083091589100 92
8869777334057989223757123205173138380516118857174740214099086295 38
4949813414961390583006484828366251641698878886861622473875043541 29
6715924680792966385708621455585356421262771540106970469365341875 15
6138520734912544638578966797289597962469051103747155903601589207 2
9197102708140312965847886489440068663867715549446386532760524892 43
0834972859498718276335367239421170607472995793566072822298086521 70
9615729730892199505494836925658929084474631567530324168641742191 05
5340663256048771108931755046302577194732652665611947352246929731 73
6221604666980419711338732866985964467946508898014232837277900837 75
1346793199817323507121098078220476850258841673077031702907505518 87
2740561887550800149867128767210259042824195830771809340351395864 77
9282382000247452477232289914094153217771285226347975833685018034 41
9502909554112443205518227257388704160566910152840771093129164424 40
4285704914351415612572152821974877540955388420590874322770184179 22
4559797954857947090223775361372395474445078879605070870397592231 30
6104583265539118261051228373336193873022933584134682552838358374 3
2898404272274580776105593096556016732728689756778286096384312122 69
3295992427482902181812563301210394724112874802316259561246794086 91
4019621867625011828578781369431503633029916842978305195836816821 58
4373597385269884011543591975557099291045170398354112341886961319 08
4843794952445850008133602798993077578542126569413082933201619465 15
0702221531639840837894733654008318771018000842361095786954717115 8
0257038605190973078239824390403308275283147368726183870747475013 03
2061893100760419744651042171771133607652739649050156381736752489 87
5981862430990359983696350731804984105380838669642412950924639632 97
9775674793945222828033847029864480178760939489292186533179187695 09
7591257812017665602105977451605562434734197584593851282863775712 61
9552679037470324941830547463860645530461742629359965208580547072 83
3148362852636323318912392817689633616271301701300733097823059199 95
8065274893239969401739271741219049178172314561891275149598163924 04
0558142779079524180611362379740136616164385749234488357571837751 94
9705291824984977614566435156431019074189714405430099632745038888 85
4398616966523978235148194865600172101217771261147578878417948072 36
0308710760855531060198792690229123213131777534653173263294967724 05
4355679754865843515201813755369139241644826433743762894707737183 03
5198395852815750041113287473942574457673642843642199367092178812 85
4455031766595269499960525907050241940423504961430398205551054011 43
4968165703959381865673597623304696348428080520368835369618980701 5
3062731797693414158996907700413347370531041700905873936115679744 42
1581552318559778503179004339358333068549086243449387894114427664 88
6393763177875646399184013944026804677152103685734075189255945234 22
9594590264547490601202578764145733766978493347574566733427939298 03
6161761191147586523295622468154793461853471815100202682770710454 61
1954473728587420595754869506532180530937027980492952388784710933 59
6537980800210712534296380497417375165958760005756124663572016401 30
2813094681140665880948839321391000736800887434614866973618704575 34
3839193252010846280690416953518226191265996267844273981159991045 95
4248553039673782000277264832667253396787591103868669138279202766 293
9046451439984209955212983102615806955064148154960503110649900004 32
9836407195898907172957004246072217719434318437724669971605332746 40
1721362184262263987029511074905351759084397604301535065761532756 50
7612847298451631076363005595480608571361150850074839353327775214 65
2496065273518356690000769318588735064411343035591391900833315411 29
3327759274642811048235844206078768963356691668849177000880244503 81
8509836966002837340686739561284441063454632170289904332975160440 05
2510585735993764583588312329705892516503128000813007352253847563 99
3047470755378978162164930244894780420716716738589319625740419045 93
2677225098357117635901874119875467477193497848725848124310529079 74
```

57050371268758617699718001899524145763476050513379496499006799189 7
342927648992642624385231716833791266701864564242254135130741009631
400968922426311546034367882234310192905911948290166593114775711157
19709485842239426737871026796837136847633936108581052607740982496 2
935484004978589584976384277330999496291118281084455959235749406 15
5680760713584017084952794604281533633724800847521649724978385318 91
30714623460851892085438023761112125583148217593853985793569384341 6
4477678649274551226633722390200821208069115406547695525473694946 46
891699897614637499017787076852022196072357176838862774251972575257
54154598069204229968547017031451700576483878086598543897903766534 0
75137043420981496326496838708983079866413490013375302241040425232 8
37256772209524233586031199463426570482533650248916142635082198845 7
4886198235417144652244938192071286517204487884232190058072175931 92
9583859634350540587852400908497577931637129658135437535749465860 11
4342194896828662777396720306627647045336261865727219949014480984 48
191883902594849516319598999314623216108824274009936069880781123283
0627956156167038804907484329987376434483055908691255081860164859 69
29585766172692588344008728874736594460769038347531416980204738024 9
184203567701516941149212813136758698787240596064977481636628030711
60382394299325831153041189781596298560507564488240699398976584652 0
793514411529244431077513909851503625368853818331557779400186661038
4198594193588134513087555269435234838458333282127986350323103414 08
745626246149487480630603790622259442815783585843806095490630598327
96347916722845185693932325099591413244383842817871014869757182580 8
3969957746736720288957181965647147857152911023502644311900032682 13
5013033251038104630765415735305625309727717700231471171799713708 676
671053841983174154137023718193443366358243316337810601872798827833 2
06931262704077367421320042211361510322828403308495494947289131260 0
341055619836505674161783344990459855934728006565954582729328138950
9392364109227783144812622385885316578563461486049104220075338075 69
99838439155362425853206576920234395659002250436246455404422928475 2
6335718269024865993489483718282700893807833724692175843720256910 08
199930780531120921533529264381945843463855946080336801572697905527
78498192408100802228005116695437180365670883049977716101763700381 6
013471989704697419569597549878393631610520154123670224547549731523
34939703961659632613626542279297299441095383090803834592672565727 1
535751657422432693575106056672403123173068334261049429778130077702
35560150399784978642590868387888741035902774415633507540726846225 87
318528129941857301708683205617894121961389190830115521498293680418
26304232249940509371090821202889010555638533140960516279436533754 3
641743602086266243018534848048476492355660795350738708041240772928
658912755609099619626864985822660416221651085445666845470227041797
21197673792196708236875724253911784831218638044521922041368798248 4
52539606452899715556578254841317814318517307418687113255071743435 4
8822484715373923408844072269294709052993872999221591346163069784 67
4160369685248366607797640541623333180201699419897693242610726813 38
86333732833483666118059197222779830032644881989992008579231147657 8
995665250603494911442718047635896435209252703310945439721172607292
40284247419176861136371446259910936558366517980452689775618797539 5
201109982599045106750956914720559270907595376540158484003806024968
63559755168243532000221768605541409194714259331359799596200631294 0
980181334688177929027476004409993194144200679953934097440646144020
330494784948733636587771867908222481551057616643555154475690365164
766732104312305405396928858066668449442090415364155918372561891 01
482721775592563107452127034166867917701341563579920506955923132121
9658894588855789707875754039411322220246088784386534012636313466 44
665335286022409659947149354888173009331497527893146274925765269142
620323941645714268687022936535770234733462611276153942453565407 41
418283091451402314689424157015845376402454775520891169485102862 60
807847636472448065173108922385862149549249027705316477904540133797

```
99356886105060187997146258413674417110788204252469129212216 8317841
63225432588236788369076241577752600781857965036964791204214 6143173
02186318116819306081152182854851097703252661392359456696901 1313558
32921069524316035358111469224774298739587023402476415765988 7810789
34785243689529732762990004238118760126881846152123738673198 1025547
03301150915648642986721516284519097641822205384655168936942 3750518
97465079535162167948959699634889249144220756591618774387143 8188574
71529167857300533082218399524128803259758285108192880955661 6259613
85755312015004455407468114445376501357778495695408852574179 0300118
91855535802050962315638521593833980671749722832967934384940 1013039
34671545835340863229328893179348475806031186202274555752259 9787847
82904827243382742203968438730509748128375418407124496764356 3992612
62349347877258494141496574889242542629954889056769459814723 3645485
96536603903554556975925129911374755929397492099118330210606 1702098
39921168614611146578780170211808921608569240060161448989191 5754866
65450752324860187818784292750897316584106947295340447325312 1939866
02353528520127382294842847475347906337505801748874374986789 8923935
12548597513803560488715443744244407688211941835547079518324 4534857
38607505736659637523070226904023707682329671714178393240071 1866420
78806246399365298083997443465964327161065880576311552969408 9043622
10127921192865394508543302775363566164740360281031081631836 1099329
74484901011301047623963427993137500309452654941642548290461 5574738
68122424632873343356342740857774734270650186545336150375731 1021406
76819998151078193602865441267290877593224727277428499395716 3163875
31555390368737077423045037310420259499855807275963857188327 7375904
53635623026931660536171079680323809496221639258159850327657 5882907
51874847062226774518620471920327399104018357195589099846909 2411349
03985875826131552723471412078220994964575261876976826643513 4993907
75913809151951448803007971534185606582985320544897255246965 6123818
63898505414138024114814398811536099586115793726669737115234 1153364
80390775139643049004341502492042567073921449643330770184840 3581438
71249943788138180667183295438197074155809068496790242480223 5031388
22445179054236309853503433519588519927445740614352612416831 9510014
66414987924994660993357609567985913759034801679151176510247 7586884
68912460705445352580299106368621063722300482310594104517596 5443110
29214035972470742950800786971159602998855277229731926567362 7772628
85206911775333663587332123720633132395220375490716306396982 2476493
45656332163353430041332950214679967647440625041314839591259 1071243
58469079595365393057792187689849465103426148665930162928956 1910823
67435389865506295676823236091517523111096417004664465575668 5077705
25492791295164359316928629882000610982611689141459977708747 35202464
14022421276020036428325311863271208988176787192768944152338 9386472
63403782258550451856438874980756202449712427883773360684927 2971246
55211308215022662621762752532830215187230139166578273926248 4965111
67823249687891633157073245042638211294729596057322776646558 0965548
52727113273265391648702885799459548666549641801350565821225 2730523
50807550208314685567566664508082000355605931983395836006206 6424418
00968553052416321344350488163912035902142608674967710835757 3103948
46092917436312400812588353391967170574900058276065458186539 4064298
97370311843862787428805477274811897721936920987062281885807 5227357
92162363575838554636460790847076858025458731170539901230173 9033694
77622864538756290892026673539845488003394949080619817743375 7569030
55950179407834116423533847105378834896332824671323606414408 7810413
84899899232540693698429516133028099313037301176069858394269 8541942
63731799974226057705575065843541786408085791052808960145257 9385259
43988735082224907088325119333855664711213429901820169886648 79326
95439181148529637527316833741349433939073757377741031337857 9935571
11644513996078632552171725208547824611713641087756194760868 1783653
63675330261107375368054953977397751095735224481930765697164 4656119
23697004281202006620365626130839409554025435303375418414730 8602003
```

168567708234631653236816151093651476251094135098078244048647389791
234285223437954092134301119089678518545094978471015976570184169066
561081216824212819842450838696672751755390084670689910041644358058
856732162875959948639143792481504516199481081439190277341936083078
773374997977062345236647761205698205062279975899512119610436244540
597489723671645107546672555613141068861364559730630902448746637407
472267989658395856967370421653053742890658423195799941072412450916
693094728417085654349866999215354143641009738303696214514003574829
095877774442338573347121772047118715509379685516672873105665929880
319527510237507132610179239216709748815626905064950329041558128047
450712386054483980823105379364401308614006835255140008999886935850
665637533475189181207579099801050892863000171669882940666070359279
396845102403504944836421053884779708076741449893569082161466440792
526977386152765496832718250690662518327135056010171647428314292342
169190700678204039428473922095213174895368303516146350387988204949
665944095809175548273415404904014290726736009588057530305660956769
645503439553787081790204084169657084022375858972295843282825725805
813836201400029135404409897504973531468434192424458198435358506 13
596180267657693754785674422861153256932889978370601703240405017712
563294079857572642287643564724693578012095210099816174770029724 3873
700709968865478380951410923058323918458142783750491355619795949 74
690036976282145871689998027403288678819732537777379694527525070292
153157679338436525758723736057779451279041841449021228974116067045
146232310289991054142962321542345819435457660099858820740081 1878
890618057995062594917798118650104784374570414958456112671918702904
300978264161144696891278557436051235715358265146731935724896101306
913687445347661393602875789933255767610036195011756008464006723668
520008186909528293416453839697250871598205411175752144398348553337
299375240276518196339706733610622016721279424210029160992248693366
449722698776530808777070802666883560598069996351855018825484223907
936715663701595220357371092296199428153396098016927499613015600436
297684884620854892767809211214421850436690337876955415212737282018
657271501190963739049327312698214423682171298880234163307521460775
018888103499334279673855630164805106087709853360171601611180286291
028850644374731537523182744776417496709554534194515670600028297060
480534632236075225456635943915902544092562899827204523669342923137
678703327951683827541232036902591068157558753083598866150017312353
308787272404255481894923459802650645484082098802578602581516824741
818207330244900566357300137833651789001472307644975005340118162447
767618466624156393241745471876058069570928797054353441735224332859
683145795395630624082837483576061884164147263153765843049681313919
914016777852115205934662034852500324302408324580767820903 6342921628
613682990692608682224463585446619301129918063961430077953346856403
996901368685581463720355005497768089787857207740010213589586145337
981171012374933876713479679469193997469593240873825714260735498641
078459603588517926298406955605571049658378888227770096401183744369
697166992250351127926598296012128931384844011721846800089532389763
050638020311842291406582220104898978092973103719171185193322515220
156650739092266407079310142971212431455311785955084947092427616379
526059925768014133223956395628679175171562911646029217894542409281
885618601413228953410981488720546424983816963414567201757140396274
064445947371619506226538897596351239323038021143698337017673352983
989912466388559596441285873863002945432289343412071980080582017506
804136142973636737925716048645000305206382027942831163779166375957
637350767943751975964421166480865511996367991131773425607830892961
121056556345239522346387399303198864327068228298893109923902314114
626080038215174911995734299984147204410456782379125309253299969108
691207654563985436256270024818245322465714168054746734711198703549
742519045011825912545125455536853463400151060219825263524891078423
381647145787224939741190957646890748428653274410289733514875543768

9771751114258324947416498724467895170045953881219159051242738511689
089406860362203387486877613425624926273569487826439058051268218930
875099766658762062128733989815982235909717767623038283008006861 6145
314231638080032441347836633756469916776338318120382781734012381008
958525108402711295261741948202511602448916336724120289751460759311
757578151472884826355932982411697971757665414022889828077503769697
390439430371570323077364938470539339194701231228287973461463537753
830947722117283546678848575482318538662753441291102890335074596800
444688740268232354069437742331377445855306020451324431645786466062
140787642753855068633958641685305668934370890699171705019278708782
017781457986926763480460363483404077235936465262952288902090843070
925083028155197865777847629083777957288122109014852786401195604833
553986478800202482785651175410806839903823649380821802015205160 3850
949036872173665542949745557747454602496829999830203583681866638233
271322348822955262607745545884993651501546672853815497741411358 7875
309980691156715461968506333389203343854330972379403911638562518636
168550779031696784788371613674051747581406875004098593227675237007
354751966923387431930646520780749637634775296052844443315533299479
043046913307362672720083810014190567437974980077617609486237 0013
065366158000828506897375813297650547024835313340176538277503306 3624
971486148051505486386086885320576488451440063258837662285802702213
869874557811829409161097289598704742799890062350257111955784695352
185433320690967725505807998711151458902271219559342875020486240780
170976482774198703346674108292045596673608156672848615964619 33462
118580996647036454913068068102894047758882933504574347967527402561
742010289272473150741967767414678422059193356722938929205273580984
424215327863996289914367780898457064771120127660206692338317577145
248356174733788035320274087196764874303216274194270839290228046957
503686789879042905964799117208489129443150056679737956123876221634
727984770657975809742262956411958950532320160007218142205018788768
844146041653459109667899104568630101461405442738716309629649198063
486131963386255198466715271791471637984454355116643854829377160212
253512804838197322803194660732744514828201736907090745245428 41210
888091102535692840243676537724586701242953742582640450070352401661
467456810969780456506645430176859379546386109814066461640895686507
781720241583903918057330515527475762333463072594481798752055458048
892822416725948110969586447977927738652845623129600398126472 88338
259684066900229838881526091144712031822657928913107068934971533 8542
688003735198672547875366584474647351348856641209797157376412306204
683041596532006680523766087314387596487922778932973810516950995096
198592313503021669793782071363623119978188346933713920618298899823
732447699351319618262851433666523086981630442766245573902461323411
293573556434105106523282411430177513427528878116212207957041481626
987953059295776490608080156003772043256393730427184954111357752369
836967428887256613089405810190587256008125095152880421152301772765
285349770211880740700533319909973481918863616704778054708334487822
594059293107402663545833729223911092748142635001709857478780994258
698757566235493743237444655673965276483882082104550549852983853413
275395404744261255330521964708411293524340829486735872152675700779
248465997790926428591538689947326851639175133163936798301967914974
429844586902086410510848889070963507665707591037666735539450968931
839099052089565913011157612197765501978358789146919004586231875276
303708805993477974959085893835594544138323794871319367363700673837
473159360202455487932230794308231033666747527516698221934055262968
076143253368113158721572445679501020074279834415099927732232884531
897162284830244873042932232548731692239194172748932674062736905434
183715722744234359899809163719564491297100327925628039609471573514
604351887528361202655241238817639322618846223018667693939222496648
080856187327277099090214617710771251700332670253153754546110548892
5978210297163514621118675683966696127089426188971919866665389636114

44565727931383871572861224418225727090021589768406073067140091561 2
80971733639229789395852513176807702810396525434905217823696398288 6
31662312527702995649440494543591527949430797770359248863940253398 3
18460419904705982169332239329395755407575847980226715766749549430 1
47987370690491500275758299230448551127067019349045137513921311582 1
73364408906705381696701936993613436814704386680544285704066483073 6
90951808235821687425970177808305420457828034879597387086065453613 3
01579394552719747248209549943753679427722215244800883374440828703 8
48151232654123979439298111033457322755987041930338468971027385822 4
07297116790829807180329529767873936855542686288869914541132711881 7
73342924408342998184536809780388405242379834013649325791680568865 4
82609310122698202368988539774627447045480243320305733756317797501 8
59500678905392053400299344836179546014956063788800534122792745536 7
20229367955055022340312430413445839946061806930210020189720306636 5
34416703322585205697946278629843872874407558758998402476940149495 5
48852338166814861503228532007097786658852838837035903009201170676 3
75172146829358969006269458644125573598157949350691991815471079903 6
85414294812630391668017296862474785089906356383553925880361125031 1
99153260553781055666367683855724461721693305348781487637763923662 2
46505776031424672451270420494455244464633289379587386855874944775 1
53851113455740186359442718802565996776102223241588259134526742487 1
17988364264284542195950327061796829928020857738139055298406894565 6
08957366808415365751975247167155222933262132926313139902810699553 0
65891952795732002274096379916277438639194446094482646279901850425 1
61094389398293091230720867436126525872743103052116176130552209492 3
79551106685634422014506747029686345885572677125505496052273962799
76463091889545838984853832783574430095614820470814716727892981918 4
12811301017905148683654974735295524745060418537467835964435230521 2
72327848699376620024374365785050678937782429635147479062538213404 9
07350297132884764308462817541847928836931764160986908302873476599 9
62970962867970490779046336943737983903375300684628918501995785224 7
31784047874048578866924601666304802637673686394771725143247356878 1
31213244212965210211044810001041072893153689760215482362845250407 5
02361807141224522039876800984019887856994771102330471429702040881 18
25160619919518253342236741245810569039200837240501226289267877847 2
91983859411435606477744841701628951196044793976186880026602199949 8
38811776488791976728297988011555867721352544142881592881075900886 6
62007610497760258958679766386542993449244127440609506694191053285 7
83427070308912419127411666165256712220917005854415623538464921218 86
57515199489727703744864458721796268948461135591400691847581501511 3
92747389941708647706380256877816292190254395805027988588592057384 5
66800517739983817667803202932278565937500731326599621444713933218 5
21357814677209362087891434776579000069113342493622948786952272207 4
21355534186313998362274062497579925852539849540030920909421677082 9
61792640376499088472394670524962255230337449488379952541776824375 9
12507735616616867919912788854359236182730884205689689090935383683 0
21073193138675991981558339698031118897697817909747341757898389075 6
14485767836476057826982081459478570857554285811998421057138899709 8
40447035071048810818100433502947282412838676260289698974893585377 0
45700360385871632364319801543474623116064732869529911176661722558 2
56323879934657200795908589045446670524625380087948600389476753597 0
83750778106085486319563522574596638661464018964287201742504228315 7
41108190563816992858037902335357433759403327947473332433602029263 8
01240243319310539908190357820005445818950476925409539722486438248 0
45202953530516155743995417309241414345116442426741929520297357314 5
59062040209567735789607289023077857331856300620471439249103647465 0
77168413852407113386928572253762252841906839199289013238115787457 6
92385727069617937862718574212433702621655914774695900788792484602
30493563522403722909393217320192597086752474411329047847170653083 9
41589663462034892632418884122958799989862787624426159615226992805 3

70213282222630713684055456576459961682187589500337811849420573416786583848191954742227549636249738910380237458966418121870386094639190621513883419631439535929888784221611150534456234098699669232617661892095371795477634508296939964140839068569133713564307944882926229282988152375159971685707069661970992513218167351010945039363402542000366540861617272473445084462127806251178632209721942879862632165972924205817021283648306230567012961322988508485830066667687900997605353864465487274717935764054163238156318918102413475950263804970590000375649678380980760526147620531977504970508091728708875750435770362002098971831862205235278481331524526930005807861729402201586891050290536278376207607457097136022452560748589193939991148392552332618739296910568311251157208333305676116046352542700668430295650983908273009781272413458822045356929436704571091384986438226953626992638203213424281379375822131006644431116767766420674300477677579555511423919511834316028144754395250967474556123954869083166244211240524588221743374368790860825020036254045014208186744167185753788767552068397082386471252779005560127384437048321849871403956745705093857926415224295737760631941879341981119833410473400595268123466765297976901732094944931027318989978401335493384820573362279178995083299398421910299544574202351691284314273033962814055529820527196582522494915139550656158553433968630412731851809758506962365323529512090289752881085017750256371653835356580654433590229299525010845569616763690334268324885500636252602625020925294577102418079674700982112009308936674128311429436810127931055159470151571743405091399660734340501250204041125938246014427672121210409867453598903316618553343269798752577126329667319805810077040292983243748195915696404254101701282623575910512813381727016759568798413418788503250888996412121766365664616932664594045357745994760523190786781118423796152053418867567464285407581293771783835757066426830047205903167755792087082664467596479659497412838515732764192368057564565727742709008995184089230933956947898106159664497616617870398425861750012074313562905685291479417586977007769704718529619380882630357569817265920674990467455960458457538179034518371641563821544934502505983098522956479864471253348396267108703155606403663735592134349808263221246070225655821164317906667228752295225749329657069748855408879623116435545854922395110985699936712630825658503640976294441502069712129899223518110336623883085504015793292008567046608457445091714320331062806437175680896619155070473967978198621362052133140338140873157330973755756844788245967902854667172699435748599267852201040821564596210111232550497091384565147629751908913407919135802138350810153243718546446843484951816107194456333545131658734709490412077464819205579904973208230514677255320425691958978888649237514710936682280077163009242345284631994359872085140381804895783605724820027260779541951581343359668094771872520427378456308643117223054473831482113390156778726586763249956524484685963448322775592626432011608806304009389227505624941035378554316396650840485298098103530286235361708503979963924143581200952382783723173994267774806500338081032344193657863333484893337586476456877051018854918718192635581654009050037049024199631396365838641215709998619385787359880322105023214541834807492669967354244388111802916839157380238751531815228204922684888027069562620766197868555501379872180428661553681186440834431629700595640001304320867031402894255916784067711988767217032132512686175893454391768724212123691365139969067261384627922647736038252003490060405051624818159785649247331467514074438834746426666124137211324843716065494834950722349280454198379797401466930108350848023207015129620452430799749320891954366795717436783611612351998502188446177594576766452415429947295317847189659333208012423740018312471825094538908636751607422451734258633535351019188779132925022582791515128464222671610483240336122004571640616745582599195171117562396957489155119917011212490484814373493277278763631835129751800954116902148

```
6596449774667476661285882622782703253014882587629199832567081661996
2922322213630553721037121785949518448731070227067970904204032978921
2309455854918880149336199118142372845845273698191040008590182985392
0261884283196625814213130410331644406097683404422756443777972254668
0772304909080164339420464321142338172844966230190241428387370127494
6566392401286559950568682041042315877416341861476097323730040510
7089022611758621110708480683648144343051615646472793004592898633279
0925906349566938688956907810258867450268390980064156286632221045571
7360242861019896727017056807750088447854445461236039764657334257464
3706682062197428475555603444790846518940911437107973373146457113
9410708885598798930721636910996244093076861052695667904781366147538
7375998565811620825758401210962658029626133733824794921798291167506
7682691119495600077472179652498648628890078875924699489077366046
0367623590963494482863039574270739206737192466630361757798192963833
2125832110441951284978931079373945297214357010578199522005226636578
9602354687324356801352280013667801977391306946426152643885718129
9809201466396855659051521870196790727562327130794703671668611973009
5792994900903149950086751509244785038683387934796607321516849433
2921577700951165353507463280567418340649102472454113161825187426012
4849457273534746624399485947312595949668205550134946833026521695963
7501361618267371744166746466309794113737729349436352636202113964
0994607928859825558314872532354282302252703074863539900261787029458
4063452467347406009494706038618289810693386030791153553433942471950
6054465400376842441329373388168766121488914962081673246768244520
6036380250954575255236836918544100388449166924563381162315854970666
5370524001027692271863448390900535565670193496736598581012550756470
7775049806575680195790866011077650220883047976386944883290950759772
4400816375403729789307959354179489932958867375132290690666884621
6679970836616920523524901565610887754199072467124350436770525071736
5739860640935371967706615655275206316568124311962318506285585299098
0090419934769217766663904103249521625100416700622590611538716983
5852272873976272500585910667294953326009056981910893734743337129238
9173881524078617450522339518987116006848178752665577057835837953
0716794652979506421713172086411883335072583220583959399923929646758
2473088317389914393478110241919013332177149350673277894525520470
0060961350041307724746142738565924398941005572839766386849024801846
6502901701632958773355829701939897321000184241118461607899102436
6838100503998957942383452559824623538226584049165661977626725693545
1604128545563401499426679352777553374307455896994661246792096935
4771673951606126240213932176183236659016158126299541934792692363206
2241869466457851692712132825531603898889891884691855946381806652333
2510652895741612139569052106656422412400336845646626056490781406
8119420830742739636416607810339130480895993557621801783920250628014
6804665074512576017420828913895866630468313779585506458847892594546
3525219262965794408335520289252769314896338747922561445659064201143
1150455231506104430257658547005054845999494792080180869058838771
7011473336398157637132347556437401463652481574114446816238185909409
7484174964306286130242735570458014610270118610913034564071261294667
5595353908186058690502298962181779232117395956881318770641888327705
4626971557197008546361426693592286973216348301451843581899086632
7213734403438973597243939861522092353622712383472237617882938720997
7321479659889946471031472333143171417642877852047682778925630085229
7658614012145984489888191386409383517408784188176091320974731659
9707347468844769055593215802978025530317358760425494773969444007812
6945721246602308469922796737515456533201386807980055372427792422
1397477545703483309109343950322909915456540082940839368580655272955
9915261947569036792297548919230672963203567640883894286266898671
5919243546121088199928147562003652795896430840387312003149206678944
1885446508391037304908117940675015202411183046974242278649288435945
0196059917048114428189331543811180881057326042380543517247138972
```

```
47661696992424896946086274426036106589442003069762602551223 6338263
92334168197268259985392258545424203272286815099024263283034 41770509
60907446526501806540478519605694854589933511415461902096305 1669962
37891456052072943260142370459327275022394292102235987761698 0157169
33633001190402719786029150318244389128292047781848343456703 8741459
54036037275442806185858440476251235388275744232863434212783 2715555
71710410594251381961366227944632872704199302655015498368785 7378351
91185446734221361907606540578427254270890155157102320876900 7560739
73463552600696002674272164749393342256703878674729965420262 15208615
61968572696793186259680356995151448453904664562647171563754 4018215
55262186743014353974633625846604191986167588505713817308561 1545870
65476350350747913103010562980660972132787363827701815523800 0617373
98091177609653663119526670856169831207891665962284070053170 7343503
34470081057295159159161280690788182400486231483442808251514 0548593
14354114677646162411571453248854392404608837641298928009177 4975633
40616687019675556575433937153581721329531273418259482175782 8484816
45409405344333887405293443974809141062689503726520361239419 5124969
31638780699753782238880645165077174867092529671407829692790 2975711
23139052040138376020643539781302226848277655939103866295305 3285185
71651013283355663867762521816677754928784295699267762705580 2753988
40242279764243858819216217540709759805479772487816596525249 64123504
85690423312100279206177816312743191583552243312883311841511 588868297
80560353106500174461323288171903574570382560578089195204104 393
02408340630608528062014983412188427623871113652286986927369 5978729
30639970261341757355611658091710804148321631400138019953746 0249500
35432209820834524977107101043978605310043362758746545114142 5984662
49952234715805070775559629358321371949231739187170625293696 5670386
70256257720677765542373956528510386791877022940393499902329 4137699
94451499542698793848216472632829252607105520152441096482726 2553419
94830439324393558451061398766312756800435330385923918272138 5040618
57982278834484858596380368905713046673346942249391310888385 67116800
32756992633201315769762705413556696256888635932728748456513 9164015
91263387388415894703414163708887876364401560406301539396552 63849690
10356668213857752093832275180255457612075838463648289651730 83144741
67561770418764922444242894048519097199484023415878805393152 4958606
24104197475021415161243350553543375398978514975992649647194 4721644
41947305962128991299187213708688195080900832194311374497595 953268
60343445642422411354860279244202642885926028057511718943784 3786349
65306171564261072472260821543258692820073630509083103886729 6022733
92627592251096726022560486608604723141115083463639782432364 6233780
67895791188872098224600742557614746164422033310942931332252 5745905
50599153181358658736404505079020098497692230029179398598996 7285319
50418281765183339681843347728386143266395565169869928342384 4435873
99548616010634533373609746526703166538680496929626183716520 6373459
82586250436337598026475113552977109489605334759944896857050 1089726
63322561969456649026498819086738431065273908197503730632040 7242801
16245538057405814293132726402931248978488729369434053160965 7626497
98514981113453913895180402154045801461365725915746758046945 8825846
51038689069718709309143301896806965237612203865428789520024 8131142
04381623759153992668083092286553895900411169626971932434008 7152972
26870670579848097677258229095727797377765501484473610015924 1236645
48670905396317361400688323324332585344510373470700968246907 3837389
05047610602876177288729308829810393792229526494853689621655 8046791
24452959336857030672277474285491150912835849112854857595485 4235068
25692427684258671039232780138600953941559584901475210496354 5810902
14905798039046779677669665822869267173668324368487192556192 4133616
07626265260468447892355145265468040068200291147592992811809 1240602
54436061914202827168740836555969004302072810366120126895375 90430564
62621571530272377584024447463615220246157234971254235066058 8850417
55903268874707539563338010268732078937130812020285477409917 7789459
```

```
00043196167970073379874938417647402971305608475762700377987758942 7
50951550234914317340735064365730926443159919028371988860313749264 2
63364281977176246068119979996877866771826370594514718469295473940 9
34257095734351200047624755788863397101213296106745199934090261456 7
43626426131402388783057910905891097310681487988555289376150648713 1
06297115658780766377749661083856962548432022456132684191548924936 4
35208804679677741881636344149062735158060230313435240158694633005 2
32745904974955110495016281794800875648009677070893157187525469770 0
36197432590214457737448929422029713283521350714230556878075595371 5
68933763481553485215949426120906579076520158081147657173066921304 9
78897895009415535885758988625910703257299228354966815633759445609 6
35171954727248635988436757659627212025020578915566353076085383183 5
35518184892287985014549861527631506448013028719863876162190248190 6
11218583803610052566802781341565530891540973223321774078922189063
61074467132539785073913068164260330673042810818964678481912430669 4
76397978065356438636935274953801251195714206531875724714777732828 4
83109764022320566654324672715268247689241090492243101635508231894 1
64348671291459244243239053074240421714704371269469215853972069951 6
85553946635555787943966823896704389868784729071443797450946484518 0
57607091197148731713309682734286931988886868493679255218748555569 1
13217499296641329188610168617558006163682254399261469985988354499 5
32892858985258569331239554444565318707502823639997310410641432211
30386807870526258704877176700238112453387367854931598808035628764 5
68513840369868250243803883432518885231410266058223353313623887099
89497821004828214822071110499222390420965097797369052725257839631 9
92249285457307596766139855546871511519783919382885888427730803723 4
11622776916448062843584525846918695731445120745977697087662064116 8
51992450578103667398874259349056132395075081507901006802545994293 2
86970833315220436059781182803378153954206193617391007798595217022 2
78397637355458259412699887963544455183370997340974266765511167246 3
06416964394650125335850680942444414128967901926840195553093544232 1
14277799121999371538718019244210892236065992201423864327252778212
83242194147568278211192767965790200466573166665295928099984908377 1
38000558357841402156402328134511528154359350936515700244709913126 9
27480065568807784166802953771106505057901167476446746400075190788 7
93455349849176022141169567011651870355971895853101229810817909276 4
19779934867830385798901932546572803518509712712130156677574907339 5
12074160136558292676943845175541757444768071401175613574824066623 9
25987820507304418338703012736896410105676845299027740614598818284 0
27997271520568635288210790464224327049337095098882694132915597029 7
33373746659031683090131446963631823108788462518970805451007683043 0
31200428117735253593414673388959249687183708543784982394070089508 5
98694412621733285776914124434661058029466963026485043902300308177 6
50207375027285955719611095476238086993588140998647354718953284985 6
07810886003195875402385018764796948180300092018821941816698920554 0
66373787863552473744067328274075160321578878841986181977510964073 7
87008897297809148904552975831890102164989652728285727696398969804 8
24907221452323453581677050667150966404745353703911048390183486066
17183134035647073445015212732087624977928849283856613786910675337 5
21591898245243115630504978064268797664181671102890666411013803832 5
39031186909681365648033264976721162286514414037881822313164870450 0
16018737681263104566490601041519340622957275340519935219152089750 7
38185093537413193050702605501450163607350840258992321031195932653 3
54075863526108100626600479999770738258496503895622885481046580044 7
01536640740319120313893074982655935843945974004613750946542163606 1
83769433316178689301791875827650481044847291233859546438915369479 8
84029379030966747793025835330740808871118523146617128368429532238 2
16579506731709329853669660306942656132312404762120337556715561531 7
13334707711510960605522712187695406697486667976969674207383749963 6
93504728456403828922381508328411771493362095812221309298610100305 8
```

63580565840214312375308708602647055332538996183516694635641043 8277
63374899013723481079257265890470979464501366517898706168206629 0764
33589688451585142952294286593627004430004988867133454795492825 5539
16545912217725662320645970014826144764165181124955172840952660 6368
28665673168744837975823146598070773189998068225907688498817925 5883
26760014969768362589690262525423996355379661467977991471214485 0083
97763338728610923750173023814577426170098703463935212004660004 0254
12758404751387907673528176783332519969875867814704610862126674 4892
01068249049204311356045810581699603498249043148930434631761180 4387
20539689263486712216303178174752299611689206752278113931215673 0164
61913365290842713240042656136276233207686060239816673288692287 6667
97984809179378643543047958453902085481576600473722547594058000 0968
82324308011787645013086746146357792930229064818563898420384570 1905
10082774954252827896972073194668331426418691263830281593130172 4533
77994977334596510729806649911027352997359031492599399317878078 7747
14659072152702350403084575607474462126132973789242856101360693 6966
52146215533685653714560437179421520848005561095534281992976481 9536
04213296470680396397757658079989179618050642819083929225325198 6774
97212531463013952467222863082521897048936124516292959464400037 193
65011409143700697407184318878268127815675239964044066736868279 563
91767711481083308863739097862001061218128793242214002968388894 0762
98286399533136760159093652184531334613337682760543612004036925 46848
45351045545505434303119208650826418045286544897784927492857645 2324
12978756654036795440777688699074507947501882729649279806717727 2121
77458164862326219503303053598598688317954160390056185834710845 685185
21056481319059245820750419721734954421695100440203064590126283 5463
95607816216517275665539686051995691575684656991776700458812931 3847
04803753462527500263851424510742290132552820499462165228634735 2003
06720728570064955889599909071077110520332204363948043148770115 039
98744683180675004792364853044092671117644597319720289474906521 4443
64812826944242896266215346176147333348457111190698157560003738 1738
63482871420072225184512840993887753146449412337082626764644395 7292
34785424847532948599276860816831807663790738967564632433847670 4335
41924667816466258539683030178624440974525231325986252041204699 8161
19415381366309717564157201189680146173811880652142307258849376 4805
11661509360310027199046726095607358246313522792585415511654420 9681
01768041565694581683191143865405806344852337038997124023163350 7200
90957829585481260754397103927663091323948281586989329093355250 1877
95510844381331977315702194604531485586903876103246697613367520 2219
64796859487113803057537343582477513620999379293524033990318511 7606
91252827655352592113077604710135355172015245545744821265950297 0017
27371119613868989840266514480231172746437082073206214709672055 316338
11295325909982438530491170010968833623879636452712774942243472 2164
90732912341213322455035842456519343765282533293556628503624561 6949
13164130863689432605811438255876411544028777641246149957211047 4762
50706528944647822954805169769244943677253474612761813606464884 1854
39153349315767241797892680048165085421400817314894852734019045 4761
73979794417987091090052305459992954707299529675404320449513408 7336
53767148396518024829337631899740668424763137994698673922929058 3084
18182058981944942932117795382053185885535149244478616578266135 5287
84707542093442532329948593780377359467426810080402378519176025 3998
53831952222781459515648226986223224719397843347010711592368889 44184
00882503192647092556462613249195549067630871002916519146167156 3748
58117008643637937575716208848968288008884802286078053066060372 6555
84276511052664916021450431333927432627276710551395984579811474 9902
96336827414309135989156016514435535862881938587557516726014672 1240
47332695886919280324130326822425228067987731829520131545933888 2825
44073458213432669938769392892108176670808744536596657165780778 8614
27305881105313479114703008232896654696228614112383321855224924 7958
05762130867667729968426113165073483756774290603672428999214718 89341

145827296086070141436678792460183879850283106647187241154676169185
264113320962028630978948898855337770998564729631066965693273667666
700503887441944218955683950074721817181806638639601259449866657247
546730540170675296003062694253903784736249583773298698495801307154
939176793609080619475647776999697629265367032565911763139893783755
564253868746412300021201861016820899183903112954777544974654925428
974251180552171438058931073183062172415544772544620002332425110254
245824177336539231601633692449465371585086382777303000918494206721
754930409337055994531442020594687495285935322270322085451357791082
713208645468386819048757996107343149306297851570406436900482702703
669576712358301551941010801482965808320450572580198141874604630866
877528792694819660726325968583636454561111895221153923502915772378
651386961690191245582226493103664911735923626339561938890555560124
675828775878323498658229203373810440208574623154419200874437796989
362176977389521431709173165908711746071319179270859359040365716952
773635667586702600571508387376859977222429876994699168211048543870
656677161475843525089165381356043088973407797594630292632994777906
928959190931772637506200244902559385151848395031417840252549499334
814952902411323885646244936325651674478297304105996884388275834908
734819190646738237894904464721807777433344524714218229848317232449
548591401059047309949578390658067575992724520628201410604011166767
574390944215949594881818773098588778843958445778183823436256431827
897104919562286679363234366142089238961145425450590306847768714204
362552372727773448265126562579692438924129812872415611125890310779
250241224035943512298823642008315580477587987740850276526411848860
230719977544685341364781369797061677420093246053524207741006086953
202828736191589995392232012069653063884107167904800674365087272457
287854030986680359774075461330821812981750553641949678552290994613
470648377742475919244484504338271544733560772219208857791684438478
817052091809002271732660437036835468975388733970150600978901742598
659628020748442614543946849659379198672242860820642823831758708342
555222226773700183364965539384429633182355268523899048721719100811
075401231125436262929486722845984159621646652298403507395480780869
887075613758113154674485243366172986697831316196628990970564016399
609964898212189639681682347032603941836381045794977082774836641299
246338551399224124718989929541166828695606138110132911339882880625
793508606435773866351470199952791052213732922323844551671558341162
296591378006787688111019871375956274444967652116755086539376947435
044442326909592745795086478066254658089743468130663071968223281686
133221943131513563035163544229107650808264231912428647965173 91613
596682951317904325064343109073254925091193475104098655637899892688
402744237343106054236817065167634853454629036399847189578474885181
886466990899289799329814112688087038230200921599389623040950 79698
282453485834402579341392407491134492253680052699557646110097254596
837588576146034785232721943491862750347816455037539784017413658215
095677605639819392083505565611838455605856085392551743855245451836
153173146211683751575723623737944461467494032436684086729472106741
534353029078066920208965674737192885657742964381973077817789342701
997487146823839953789912761474962646108691426860883803393660 13614
455904183237129827110549616123113929032069596506052017745286931039
019314638174294407672842725557757696719586295458877845802781931069
120446308501408159893347236166549668042356721085239384888550286863
929373416712318138661307405859238304545465348609222150878073181114
692151631675911194806415998048770931136725212452813708397668410931
169456611453429497100856106921233418416552144874499785942813245911
136178695675308861295552516434506240130081105111960382435573237312
505611267857804262039833905724944225056484394357452563942136805238
431873292744652425227879073218859357478026338428528022186650453750
954725837180026717886968699134197142872551864112938831053833648748
566571809376997249728034581245947825555699307057513887557477520690

```
02574120251128005028207673932279499455773316150873336908615344077 1
89685430870556517583432238947822105004658618546670584739393799628
87147676466403570737061965256995627007819728322503719967137162405 1
95232753669863529188457270525626798013131649774320299329127362968 6
96229224491027008521737871528662026386346070094468344592217659453 1
77501206201399511557929395370521942516518669215284037890510497178 8
91545822741327235394753434517591875285493032123040703204031259245 7
27899340793808055802178570955447877105028276983142825297191518567 7
48721092763504581912730137439257310908847002019547381402110715251
77064303776713322392713482064641959265455399441051351349598813791 4
66158984353300626099225436031174572331589857619452120193249200711 4
83294854491085206578039070745010546771795038037758174690440612501 7
15331374240873606789883443782330044286057085340229479809795740296 7
09373409220007736840355422411585065991843272318038672312326825555 2
05449621350328013208371194929093792438162786991801268073118945255 2
18262828832397718225484753078602820831730863333046907960579525095 4
63079445726494225236862220124674317197023848415976846975116679648 0
73236235382167617953648731167666919991399762756682669726050262670 2
66542047316546316894348419736225077369187176190444669094975950080 0
85090738370942926512331327034954355556151499995329774148516081788 1
76374622645663427701378847851033542037598571143980625257802663968 2
03045702534612394305914263663082715938406681439154845288084500849 8
34804600529695168419532626344499666669308377768027640657739666676 9
48847224944482236510176640107014349464956708439986574737136407070
50822052970020568751927466721487331973996692363268814801151369806 6
44219024751948649528710434207233201737460614156839197314477223890 0
13958423008541378999568542772434696910944898795304528662704812593 1
03497003181681645007691809088921762225193333657918803621828612078 8
41266773246535904468179294860664910804373550677835808570801624829 9
75909431703133995931942161087550969510673949174075336824467432358
13280210962594775043970318821174350578316864829066146496603585845 7
96143581857163797388567422553198252622814404550105819479367723134 7
32457631171587871489303320400496504843660825423698150866659540129523
88139153630722797633474574801238990934096949864289877454551516426 7
17636137516875050025389633805283300139166744314120968213666430344 1
19521921071258054185772114069064186758873776636543297113239397241 1
72100123135944490034248090231017935805860667909253313595913704048
23189696189102018244994298099094434058420569033851326602985744478 16
86509068230298649317473527183113397486438315179406555173580996428 7
38689310124207102875446158664268921715231122601750831425215028384 1
79630368031661502225656013938566715837920023157797020602792416025 0
35435818257889618030178874175350560757051268305727111372340732704 7
30686382130936238369572853601943487507494917614345733003786865343 7
91314755428272720320476638277135725418367714101729636917677727870 1
79081940904663341299330020720613838804969918023625930997763805324 3
92761802338959463000386767529728231673532463925460285230304774301 3
53398181537787499864027609414679015235229522505903501643074597017 86
69109904446069634242273385618514715287123280520584687105943804754 2
00971302386621157673644066987758354030616928078292200769495668505 0
38866665833063502229981529331540048738163506451544842915504925820 0
20790606371565091107790316741009390708727069063975633112370646951 5
86419056549198397110513919224710466560866623897511882397701168357 9
21354612466335911735509735090001766671535242787568203803337094166 8
74789020252020724797517877304373708246910660949095508295051807317 4
05560093128791556999240649952176490910359213419864409426534600402 4
36328520625210919895971051865724746829401306868136904615810241109 7
70414590941110382210072574316276339934387412532424995304540420356 5
32292735963168737751225504834216634216632183333024702907770333856071078
43379337652229661666849341462860711208145254292430327887230479834 08
19059910859549373970597510289752151010417301489184265212469897118 5
```

```
0075382663766868970221250273635130663609364442340398593565445276408
5826982218037163446632590554088425570667004613089245810185875194
0368707217145803573580142875177376843054766215052944244582270693
4404285984207659454122394540388042423301043810983472457944553596
5683426401542137934874637989176712082341454691692266672545776017
4257160854135451126174935059468560547636229413150275989599050439
1058227434802608273305175147602915571747087705038822398751762270
3338576434896615216318957300989523320116321735028201167511299273
0713267260052876301880209759346127721596729173496648623783161743
6698041206647192144982268940549308611218565120221443551567890925
6823648380887101752810667955325488424312772881023409104484877439
2995391213581614290980255723943439667559560758994049157336853813
8485821194616790874618789468917988728073168351633785190542334451
8248669267361534024208813908880103348404616886368935966227882776
5261338363667423089498630846433930887717943234634062981180337689
7513646695327446763957091745737213162896124476827467343732398045
5436842248017638163567461392617499658893301926273138137487507561
8907016016819398381833062519689517451289941907527046727329692293
5953998363055988264640015393361046518243848418644340506990242670
1079181071096042458403761378768855816569136065284644177298806077
1030087940515750336113812226024878270571113554610841551811336586
7665845331703285468672719707272721512796379122858837762099273481
4970673946468927241281883134565204057923164885692160137208928613
4473426101059903100725366969935237359151118349747594798699204439
7520066426724008621638152517329007924359669989744997233030632552
0062156951309316215427296522530617447221630071305528637574020211
4720249079371840943854740066644716775821379218855935534031571755
3716666934979212292480321128715473978834652343744036540957345349
1414078153022245108020926079742114592462841159740480168620632867
5902381097938381683518546489725876502698669718892953988325081657
7696882466611792524696491737938204467460275559792750317244666262
6013753478392190765884307586727820494632774090443878033892979962
1975629622956104893898813697955118480416013976178369887086149964
7728341383786320489573528050455130070143900237690323087293304816
6664955859396528273676941099705894743530522075381223813250875719
2351201827463333227071093518595999584547289523563336638487047901
0738802947139098253869207712204802514484754304450282184833752878
4798863620014060050602375537333761814074411572462445381039698689
8724381424542096139320665187875639892242498131165994297591350919
5340484827275541776814752214116974062306806481580107104957032667
4977387911481356811675388165263505343649346920656900414621285026
8945370256457659257641544995550425200768292285402684848821037758
2658272051604775099082983613747267020457185218704330828626591204
1773638048244285257764205429920263295075043415638598765292143537
1845023064204079417138883787202239688761740037666840626100645725
2605782281291941693177905522098434391665222911808391670932190206
1304618585498359323498019845209533531758060111290562282922755826
5982928535846974653831196307119831657630152116913337200137791294
8917922301347225225247473756900574856666061257706021688768511658
8379852059859316773387454779247707927344815280589952594396913916
7638414576290186986787941051757187759972671728169822526273551128
7567216987809198213187702896074849207721557371485647308546497862
3328333963832378040576697733833603268100541514270012532071467324
1048034729453543920827712217887260532934636518758101976419188915
2949566795747404511049773474943248506828947966749926559273258764
3042015726811192978460728917955940221027046405024976149346487114
0499627614553552667788279142649776542482335287316825705963332807
8895991892995452343954045465374301864913114376688338339445607532
6397410175661508429415889603815939582832411810569971875905968161
9913377993897724274417072000434066333837728122999571755383127863073
```

```
0022809883489561682808931695683545551324577483733123972275567066 28
2680425250706519034521598768766338686885475102438567938092676086 3999
2391758160820357258286254850262224949923619485197436100921397397 896
7441624593053374011479751485643454373832413068379170359394589118 89
9342636365375754660433041008496252652519693254053371979694432399 34
1242833623743685241861012826065760655719331665920764132128617478 19
6809716699681389916393418918114881411844674684122239536295701823 56
1131765372743398900705123479549138222118136233764769515556531866 39
4092307019583820869148825954971040053577444462987159991618620091 56
6403354080749330788638860254347857990311755095098947121462329832 92
6333359235919945887713506786976638662484357837865067938227689091
8006381696567371882369918686939068433740809452366747151698150561 95
8222617899913208638901003573170822283936343420582992323404675682 3
2240476982185560240834965972239456381655205890279446276251674570 44
6455971118067842646398263084395704692169766701670215418397145240 68
1110789609644310701442296183382904931581380238792277496024418664 15
5382819216810291565131290804299399696192432099719627419607662716 91
0301271872301833584434780743585523375310345244334532861121137724 79
0092230441141506413703544342927938289886482755757742895298374946 44
5349656978907423650419308452572428297530865540392109528290871989 6
7819941805880723151288004778933700441079967103609490931630729705 9
6514516912965130412762335803179181430995665227617872649703672296 29
1436170506281894033124246861055454477340147387515998715732263103 89
1157295762792180287732580177590524434906478724180138376002991941 73
9011398081634179620987076543713259691226636681943306224933160951 88
6283956994122985921859565557152945505056317527222147818646334089 22
0457310012973042710988354030118569430346400513396190150521645775 50
7348484099065642541407584518477476457307346262381751438952504617 26
2982223617168549605962110759283269633443335265856333544879111389 05
5958670473875187046485691347101447554308818507039972297703761359 23
2359456820422343519934979651257901547179567290602399599066717942 84
6550228106757432851464652006047507875125920951856360269185104769 17
1375995048932563425427367088838745632325854928543063219163261101 16
6729320058060045041439291897024106824502794959244109661199853198 75
4538220657518497718021750505785127344984553653063163827026207567 1
8794630549650859549569097643398421646068787545864844329424734429 58
2281312792385251180334411255892310125962504169001222239918088657 9
9805586689731942081163805298568487405866685243797310892184678803 766
5188295139735076727642948822973337056359720180026399398861741899 13
4721514715637971582639166798052724084646518077450034483315173016 89
5454533358724974866955335575722388416318700559138458547868514743 66
4079947258587337230540592956725114877515180339226189740653287165 59
2330489788698832246738662259147727933140215802141484212433911472 45
0473914585344356214062677622710015979947538854011644327086991914 82
9632222694606103956513394655310443224776025382142069367526190520 58
6029183583048598589382309413833406259383378926278691827708629127 56
2234618177034798997585347336575979495244598309373452079394184901 56
2946063902051109629313762555648811700067290105213329436196343557 56
5205874865462197022770257800234904767961021543613042220341829302 62
5551462594507274561756713825460926410769116417331696926056384740 70
3798758702504320932490608917175324264480851417853993150267101884 76
0058144693627018038780184088912249442433852688414915712146227212 86
1408921798329455889041348529702130909303506827346025777859732498 54
4842801260694579127310722588403494176709909890060342887170847226 63
8260070659757813581592143455758946216628128052567057344767641940 87
5104862305884921533759373001831314491962014750676751363278098437
2621148096258603763948944155147470967347897278177434243146264 68
5938265870717728358196819226976211128633800122543744153785398002 23
7634493729950037985247233334137642648824869617405040988744166201 03
8608708808831237763353894118371711564948059841113381287608763807 50
```

```
38570051581429567542342748648906687970292389416672185937468415715
21878848625163332816222489469651897811447139338794797333857917862
79798813572796361571877901823404250821515443804241786299249286061
55270485997587604447230667690819672470230803243559326729408631834 6
36555045954374126872975418147626373698037209894289604333711920907 6
49573270449301297552663255591806050581149941615081396749207672807 3
17456919402520932320549221763070181259324214779344251544414606080 3
08818244513746625604745594681802664921837719096756115445775167839
41259815706809327189321212915312715282022303391823475806108181527 4
58158892499031022571076986970127404900256839581003827176030383005 1
01589683252385744237509013911122859975826193948986082161267827077 8
03165487121361163865379074699106388243578499993725798094492177759 8
68388171827487475128657914981562144641316822378973957798994424877 1
15631392183036125122864292879217343154603283082873453376536935381 2
73548282778944953216045851723572866436122776083483624277882200134 4
91597351553180083325493175659020071979430112131118492333278202654 9
78682492974484336641817897812249547245247755222325104027862792404 4
59043636953779716310899131880947330879069564994843185782811641939 6
24807699868196594025848417994555481052696390629261943825422058471 0
13164736706604061386360246350196460634970638789100238068283689301 2
01956892488501040347915801993586958345751295204201582228827205474 5
76973546869944344191427194272229346483733653931702894999573750642 02
77304224775648213167488881913580598811189069298277777248237917937 6
19470902078275303270570520985083499959589483285018758581502087218 1
29971874226208795889244467567768743954407944678892262569824618983 2
58709647053901777502233094158583513031210152715670577529453759078 6
53057364656616027427898866141632659935592793628054621339639663329 9
19068656217679737076798489863919651497989304836809348435288021489 9
20338904832397804299385444706309630904916727421586856015014113775
93302871827123901747428815698573184126603733885043836936504815659
61712491916871028552623724234744222268748988830741217650544648466 7
92714520712187389504084629697754907799659133275434181740049704939 5
85189189401977456229074913105488223812375249042785857656355010648 6
31225546788727906719452177692368682929226995242454952291402881592 8
01969170499466275104291275558947304132411651537022618818472936726 0
21184904201904349221391678754955686104274744030459149464546801808 0
30465086669482133080705192150931660063367504416138109744075369211 95
00573442768488318639453562911453134642528565188395482622302334638 7
10097513934593578693622251608190522978101513080732037666525120008
79531369874397901767562917401748369456838652088591220856332741274 7
05011696248475373372231496748965637823660510203944765410301647095 3
32267984986802789880386455993054896677736556886137089584643256665 0
30210747952344853771765938770498457091773347516501675590013537751 9
76806180571006569612400084901009385805005483001833311060060185682 7
04123487963089707314490447939733846236615092180730549664127940464 7
02688341757329705615035421509711297787661137628526345891112430002 9
13667436382394280007474095780374296995349907020635678839388678433
35136547261190154557442521035559120105343760865755516615891221034 3
62476163452284914792782626638469818753013238468438384819898225912 9
43196540130309103584956556886042339929376419727753778129757657675 1
10422067469250826111631348342175239471485083689488569079813732858 7
70354384154566806582243641642314639385420361939222129549008319033 8
86704663690954916398029276092923773318171162645862228801632493622 5
94596845813160900455752491644127351538119202888181105177878340704 00
92717305000915331566982178326362307410274780898351826896342116943 5
31182366857113501820728813552528398665812848871771633908035872326 0
08868597228422581112521579584861191961634143487966215279633842959 8
19315830407846930437187705989598100502068069666580058933713083840
66896578841358638906377088671403403997388516927969025585844724887 4
52416246492782534577943365986524544875153599725586890231449061046 8
```

```
750552843715538971359887915282925828255050120835419120587880510395
496202435877418284809042834012623317913548824250786075509868266022
155690538616721739707016552951826185149785155991687413480771803539
647696253126379133931446691551775242879398376583996804906330680176
854652833690811941784649343624382999817942041651225155281648662278
969241316967283175771906598106614920189094814966766193502463479471
325657146836099798366877167117645768363846679320507604568423013645
607467711282291454410457836229548419005716460338365284954362664397
313112157990900481873735514627240002040528376748055046017505262238
339122922205278208826476178433341282467090434281555799199936787059
242446539841415248337185030982365045703951871832547505889146921972
787747493476137431476334074432723672357889683072187178156069992366
992607880234726582584111858536785494634704089537920077406163245285
854832012427796206571255740619001540568666359355115191485907385418
336738943687771213576680356395554954319677643261844785458336548591
964475108085868241702442533525220358497647116816247683995362447 6793
908223428292953081300906689380635274420004363883387069849021712445
883967870819545956703139050196133302424742867545568883597989025769
892006589434200237275867258803211692625170144931468664680549000789
632227031773306101313017187029864949930259177391147857049708705092
800940844098301417159418161489959727551004889962361242184667921740
402557935757841830411854809829851398218698285647664429028480990292
829922586242989685205259864405231722786608191694277437470575771880
761080178170144412923472681276169061375346156304594764959765848582
440358799678373308638092246773164888660137582230112646771435 88269
322052862836006846606417608669812058388966448423634127830430019766
438574724055586785414244526663556478233814748800205471190740096934
076623723196500713688019886856632248183354617539427810528634045487
684150148177861122004990204474591552400838546972384386526846236050
165034745895856420487076738103344998421391627255092008166428318460
105347025818298005342090247498511881750062670884412379523195042061
420729591780120441108210043139323316050122661604736720973055753935
978395931391217709166253148727803743200032297288350128452632046447
526458249676953834093331796811966577917278829965391034981801394378
419558229713114826523637606339828606342962398279776496446686968629
837267017460877965733397537813876632044627737591511611802806674852
175256924858603718711077225232573649510157221247061001841261126143
113020371508764915462560950255744788727300241807695775894301097619
588539241071395051108671826885921799146347036185965741416975160347
050238400013203497201690750143454481840785001959616555552021750376
614208353600030855422651628313900089755352930457979279905154554170 5
586820451135637146704411942761822983730791428614295864767934829032
395722437781831378383090974524463284891032578685455840364603505384
439735356539103424668901769280812754245712647903794048222416972827
054629989166657013996978232508098930619305304874846505714394551522
130923050261480079978803377249805525957401726449051295118891778545
751052809268213446972071235540578850789568719093781853355007751043
065888274679607849931894474395039999440818345801504830345039967155
327137758714638778072081624941500103065127483383647834568784236910
304330175565872230771194207625044108524387177687336893154848960604
981667626690323755496624275319348817481407832457316079552872676837
126118081584623942044542980533497736915195469050498168948792939639
968752324908065516345216962542132037649917330724161582043172409779
780270130087004141876079184028490145991177142665209285453570636243
160765058847376985987107871715794338130707580432717296927019986816
588918070391000277603084358151804591335015658811959424430718750790
126976529417234363240416807336041725105880539857167060382979606033
118535820247244763710051832406308449237674456287167496676490605645
124991855267272564935779149936017213357891928354146196788689475497
941118320386604193488760428799842688266413662338773747948405362154 4
```

```
989893738093944672955532085964288239560931926733241512640389426756
729012135781521056356502450634808667200331532150992186650264551465
520579140434331418931744145838420034720114527627475566386463637199
154891556164136320487255709853796453133178054994657384417413499154
230689959953012234156670160551788060454883096589803517762589040817
303354734957965719804431576804129356102302309380869894586449484552
998450548608120087719720164982137383165934861750474894264984687492
483015557556088823382468631822938489826958743128999273455955791027
777823999739474079071706240114696142722460394052244672743800904788
068009204257381456815446138608539893580229084144989604154179801310
601766049783453641531633197224496825336807416715398997449480537924
789476725847323987207948415076975129577918946249822894144335476631
853499993586673581654758791933439905537468403042134167767371913878
776319919673229649624979546496568363762555201482210355676498158036
856868528035439933206733394042646651584569953881912796585701315163
723035971515092329099272112888339073900630268586743036381664750134
279032424692734424028488879997900400655379849337708746593010508071
048941423036863903046826850163688100183578933799329677697771516244
432831505342567167908835992414207504466339326272476704203597 84332
815578267651026423239188433779193494885924473013410522507517984038
779077812791771316017378485164931516371038572541771694565531677933
276154956643657568533225261023257517130919222912037868211070106815
689349821999351886051472789723308956152755259112847529248525450462
215444218657403751132591100938338672834069400299224989263821091937
119486618171512760816517655776727261418602198384922131227759032397
499396244564780586620774443935110169653881841453023655559151086374
832503926500145515591650607843333853192817359581558458366214501953
701708825489775034719546868729080417781621620256162930998363284038
401945381868215663424101427934990548626635944994402509452009183025
520397214084740691796595681301659021732590209567459603718940073653
349450183082615318467364198037639325330286271045821606050861605871
702342043549305405651136441092980288760148311068004486037089144269
050402978340847669181085254217115625518889434756998384372769449941
429708519559026365592385928030262310269084975919716402328847731383
904355979530301955847218432375525820627416619828984383021062971782
915921569843316111053018858411779987507490214559725971447090225625
560205015783836145588178379268453835346710734530611392405820451156
727209613739119106945585108712014423488743684572259093038880828221
033831988164490376440625650760124548099875097829844098463810310411
827722681572136732476956946590412064482082337446955994162810986211
350871573563474400228807960266077457333640152509336919698130463835
087376041843648153203775053730654315844428577432972215583088816557
325209144771385341671813443899989727793495960845508696953243213134
378088859951426680407416510593079653252371864645999929929 4239832237
197536799832552861852357326487871009326795792414996495180687852579
182314761872393062952159868693737823729092141976803024842337931249
212091294171389210146784397236873930708166013915286514957866192992
916443911890321162611480909866002685330394585625087315335985026307
484834690610544701424902305482929966185089428901525947881338446026
002800961728992571497477705894917567289921149108269974117774785022
845859480960904624945983915777808117071800347130290630116202830820
514038486414156962849536879123574505379395082788818907286254490746
622170778380512001880953938635251023490891476271813372322525910660
537642114052353192613420140635060630018743284514593473055345970421
994754697570192913054134089336885942483002748062704664044357516116
741527649515523702214487698862684659580526666008884645549681801107
100990936424888996698200975758346057385378803447003420408960409947
924574316693818153869606263715503786284505705418845096469632896205
705629234882625133792545103644485308087253873094400735034798170646
065234750378998241104659747174811552398903123634887457263482462 7476
```

```
0957517438198528192919890889743292656407594333165862841246724 01023
5655690481117569753737086162435839678751877009847840583691976 50305
0929709154181824064011312900104279320543904411739245556173621 92512
0830998456641624805764707386059954205257697741047330195414833 61533
4240073481508862892579364150423402253408493704434164662057026 13588
5413312616712157351019987224242692783799237163539075049774326 85487
4338248930262067273633607793776075367700462927382559310066025 04206
9025358331972614728631742206538472465520630618072151259282154 75925
7828854706800082211190153120865951225651470830249538010652748 00286
6287127462213996458324577131012078227598763520822289382690288 31949
4077772543020190149164010198782913990163247907191802436434625 17701
0696153522120288789546337752002232788315679671189938532933536 59394
9903522938296260379087717906949906119125422332681040549797248 52595
9491095256230147834898201045071956735116137525642503562293164 46254
9940499533763935606957000818997615661104639996968504159253857 13482
5988136895394133983051248427298333256580001030513153509680790 99652
5737825476708965111939364287896837275776845935078123417746292 36704
6159317039902885813378987545460643406593881863986652278567241 85430
6304162097294686920039610154195508035983774355888736941366897 25746
8094724417601896962815706914230959157670097594765584986670528 92464
2208889780000051543895753363327777800486387310585423960835484 6698
0252413564183341290213942664070448114213838095809621677599456 91040
5908669812865187486504263380538768215603218337307404430486350 82416
9945134584245177238652032319352003662291015738060744545025072 91141
7045165082053002528350760305671802176735610289235899758448668 71602
1543775796514290099151441658535552932583816382863657600898384 37476
7972804092752500822022496907611955017332562673284383738858246 64039
8155943521281168905998147088079444191830776585591804084943131 42557
0892885292490987354315360737039483562192855283379008927965446 63911
2193655696471632750812596369829675456851024862840308743355397 96624
8219041742025273544284231772764873707816328369163501082664202 07972
8533408734892267083142585499141101668403192792861404454428634 82258
2245094365804777222659499237653291515384749554204367363315762 70573
5678725541454831075538771684911434311272502004294746902048758 57059
6687893397307873667114082913580284627017952455906619170558647 32170
3941461655344148026161983502248364247934147172530412891534045 06474
4165650827253312325107593688256197051598562206270940238792736 93273
5308414749363386465028182764873700358561355235930913221343165 81241
4043980308595800535246967078265215553707022012926927272437490 25354
3237886194196165495204872062313270142581298949265471709614557 02829
9558298867419571736741057610922022553977805597734427686847839 06964
7410872028277529928764215738289521294800436150585533096116495 90619
3682026407171960806067462055669329877169852302061833592941970 280514
5813146215349365031561472101845006263794606002399121849082497 41755
5068774270757804295808070943900407287324744751598188023372941 22233
1484650358615121522201789113078995687744410716633658650244860 8390
3810132989270744792294902280489683540054540911703858573528426 68460
6998888741032312047014616915777375942379415881681124391095728 79679
9194031264849829862167524251390830132640076459710216683617362 96972
3910161601199354977495578697924052967290339315423916841415011 60806
7266878009266768052142837066934937284375792362567511043304698 24870
5826533764318544358591576527547143088579023160078598029502306 19709
0331070249241089133246516459945577089213257309984201050373015 38751
4611360110091962764603374961379446922186807855008948112724563 39897
8174763599111751479053642604525866534522359813087995745859244 92013
2041250396245632898780738137300671333300157557860858357780856 08286
1897468967084432406545132530038447083814543263528092742538473 86546
7557768770921919443785539625679013158885854197607848999708990 36352
0503789891086036369955901923376354046672384146217564466502097 366846
3352392346791941626672407983689793122757229499336289156376997 78865
```

```
0186402026960957051472993199633682591151266364470446006662126630
5128987845373204969951514137672797524345640551113604444560570801
4391185704705551738082865844736344141269868874958323479530603008
0325938606946185192001064291936582184544154569366810563204835715
7376463453903642629326676175338725252390385646775180513862261917
7324726240430645977603498706002358299230855770510900937417657434
7234925033940769721939086394296353887953890077489238368716607543
7942468997550504370854674817201596480839520436756430205010743705
5683132390425989129984999194782488190441967483941357124233783961
0998535432090214777949407571597296529763943172014394665151269249
1710717132695262522959121885714101246179801201017003373026585433
2662620691809426744276510874184400109206087727293789214761939035
0365130044247570121077615324575337537932674660637337809050695406
4179821664751305114044041172224497467581365660502052712808542381
5600160900808397842671113319856647746494420286983443413412020554
0302167020706789905604381470527089684832528842036857290992582349
1314666403874837515280900031527378130409655876389125304801439432
2207014094820762814904892695938758691553407192874709934425571090
0740763511846545224504419515115987476939659188800110992980154450
7041710299465222387243055683513501184977807570474848987035685687
1949820400522174723349473993497727469824109341518595267471104314
1809613158168468104147309290638846502758460988106933586692975700
5223706138174156013412902959349500968895238872190138393250697294
1999660286924622629318420511417694182206625429675259094629998783
3289804498300658221335379271406379364171313038669482311254011813
6356859144672825755580045257704048842595786077781317362719134299
0130005401734461747215149245580177483422619153971816337186537933
5446215316379703670602825672566270135330091550635944777201147921
6706681799179017047734364737033727532070069701205279236308965927
7054283852391066513689293899265767083942549073583356027336210318
8573228572464433004326926111110993952871413424318914113830839211
8310370546891844208457461777507258978602892293574951736711613084
4543570385729566186163808068306229746817171870218007459461772952
6984763294634432039297316704464537018388327643218889474056136085
2201270428440837384688948541975982154093057791543353898108770610
1070856010228667643576010007267497000733971131265592126331425883
7835377702076498886096464641457039869295589796195725997055607876
3286853834355609229110975111154384005998475773515480725390190606
7493015919320746932120514781905775850810871423797515884535608859
4733473011309162280042796635906170090700849507770734797861569646
5158173734200298419684926713897891555151984595366991488689567598
0818516350342114722084003012061996960062758479518416994021128004
2278935681020933230531599631285929389855871216914299873564201472
6157644249183401203837157999199511857894126994751983237780105424
6057648509907567962411263858013361900172365430560200562847201023
0925669356755876814529218598230336899995283290725978652460714532
6177752233623459644476149928164662978227479423137187503838742074
0071007156689387998214744302780843080105194017792276766532690658
2578450177693646563095978100075798285671399521890063690776077162
5809428370526276431012616584811391213668049991415537488565844096
1604763805823938975045328084381574719119868817213961258775074981
8324471845267388935465162389140126020703564823958601290322986038
2457746906264601132615902587229120684243235099273918484604114973
9571296411007841269991098329855314068051345372166147257581810522
9558809309758604044706407461552885097686347752037783651276976107
9897393283159385895071143078093806554564587709547004039301354687
0378080386321234085682087258197985228833910808224938998393965092
2161926033302555735080560627287827880489106739537611030952118379
9496437827568688758333370714672286159960591050358021863591887284
3895486178146130771180910812490849963162278832846575594253052046
4121396892260883743292735410371752618280291015326178949602481000
3701934187080228682050970900794298847144241821877
```

```
3768861843099810951590647162827639710855739232366299543369517 95667
2221617776145901220205021035647523447654091913908169541515248 42248
5992665864025102207427896073210875763593502255936763960146221 97242
9449685875680715070495591512163060627112112147489087979762622 84575
9965671927541321765027243282986873623175711999387927322986125 70059
8328733156978971662705909319732438794657985417750533946094721 99873
1780076716008206327846955952410960248863270807526834723045488 79506
7255861137014841542863332421274439685755350150915471135077423 47621
1781073221684164922782659200940617589259055238051584905418035 56545
7466264227832969866348660856717305131101579037003704609187052 81784
4553570948431848180000588203687460311840948245631457630975980 52214
0078822087503759652946110817044769871467128780465285500802105 04635
6816350779613241577274287679325546354604364832185302674937990 14670
3631390890118950746757210014971430598121993512673187315236624 59656
2910797970414806744210490101788887887261270617678435745421233 64529
8317410254996844673829770084188687631047878303180932108086182 99839
7284890065777065751653951537411567904286097840645757874659646 81082
4204777741142680858377161880557493517357845565300467222455036 9475
1148940571600444353023813572700661888425561330275022591075185 94569
3009423418819844949183555322524954609776026057849839448713691 28757
8486983393698394178016707891959247335127280754594115094987114 18142
6639827788836286964575221168575815912098922701121833525651729 3636
7621051932145009481404459126153274594510066979530314227085468 8739
1845817707290090378192697966247706585908164916708046212513888 70006
2931345870802880778924634728510971040869237061458003159590935 54156
7273326471335701973050993681863303073416919477744045409248412 07576
9458540688972034242180656273178897101782330022881383030353347 15210
5705768345809461933541177649998178171646552127770177116749936 06360
0424629195518765570888626055565459333578225336007892478936520 20679
3818083966374149942900186449247053460893792649105889062174647 56885
2477610239130630994309231488605391389049826954837716137543862 19379
4039609277191637559200612770437952535183439342700726037376183 38356
8538642646126005611698858934360040588872761478162111107389661 98852
2624011337894276219449124450520915565217032119604906543673920 41337
5195070026265821556514581869106925011332498767162533806363036 45961
0957091311398259770990380773298415126534060016152134848287892 52628
9435353590534938211246282063327378870394866322265017931364783 65149
0067651121214320586291954220145643554300407045474345870123234 22381
4773776360878549099119148357397723103112281690277386316348735 51639
9292176593575116018557216638140780130962613880503904422655522 27641
5968483995298420018665228263060575946232734683844343402962736 67178
9492433216439317725287773415081338385853182185817256745106229 85125
6358434372125380670846714651692367965790403036522208007385996 8143
8723603854470527078704964660357633278591373682630015911138178 69243
5824228101494387937081812728199615111956310742022573428936818 203032
7220968364380917860869709364474866301850900740788083447564412 03410
7786671707299718360464480693852092838372196625798822505776417 76077
8783803749834016208600260890715750689080315245208780621186888 40341
3023003730962213733207299288667016614405756359109288763062635 66935
0361903008975656656848860197239142264480984704230631121915838 13865
0616796296868227712562427285306849409946459954094931674087697 1089
0594124916616908686628026982932261160440478149370830454885903 53087
5453538880604490860883793660838911041176682646711227887137483 75924
5632157691880071025274168755164805110797882265156946483346898 87474
7071077067681812546150725053586929824442128683024429084353958 91702
8094287758170178130118255000187580985891511712811803681333510 93694
4673092077590986154382001162792751499546241027036830511470858 33369
8093637649277560523472276709318477247530253376460162812992938 90055
0523902526264800254984546270121626353529575643215202222778242 87246
1188269436589455812695555400747540855734127125858402433534516 21728
```

```
9009984331194409839882082711437144439209759797859787676868941083325
9417097339335747964194379299638159998163573302805024185862652508001
3867630115027790940578764150480421255386937872106226418587771821451
5478413268536516829640661424520060839841773069977823507866851728351
1425111924322045352305680360814646536531361539675629674759966353691
8578880806222970365769276341533636893568872194370103773952970964361
2817377965583710399695863923049408795918369555339659079422528031261
1304179985684053833404088638304176204084648957608525524057947338651
1633006582875507428339537403278824368035420707277650899927371651021
0023632668155914897345691221331697102040850338579600024909895438361
4087759780576422199087858999653794089950555955437344280009261103071
0066846926296462609863374420905443718529588201847267175125248638761
7707450171051792380846070013424766647239219975538451981016397051481
5419675882856293441581476287627010085377604997329539279110539486341
3985945813047621834505053416466258244874752156047538835823429124891
2308480766651796693016420104677551510351064107042908041918607209151
9768365379810185679250372508658114949539466335598325855515687066521
4463383436682190219165330673824020630512990746654796488731220130131
6817640290334543866785097007002369367791025515142481688418705147271
8570587424877122386614774241774807206682372847124127648006318918501
2112160006797043011622831201695279226813260006865728793161219294819
2324325615493944784094893939385527493843942726131517119199443701071
6180565894582083712166376364148369988179742302155962392238131825671
0065112232260390556037504635973535913428368195442937563318337371131
0556464719467372684891275343243157237360389421588479569701290408761
7019187110350946987845763085124879448688695106177415556360159691091
3285528674220123419164354764782653034995563023316482734073650081821
7601507051683661391678262373833937833129722342491273356193266356781
0150259434035590504017728405820183692964589843146583563761703695641
4418759924821083933350364504161803340849583010869561202233272443911
7201709839513342256017193103398791871310389329311403423165985834731
4453121092956885878161678560563666939604643271798915451420119927311
5128127626804090576754887898834122405698296212396515937185820107211
0044425687770540957250167222187774199239699742424548052508007003651
1772294986499015445827627842205454236681256909000930922904684104621
8400055085367820839412438054749831235566503163972667292103672021931
5794455804376702009827790201905035677344403077045282287879426902321
0872463084029566008831024628976509266216131696886680701451369244451
4072753326002088666025577523713540158210904470956072117760677215721
2517386995526448281653924962917933264624545569307017521057868831181
8639761543759181843490160750155518194291214939314650934369590045431
4118215937035816748932184209235535796164170970710266577459163474351
0305938521291757951909487896889365699646366261155514221764701176091
1427636235413560144469873087108825559401557702933307168065071768261
5890615153902280501687194706029899966694289139651617004820858098670
2419163348746485156366605601880505274050937710995639730830467828361
2975188216886626775577351054940008852827265246421586600057407722841
0665694825813355725609743692591190583410232952175787279807299238961
1768273980489549060588528596126904989942603530302926581580788812101
3826958330327474605471748053286310264149730959824399131579369073111
4451472512334249467338144789666069844796105753759084310553203930571
7292020189606524270856590700226744022102298415563365146673692213631
1619955273184795179743930272917536976022597523229896679719920382541
1633115497231881323909486824721939039073279232089220649021092086101
6215321529644275855956270680031474536194530646673376185166884351401
0545795955163903609083185340241724602647827855106535046721815686941
5020859001363203945542104889770169032217552535272034390120254058951
3732598054682351907224679120255482650001689498736048527209759670501
9429392994382675498790441922020428184196544981294172220124164084071
1081140802392027387102776545344158962040509452981039063143471483941
```

298993092566994451580883764470277249704281537541403733840876233560
031721798735665881200677024800524342584100572727816789918729011057
834895163277454345624608173512135045447559177232251166417870394921
004225707428479231963345522858074444758722836078505271378413 00142
294036587843611314811680761105150102367765805645762827001787064872
226313140597231528510942673048784618457230651930451793553917793697
359623433870925449008961642842149146255376032884948724943769816530
469359466083274925042127109188608399801715154748403227072750115912
530976139895546696390166363555160537362970299815348192612842226610
182774683861357990677908850059019358855468448662500366004640278553
912949056366183740948509569093149366109815181653186745159876349361
119908775253093382708991443573161402203033942595616544717143323457
239865405966628819595720691204491916634030698106989377700043965327
488620420557843550618795370253877098255654775700944957640510307952
791557163931493399926802208813549025110769777399264405818027145845
783961661603337128495286472559121313165571856473763010746738795560
400892379474647112760904696470739028770063249138817518739416903425
383382034213674217691213904304776755624886660060102976620319670397
052305021987824734057527822757453932531064599772948587253339210901
925660614329429354449552126434701953943638707330140325025303 97277
472546499461419980823986254595947528582179058859836715879056330807
328499125745187823167124839350762840382643270941437677931361736001
450719368026294652971394005463101346480149911526496521227542 28935
443050276902585766457852642347414682457558997374870043780641270731
361148147950436769611848407918426531028930765857395165069720461744
350483812258916409833032689057266565563492848622455696404978823181
021272868515897521111420088100821937046252730416347465516080241824 0
992005928782244496463184024053349465211538933872665670779649019 7439
903340047162541428984700321280530112013434003632731926627171660529
424307383475945727607789094797110189355709756539878269266663437696
386761767794271434435095362526410409448573878429255659427 63327674
191890543372553491142659784510838554285874625364275288138204985509
077791359256761895905807549627513438728306328894275944377855018970
013423703699010917655017068820810422433433351856378595129218470442
487790172946877813388584765853418494564230843975232254584541870585
059409234154992326050823892348791518281313437540590350354503 7976246
817092565717232807711799886210977954953013288193725232292716 9599018
618537590608507830498979182821289641775828334301731744187751366997
921337252228430924201156275952762918563538268045736171489805020672
709740189126541482756075045989103061622020012315772743556 42240430
248401998611453063869488579671973350187550836684997514235742540866
098253849559901854622057493352032381704826267034038046706713 74621
391708326712612767515562422975623189478414789446792675459709808808
729968356219103003398237439845514708586348625379582933142875480363
929533391414925907410654089485188506744170725598270667221226974828
520730683817266968658817008567564758133916895408574461368344127384
007270205907335692415155145540712038484903880303635497768174718503
545608541319731920783235258646483841318702588571269070978285712935
008070417408180941579495255508833845434638992040738538734858891105
024882800816941451383286522583829223131475853858393279582256800419
969562084186912291114250311771955667846934115901095078641799292572
351091518115559836053388770090313132719059913723647158175803 5720
555228711561094158563878063265320605888387199826257917322223662895
171024037101772875686927534717342830891621136043261025209148811782
870360145311907460943380766849797124023747183637116819407958422224
628083337982026409548974113921572738171674136979584792037689323070
133524647465713195303863182456302910180873999567496200909493186369
162027016953739662436571536630394401902423601294981790089732382467
673918121028880941498049233156802859214186878360023527052429112248
629634444262945521602729740722147448764558710648809422462101 1301411

```
190325195396060498261348376385234086219075091480015231994970853151
694727038698501928627522029791444356244908799451922435782038757172
062187775443811741287385128671899199262865905912462751982438042428 9
957023242486811331701186188720776916196219551594271887685566884348
151439796060108881983822356980416952639925207581682447375244045811
533476007680308027561332184420172015933801046457449539078864611899
955059208644667312538211463439054619761576459558766054760750358596
507400214877481533842867836894622992322259311223987263072888448295
666660761383263441463728786089244854567206302227645213741807045828
841286867409874126561685924316199572602347579275045806144219060760
754325556199581406013623751864414114322281040338531568277294299871
037869484308000768631029952106540503295265312490951283916099496200
863749689581723981090104865776121683181696882151923238314111136169
247776053898873483426243576591204519525721911206259211592857521352
073191659105885887787022587663827501853008315591887037362328149521
028091985228495661341923285949419770060553999221568496883816739181
275298938488995903440000088151628572556467326194747335027428675381
776659846257771561239004488177582799815401216212284999916269 9583
222144468544064213463912148677036404516490232030051453261166125078
164395029786660592660526237431697876001206217488789426400355837520
617768923611019768849918429771130267638794179153938777070116533300
175824097665897367549129598193701229010427442705341245697932518877
369971009565147780926552856100678614307892084145110610557411488608
604673484086421486268582967458796822162830222548443750191352795020
202610144149359612375667370637069826047831172964028414500633605700
716501103684874873199136742003687704593308501922674395127735603719
567151991224576695824890708668660562270782674074825243883990945763
145010063753779106598357640663567508018186139640675765123603403648
878301711355508403455128960701152086412498303651068886072733093144
289358491030502487684554001008583473938253412926012974833189941127
657531155945164027672578643188240601860856214546096077508355587933
853737257095005863329412866233224014564909908193406736699790803 0780
199015545066987229118432137634633506623171220285324684473870366411
827888936200348658575390340343182863394985842078288804064921427677
615330895433016588494728687050732472047555062954488371484935259438
106787788625227032943525360419984476708369644737172909728630479592
793677486868975694841457254489048870193668971191106258517487329557
793225523030005673960012073235490925297786037294223616566438250025
589995885277118180372672405835787633069882581981935428261046602845
446873600640244235033834431083522364902624462087513490332142339451
226556334959101219931126252036504805344481613244105555752654791746
310455055928505250787495987112371445236237997582900837445210668455
686478684785017949976974764750855044345958122951632384967506134315
102405218605133653161850754258595933912804963416755713212746626694
688095641728477538136949190137575851277254440415630074050215028311
430625007139020129896031269837134698881154587068501564245840243587
656877721027223967123808804882099014233017199336625644479600855529
532931719823666313381204885052668978587061845463690287373091004290
601057433981498257800746166898054709421847902302609166415401823 38
632718340342150042117498852953580584229568226030651221674062264348
094739547411420721831831173447845766448031232016492561714090033762
647708642899235652258883763222100378550112616016634932765563111900
210721123609535665547068241659030675654834006288425294998305878187
593537654831324395888522727625074935402235412338975777888328424440 5
298932752222419986542344003026688327815567564813562637454153404115 6
474246005170942921822273144138149639404117665757737772501087862077 4
884296858495606345872806287431788497993385617314425795600625438251
598551453022935164176060699485935263708128786929046278545919306181
085393591021328010091651103911210812800157319608824844365665519648
522760073022388650637482824307538375087738496926527763398421164558
```

```
3077745313963423813053331981929582891473386612695394497316725667 24
1534603460999708484106268393735982125308895521803548642311994969 17
2897151929289794444670238007164934813124424599879033937231347324 58
6687210696558058479932804332926306039936655556400374529494421230 44
8095573322885458279941076204611199152242828455003457104545393131 01
7682252748741052792716813397429501166618127663001029888563284240 46
5658085431819050154985535351515705314820805566212432853357752769 129
1249663329658496690922725196860002751977895821887362023551584031 19
3888826455856523470284526677556236637750954479906781920011493859 34
9170159522681233449038887528236704296630835867686785000132149573 59
5136469365367555114035990407772861913124936222725604314527721679 91
9158872761312596425960497009290465355190143975155677551315616208 35
0720267839728545463692545632415902858725872715718510879991812076 66
5635383473741597707148636776029922729594047292575545679835496906 93
6142841057649820373386444525243917394310124491128179813052926217 26
8230970609646016928694453431004893328674167729086830679197445415 43
6147294796201406551823243126264115376487807706005745274332070965 2
6536230261196418272651375792724995613674128791182321785174623406 55
5580956841441222771222715573580733712850285556458913623453940225 52
9260073462509914853880658749015864743928791012433893273712525118 42
7026373321910358307231861499322648857605268256581418713845628528 02
2188620076962676478925307083793607585334052211092978378396606292 93
1192046139810647664955648166053552687133516826987466489598738914 79
0064346289221826808831626073884291951862499052298653988021289300 07
0110882556020771705921847955982461286629043449303538347468796534 46
2226805529616613564169971517395946974973869472223608115479619706 80
4090128577944715903839482253094672822732427749062279281192860618 92
6284631402992356579963824833035565442997895865209904786514956861 46
6814047145007335948031313658485495656306380865177939513201358251 72
2795771893241303624462095851642788419951680920010306326369299494 22
2453864865398727691237013052995630580539446805246955098912143243 19
8184862828340428102867133606074802306809803741524901543920000930 70
7036896756755238592938197913483494374956216362332648800419037836 90
5915471887405877339535524211203332904625538035656243285634537063 20
4304929323966637755719732972839332655857664937425923654756079891 97
6982254271699061830530651234371605914513116807571635550357527711 52
9160784923762332236922863786213377607083746740110512523380609950 92
4231166459199657582349133411473454485555839916670988209545436108 26
3242798043555068446193412387800752287032101164115018576233302147 40
0573159254029107601915832655055190474918997046821238176555950166 88
9123030275007880565075800490313468421759096064432249038368512875 07
0906805263289453782747091361038137306717040775263231693602917482 70
7228826477725159569991857102591569042173636460520317684455742840 12
6932942743851495152601370118194296546188512373619599279092646621 23
9612960998024369230429945769290525398781922535646892047486171413 89
5513330876561258375201313419237018929205521797245754163563429109 68
4882374385144965310713627635237095174114709050375115859137506437 02
0190584451437525104375371696833318241056721768169415403079743536 10
3604457327070796116001430052927942123350643445568376680004304662 02
6189402108510853502703712017928323298931720462917891832884592791 38
2935066270311103281203365357199178896502270140060284635867342633 48
4038138150494636077936567471656264076108041957819840810006143491 30
9220744592482237262320768905787764897702170118871922624134924725 45
1763530786341300835263748451550565256256298609311598337141139086 35
4984924997747118178063357988669850653745661280509301853744843801 56
4603532781571612001225448254058324764372763251063618008054732049 28
9641158176234851165972958945087514490520995630502893912679083957 71
7541231467018120076263603686040275900140582867511769362349760072 25
4885708820418299321766405058326458870862673799599409099472293648 71
1363869872896176047232335275288766863689911257258787809665328887 80
```

```
151296236250527595997233120509885787330278437617474112850755731460
734437316342908026098665577494117820293702853393239824925951498603
929008458681784059001502829840148436452581636428124681610077608519
484637729698612780243409947746288433303219834358865839398020859510
542638726439073693269928371482684703889155629572690361975374229159
718057195927499315557656417855737505079613118934633771016187774306
099821854243213824672968358207459044400713861196502170724220247317
680216276175569504433940639906956998302035442848769598372929821572
606386279646279113429925129704904649058166376858495677167725715454
316449191886787925707733389084993382002862170811545195972041619889
591921963169060382158091021286890230248103934686106773181389529965
890788642975598998491279230133870472216119701690281629676613726294
629132945483098004213790123118486280645295494928155719344344954918
733919171387484090848942584848729854942532266445244776123991470666
812787566786784371059526315774565252022727510164656252247922795523
959421371667485887276659450955632490008970742506202682701277176541
777812417595017790407588062944089651966374677448657745649613869283
329892881120184307874245432265735553265782121959459994745816028150
276062718940032147391783446073827759119202210968980796759623061229
037582763606021163372172341243676462230723258360192166756433329359
970330702055891840874114664851659034442324963656326813020355381996
904852410738834444505620361377054263847387212546062369828681927966
800722289252717063234963486201308545800490120353952789842460208332
306467915766319658793329342156784380520215572985391107023812742873
459692564177666291847426228315882261559982307446859666695338069391
504238781354219526222194651355348205719272397307283547154768226795
659300496111337343536094446183284959422172643965729761154468640912
182763903066366405768042627377672639103667577893716667791118309632
673734549217038474109842181929932870402294134277708024071582704529
795334331707700424513047242848925759941829279569589963227907964826
143743884670720631936507288012420944026765301779529814686505744499
320189253169661408201033347347457233172917770023499106040440096220
881460575052511656813910215345493006800374400903367877614928045919
103145734127975125257184175473672201228966592953624045339708271462
937487249399925801450625861099419208086296844694610325740923205322
982641302460880014729134560750695730773976280294155384827273974729
643191575726074948819348493419687313623148043162678373835702377931
754770686421985634510543468853036969036217558078996084075297755665
635311024077497985395693028284079509156094430142012579542276043212
301716663551392935158862828410385890199134433649829898294376442818
092696950790299368416228764022920159955626906220456651361376774086
201153400142439897964161262181684904076865976296663668273608461936
077011505752374798678937001009516203725342619046856538188188327136
477081429300700956173748670222977225028664663601738596876680780638
583138799490240060994973905385651063893589910794337084388640783365
544785881598618384947430897788823296597524502663351362959204367591
753392860084962174180466303212415308853312108594983844476244960414
742549812729544982232465409903851767460485181616346093082293838941
345859642843570714419759466204938387461999533948067569133145079971
508658785434998905065116403720678600031274330900347190068072841949
979454189150778847802451303099888847164677849181204491191272375792
730835621039493340962416701851610581760740477389589258559431991721
968476225638043261399729804517116231917734119300979381000588869993
473304112629151768549946607102323347520391207000131425974724178283
304314948838929230145849072225316985358919739869926403694984573898
039793142185178992517031732724115886696382017886935467814148188365
998613370087528772176033832251139280903629251330785428184475952720
510979329896737952934389554082968045845488026897972871681217612045
996020483299138381131286540646213997225401984880374858605268921990
687754092914860751166716478028895820992904621897109093981218115155
```

```
2992749926636633652397724560470307541311759637017111424908947667074
1616204229775287479311552933667262457648212689637449098146859122670
9194821170057023703920962683730366548527217160897789428370771688847
6094226160707546522809720519699545722731403535501070447232716700085
0264148810480190824582235846326137968575934514049404901681330712998
1473452399430532836091711876426756828920487508680417721771189873366
9958675477100209902191832463798628477782614108937974835177886790 1
0805909285662951394705770116182129015471693379336603673388505867 8
2372589538605170312918423109041129090245936954031915712966511750 36
1562645286959738689249163726973081378948202563974300998041228670 56
9404392206376069614827651547431300928973042128161508736591295912 0
9214643488300596885878739031236709782130150055082260590421962291 43
7192877664166499541229031550816557584873441534444883789906991636 54
8974510191912173876031953161298958177122343143345831259480456056 03
5140793922909947609422519914893586815806839797919263454270623351 67
6340140048916424664510462802466370510764228275502370033465676855 44
6519875212834041108168876146384690248343360159351017065883976264 49
4724955786960820915059068051707034816575607201822980104704726079 70
7983999010195850917740484204453184993595881292111246073523569637 36
4508465460489482040392062905243492818545322269127448812828459158 46
5502612881276210782663676909472157102670633255045954233129008863 52
2122053892685046143294035569543854076881988643946033057302776529 54
8451070780301107083687212942137934578024936620072911903876251326 12
9726112117639086534702236502640593141079083914976347682976333470 438
0316361295056769907031204622031587779769901572744451792155558067 70
1708193445729741107130012699684530998999362390559668503090288571 48
9248542725965655599374036836340778028687385473803258051888022696 19
6937305280686649390676148681433788168726713905190126309289321560 25
9277540188728311235716476448927582106932360066911640646327955901 59
1985338462954360870938491544043787953368105858383397841412366142 57
5806745488790631318005966990753354904100991515678129852892154721 97
3228093831205228062118377709932460448070891195036403535126395530 1
7969102525788145403421499694828053494480834387389499446950628126 45
6362300508186365180559703065665818778592713505517363237671668552 15
9613593544273882463832796194302192545579523050244465052323615239 71
1558527516465428398484324859088541530322348532363156908925850958 24
8992924057511672197685458747093520372858445268053136263159364670 34
4007691856443151467772873914467685778560911376333635494802158149 63
4238874342885860482536785160712137109793117047031356001163668358 01
9569988185926634078290824296710009650262343767450733953847307361 29
8266422092273613639597367825201360728788124389308783162542895293 3
7974787488597852774386270224438809034759647298491200714320072355 94
7134882438331346042912914702900428058812898812769075863076219639 39
7867804380983981285369069446118601643824108954422658803945107100 33
9382208516666071670083114575541795729045253043947353359181698537 39
7838785059360864341641385723595423278980437636731900914680652900 51
2196523707757048200326126536851943559331448045208133058610232484 90
5922191238011974924881073477670309323705646712704539500710884827 84
2076598681939320823802014424313928940795861006365271215761235280 62
6664663483510343877477720510555587781511729148825540554750732124 17
0870656535655224489907705208139867738297225175199112334665353373 18
2766991445444457100640276579552095787271781587173180450260449683 42
5271845053590505086545033945541447884634833032555064653828200792 33
9827144154093282245042535830190389199788012711892736504005370821 10
8186162631736155820471854678272533977975164724738980802658228431 92
0849071762914963957409534139835769878700152850582180519588547309 02
8411218207261261642974770598348713697343934300712147462887622621 76
8509945948020970171407047458344340287492257214778163764153178353 20
3894024883310207111945996334097779583674103982188086100588432322 89
8381891978666491068698007878623844300836084820622695707935511958236
```

```
724320961214505239142763164261085938200915299459307164359811509744
471399610746727963006254946006435690761355048080395552252346783932
299234036896773294937781306487790982224320958372209877062533962694
612585787218219183027484623301161892836849192787802470304506840133
470131692923654597261273609629125903509199317183362125521597444895
929012539038729311142783505440495348885481350496182633480614510241
409882632467025746233657958488261166210852788614684855614235967594
459844314643446023143948486342022317127554289753878443235327081432
398567042852982795767230319747780184442895846935223907431983887194
656310190393429573566316924244937977151743939275568857332386904714
405538086918271588080591616051919986062574341101198989490525667143
892905505692146371145126959221108498459953465836743478430705907311
672571665359511888392634254724318347884827035739438294766160949289
529563023749577563590454037677619891874086766162887566873868109632
513218071819935454412044689631688964186378796481740820620199255831
873302314991035192530086999689309981528890965603168696185716943887
220826974646085728531135598174151752781031092330346762940184546387
026461913746580079299680507495764958132058340412485203177241844774
816802375697068929557134499602214818158217088931919664042836408913
155276361583691155296902814613831790766465878385109276442547547757
519858688445541650813928941357860834039816911022339576608888740266
573146218234401068704886921492941128865313448700896352736694896862
195127660559272142286606430638508332227432854799954728207724699782
4776615414524344118410767454750720100248709574284726427209073102
026800195766428300292894912507151121024962421653400943291943923231
118338465947839000144016093262962430714275045249507827537105771747
095136888110026552115580809691419025123849886651710280443849394358
923120210001802829913890081885045604464324007084834111220445335710
298550940225441130757503870855116921936715421340345105158485638301
086768768902753850988213495323229943997092277284204923121935807173
3180545086579329586689830311685358277864039334629926487621487905021
321358550265353951243382881314213917972048733544372623722931207641
064036112524798287093597955723584364596776199446110133027101759389
90874920962517046845145724623727497223920636085458126815932496638
525913846335426794440484699404481673685085997350647574086916713131
740094431543173396378737083889255692917939245801602438593630101094
347019647538406603130596476269469610390969582932719326897629157678
935269144353971950444651800568773183294081792620736085847881334106
436614584374512467989766162506969073758698591799608553715668432840
812904741347070159944098119247419686336244588028696923008740456350
940459976835006176373348234862290999909345604389767917342718752950
882927270094681320952245429291069233047568099874519370696699639214
654909605190101803581525925629511788816557600408579722911412437441
6373738864493025525390400038538226072626466931145500775214249393362
036976132246090046234947356540690164298044196657333106015318935089
841656371748152982806764509218479920572882353171494589597660274930
674298034056935181993001889867129075173932864956241186176402659572
864665135279178449317966278349752517634939904161363074237098229142
320244476341796202360115469735498965778581277744702532542553085702
821937612021212307951975619248074217221643242373932213915662228236
845367785339739469530432539522816591502104211183980829572978922838
642442274083462291353555852521045701137692240977847731079911988836
34438621952233096258307738020993505395968383888184971572475262019
481484924480526387544182041021511847529192604735330515134903487256
535300428558903307590135185020101822529975005126001233514568797015
788477949473620974571901924368056556852153470683443159096493184425
003391809796157290168595913920147836031414196691249865876779409827
777777030700456171857300221152948844594245524322382862559231209997
198493546238704083314580263102044171533517455171454659811995808683
5306089698989527740917440644097214771131508313828692664516064331163
```

```
9326438551408243660323109563776989269784013495935457097656176612 10
3153831701587355881845905949164143368347728551325805758356566009 52
1749941039902297065859489977873797470122894423229861793971140097 31
0200515003039131406304987374164707583525566427723205094569351020 83
9152065961716192193600119390975986276904599780142553902167520766 16
8048636730571768923244370268787604308599482103656518551214670215 85
1272108643709070503228273526123324719590920861994327617346063241 28
7752484900942537585782142022980287276758605510161550503047626089 80
2669299631965343187755080991347565797377245397335849454864810455 35
7309480361308636984060024250191628324619491771683490798627255306 31
6634106272468935923292241178224379007627571953287173839748871056 54
0577307564532239228092259486621508442865547689163851727060542391 74
3016650127834840374014655305836936230283054080695602219981509977 96
9742997156594769194468153420743243224493965493171348797648888477 86
7843820001349626097220050468886918923383826884491413835167572769 17
4546123162306877924145277892614846251797397078314863413724692890 94
3970168296180840782119927134537103411666409888554897974592480978 08
3510386679194524142984821547894852332512275463475113654289109465 96
4035502463927449861956651518575294088437124537185729065240197714 35
4715498368046642375603831338136331525396165707503296510133698544 22
0685387425717141618896259989150903989287911396089923280949722735 00
7499719614193891077908921753870669182738768386853656800438812422 42
6023484218032237791909896286832185485646668691491654379894259070 780
4834342978624721071413759793575365289590835908581758854234114354 68
9891517126357783388497100094121385068729974158099040301932196233 974
2923140602392725757528690721273584360608171423843207945190199269 89
1199532236801071798383251786248183545496537508033904651742703585 9
5775772520816613598891324971659590784376780183650616173195489794 79
8403417255823623859487350375059165037244020024383508334606535883 96
3581377975208996276010890880410317163452548585236401606026178928 37
3115972932457866257795362048205873134696749582217762220595247444 99
1754712211089511158353142818716293569338463751159831426196893985 47
9154089581471247065500505743018099282836911292513101670046597620 43
5932723914576937307759617348135623794170276266626860735894132968 14
0244554461794629229967896554351537280975090061275168299682323106 51
1063002271290592894747507904779450661410809205552779908061361301 05
5938647966560361763521798069337196234387328598765746710579778626 16
3922179491895377743575062235599458838509439794529372471075651253 35
3725151065325883470926830784806461388198157058829135235350857956 46
4509354284023581629264676269237180243436337703873925415075367026 95
4674759901383974019422836789310740142739509435283107847364342631 43
1721965061444834204387107911959560073976919024271566841916196465 30
2771319046965396901822931628205696646953681545059448164730408142 81
1064559426455935153469781689321413423981863232908174252075260098 34
6967516068934972335538878891478032509254049408363763391993149203 76
9988087240866120709926631460860972231289926917416917149007567393 35
4711805960833401727874390989117437790159926208491910319237233387 82
2764770294854503911913060375930244964062614897197043186055735466 47
3477515985400259318780992815503461655638774010808636355650908900 33
0516026259375366743370884961163072331542511137810282463955796304 92
3847600206311281131945933972733859414781188965214417266612297755 19
8086077669687930653246509148673611306098326328793152683696181758 74
9274925830802738021947456854817774180262168761888716521795388681 94
4148416320239478446147305801296703139949251677605460501982936986 59
8464730688115514688039754475630215198584289087838003536460616075 36
7155131894005973366200476851976911843914364794855198932106501980 89
9792462645128117087401443090639482879815367335304517964743338385 28
9711509180048136349852059920982260589195909686665924009719183713 80
3912194098611107081811588567053562953820542827744031876285010917 4
2692890715254632332966757312475335314033450907576076230115065043 82
```

```
7129259226926235359096870513352908825194426380640618984692893911 72
7753684741625107999411062287734542841919374519989309963921304502 85
7394476617575780321946140400090732701324725875034916050565597341 61
8242227034104419793261037342467650328381582901541249614332495848 07
3772393912478707886085251656990776398811494936675986261664038692 37
8645743515220440458274986173261936754480308830956739426853344957 81
9308204327546856240619529482056195375145942788635636149026600409 71
4841887651835831494511782895220094198429843647588360236801741159 31
0887881043860140705856400002615904553175373832819013941603939791 87
4221273470341594769929727299245554429585863170272133522122479633 451
6343363930365832390479775932788436198108560735879105385981096920 64
6492514874239312895615271140231148103314623928456069768124483063 19
0513612979898796635549293215502479357740974918373008632058710633 2
0230694749788678744502658688123255491339208379643013657294129495 03
3736227969806371445601349048687338553036245057506296870479140796 50
9149409174481353570437552938595946779432578245000266752508442653 48
0631988578495883119018361757360120730972656180503493135588375203 40
4235059092864671524639106207438219704977671841870260148162795921 05
7028155058964097193853480623028341486837688057432088572937691824 24
1297776117062652308584376628868099690166753710664313633118087440 44
6982279182390835304285959637059490312892123726210020568370757821 94
8402174233689917970909546252081448666758902800623392766597887015 61
1926291699931209277988564757955020825098543665464639513143960818 947
9441705785283625552711721168410825731846214442683753937884809844 94
7226623117060998315667143896953034657676733924584526793185678576 3
1297655793955944022239907071350263390259950683317084734771242161 99
3913046838595196183750735500774048975837053860252777074251072225 04
3867791098330614989593433687041532251953947906627114272636233372 66
7868457265143944302872875716091862608083402298523035357549152306 15
6712521733501838101998213160752551751012478378207452859367931930 81
7234959183076397443574113046229717227036418162433263211765411259 27
8434787726778278549100115670952928343531620844765138813124669017 26
7231297223027553939888133776057050637027746672934347848712050984 88
7290144305840888765598734833959194342148135500579100184631334446 71
3720409792055721882443038555298164792079359275170630486599360728 23
8340072062271526359646609055877950795948203283512067739139697321 81
6111552505787559347390867269711310796046733961498339640686759738 17
8063541635835575854654749437420511848818365748745492059634583780 55
3862410744596620607782339879182459841801713898053432643789299938 73
8936312794906657362666632728627835992045309203710253331632159181 90
6814701898173859427768080879109790111326186023584667195820079430 47
4443677915054193978429509339814457663314083690946863408116489805 12
7935520678657533704114864192975181182543411718876716490023520091 34
9067577417324208551413628299379868498513045124369963285387276777 75
3066528563734299161690070521861947104229183637101067036713785310 34
4608560467228351109828829008702295885831651328114493705161967607 45
1526013585046745198728096899226445606062572157545869823890977051 97
9723599720315465833406823557362392991841844068466238333074948660 63
5963171726743635065829614963478438145691954710029984536487422726 18
1136431923487160463153028652173496508667276322325067274450057248 87
4091615852223625878178768433242364171772556652419005916370966152 05
7214102966985961050586880103541599971500675839812648317024496596 89
7022530112011312865021864925323609744320730347296666657578205559 77
6915072597955574114684807431951323469133335917799681713394949263 68
9096194198732130847390271352675988680534674160723627843605557006 87
1653333224631478433189691636502747937715741423739094064388208688 6
1312857833974444101243244259989179257953946722483472719771158241 62
9247857069935257000880583316522664525438920498801255539074604294 0
9781140982333496916860070400989314027471155898326898630919890032 21
0381899779687846993312670173257985175167727638660400331834617769
```

```
71636799874328198163655770804768240489402961890830180259113109372
76945464517133760527983318001771440054467578972726873005503347049  4
19047784557548214896174358034754867248223747442504329756683789032  6
13752282345663451161975962154849483120844752436896056650130320704  3
70775660121401124182886783635915534093657409276878486954134814260  7
01432738673072310546816547829178407848504148629252437833414921450  4
44881978032991795074360134571108790308514800942024708304393377271  8
78723723624296205301723136459074212245309078169150053814677660817  6
38702550220593255069183597111986113508637356626184277781159668245  5
75595606839172212601432467961844116854155245484716624940961549435  5
40175661979537227672774287484603614346123495144138513918291921714  0
93093896326086945500387176624939231277609803787514116824369538021  69
00937169379509466883047083996807894487381387843272834945472297290  3
73857477483587677094706786967403393274527646845449196949484311486  7
98622174683121794962985567005058838564886273765792637370759351500  1
20851000690998661594348510868038070831590329054529428417646769261  8
38553748593679523141575321459423892033934786254640890593101468094  1
13600215322288818279640366989292508432287097938658821021233723035
39747983314734812743384023919305714179560839593229977802046801036  0
19943351746320961235468666010088028873844858864871469878838843926  3
27403806051737545769506955900364870690605703655643724223286649615  2
89501957722107109324148371809457508435353308199838113842240971114  3
46766776115499253112045046642774483291256324319005733368008428357  0
58131337710518528962742849576283202815347646262350114325889319350  0
94089422570988193565517725295978734367494291713368705995374953730
94885237654819724100280514696590602619849193488097745642591937419  4
56484612794538647195637823650507466661108120136890914280272346783  0
93472013811657611478869749955748873252147982001832901866979469431  6
90938137255081190021427750475494121423244763677919311594975981588  7
54316887909752647468136699105942531107687041440335453908900410105  3
37207801261286470730994954070198168283180350685123597282965034206  5
93831547855573149609749709461054003399445830428351631735355107901  1
23406597846972864170085160728828654849449274606578351855236776410  1
17562946884945065135019356147298968214599248791524130759662844391  4
97492273142701099841969292775271025360312191041682859810083133460  5
87277764157635563587256762577883283439065449852829468287043684571
39650907890423427169668880846128243956717717532114019384897498450  8
35186373247505750029452535871509544100578342896740808308301448957  4
41418179686824883962781009903321091902659714332205659094950535465
51738166824279589224863733912973356910021146088247561121203336570  3
83401282181480344559809196598994288439733427531030691028196828618  6
15795488126807837581413003437974091424607690984066099113594865685
01719589515268012925042501463366750932153967687317776035010658244  8
62798964086984071566617597647124879911462913379771727203441609433  8
94521165268669180753752361945660408617986403165359524823564567776  59
00456471123141563875276965015245988347659073505992575985890423989  4
26120101892021650165250621058642073507308315813915401280586092699  7
38480439526973291364423837387582529752694931423630364817439300906  9
87697474261000761221723841568424796788222350717735236929962525238  3
89078999939371569279598531529641883985057515542061842314332290805  7
80447927307000503306713777206557418367307022855216699502269381127  8
63029143879921157647614618951537501190267757321474668928407017786  0
97212512597067079327165803426702045836637255467757826704912751110  5
04528768252690207549364526808804169512023224331843171558130562876  1
53348039882903826134568945079878513149343588514053409333615844925
04985033353448967937213122234435161032083872454669056462710581246  0
08862630725118657131726133542947605382281769503805019969336461807  6
00308540821018788129303371685950916442754533757804264491358467917  7
27459558067848828711052300795145079203212200348295311039564593751  1
14550724795199232391555698871940627179088561957528855680635631702  3
```

```
90446017775352233617671776560656937561936193625957006849224843289
14428797784280102515971795248281793420502299436973398883127868943
80883694989720399631806896130045784999474716058935338285983607588
75431581274794611734717603722286489471301704393985136109701248283
52138666187821323766164406632541867297548362655250981940527735203
07469054184347681212646407002405914653832564359992898018183780155
911150639925614899522018524983531639039060743883264204279427190585
06150034694319671606393627127161369893660688471815325042660965646
70589765842943834041334489981703509585245070376250868978530336151
77297452300877119980938202182002832248196876523179627503150112089
35888075560863606686522725620255204603549818864166539665399349
75502500443517378078108911971007001845060791121408004350299097653
37370643707723379578964579544918986866109307292337337799563710281
53840313239929169288754888583645855133628103130375974465997632614
43882717791223764087996184348001682463682889243688590497716003401
58793703962881241370090033475183831782234931972553767959181463566
93037323117552118652202616801645865331479190320055615233120376847
17622683069549459935099121858352992717018662466195542650555791922
08992209374160633459662828650277061333837852659405264411830116525
59781252889062460575250825224715375493965389238843297234988721881
68358447061003774797565237579900197363932060532686440217432937203
20588840765651246479469182222312837349991621715284041684638436562
53190722159164631590859856326456538219805353601019034722959077108
17963920016920444432337635678116679844132303003566696418034085052
75830397152153094980651994807763437185234670357137521901262395972
02353666925060799714128666232858131103001826669698298230546201979
52378291902170705965370311747539032681300786845008674345498867023
17310811264627550219881281730952362182115460607012873066501954551
50366566592293173926430194942848113830924770555947397123814212744
07606962332172705941507138868919209599018903706341755049947675931
75337376098118569971789123721485265849902535671806319315337384020
83189440107358728231396447490868722744425218435179672785678857347
82070054750062157787243746950676902332419680788697462228506011432
18027874274474258213826943940866024364113471013511570979410237673
63525672346241586608470505296195793250152085456912319232852540122
29897279491167347413527379452144514843285103670961349642566871978
86170938900790214460254661278542169952541713554785958251267326901
47462776898750349813242602141873314111691875857176445527786922089
68144858209938626410651870202932905331863676087401755878042682528
69690114897591568453632370641657499860018496204462823603684213504
68057794480361348771309583721967331489812688341702069538021919198
51000865989117091573818418386151817129983877720094677561752585207
09627958271933521894050406448112945616195339327821288172005426875
94895331457835574920048576253750683097854334895374975325768954054
19326270917720364088944671279079740958300912842043577048840544083
66217245899449631293592404360359133554963973636331637286100225567
89042169516055266305107949422569185482022457225720797951722773264
95507225796927216859052186284738283592265311486688627915331568351
18137070883693020725977831132531886494737035510550466774803090523
35406486587105660488912266007994803991495803081174408981993983403
53834215098021154616930358555668625589248049454439523959372931391
16277356735319479956371954374431442002474012459576864816870877179
54141228033942753624076059793849047357085357947220635960180232829
52104486962822188162402618925670849250527459299252204382188258950
13443281728836121927095660560594844040670333886103385868787530310
20420405688138899384795781312876809832636711055677415654202366644
97668081546520442645161317135424922414930883973274980264395662125
33794005529898362196046158283593069648321811500081909673527689905
99483197194176984221322937382339372115483425291908655509741160651
32809795630560534783889506825281143609976466828233938748414719396
```

```
8899080383490762602594307447976737071051702488208144548532122245644
4681673416532262342619055983149920006092631526920678309549023422898
8838821526131068197401371957873879734470295029043544086904068667599
1743726662223946046443498655673649141541157440664028641699644473326
1799539660452142879179739077054286215250548189129033267769530091033
8453245205324654194004136473254811460661581881867990681705414030199
5492007132239772209550276731147174267311105344811387495700163677222
0871509057043833004310158164600248543330650069985693298328323954599
2053026102109782950889039463098645696631259191874253388746597252689
6493076183409810940024308757029623926399208813839199459140878344859
6734121765356007301111666681302543588829594223967244361576764802749
0763750657142900193412719368341177108074199308838731885759725805359
2608217936733206705359049015937804876474348648387186601152061511919
9452266182513202278229883188306164953410122149089701672493745603679
6666116811366356736112759995475436456578864714770635215943803385509
1118277427811933695009349415187867971154265494580438059500464302099
8551260553271598091760541104013177007377906733178746544333206033319
7633276441653016999388157768794279976658482097636321868406143700319
8516036743893683951815183806870601292380504103970937013353275407049
8164410210803232334120297294537283903703913120265145332771719631149
3203487950180294781391541671122546826020663848273235888914692377389
0285182026358946917429805606882641770648736204882730494772227608749
6194106492356912944970559887731305487387163928907281858892624323169
5028584098864301500121947127764941363103367301489878572179052835689
5307451716683762526987431280334072871201423228886512700758316580059
8267965999952281418562167606140863359418484062838153882077027722229
3830592678252397728962906072904764945075485922505119261757307770059
8298985299959184730112300780119372658949549431376486670784062295989
4815281620856386485679713104611381488123166184161795976237206018039
8071461314291055007442170556726291215549193140575807302741316998949
3623605952245074230397116217164945033827871823917213832806613160249
4165641357236975859155316736782711368636926673938874657176848442879
0765543744025862307044122889424843382134420252316883168199071799989
1048074149036011847093320907379892685141437390962505158850336180349
8057369625361223820307584006255968953151283087926222099331578021499
0241828322278905410515030826477623677216198167062597738307728127869
4152709112741041382893441746192215343148834356007502960183506993049
5377700718166001491716561761930320571310180758755112216844896947849
1778979290722696782472259954683978166420778862933606629926754833769
3526348235705661662233069049931047764600407747957903695453270777799
8624858272168266752283572442028542925176968991248392574462892159719
9321824380047324071509107225599135073235489242217914949842726513579
4371389201322232194436388369787842581184486531810135359986743720899
8877039922342888605536831598782745663342600060527852849678972526399
2974952005996252456799866909835746361108067222176228118076919206259
2648690136415328883449756442534127967768305773546443844171060884109
8441410379124720260654512279145765878485670127466835259047952537819
4324521686583109133661391025558286277239846109521509239743007268519
6357244557738311092778510640173859591975715669756936823800393043499
5643369826186041199979825744224797213206299382586938932051163113
3590272239781856293080945639812418072366115753571765771750915258629
8642449923988908451591960926233268041540202849914035293799488661
9905848379899745321982807748539649983709001189757836329221569910069
5135225164777792684573593276827610527710812619305065415546204242789
8526143461195900247532195892412870948439624366773040782829042950979
9803366980199661862491022715249014764834740909139305578551793371059
0884199913505576491688183121626684989496190484005039433180122988169
8778202001444929999014034822801007650917661050036551885174643858809
5230029184080112294431785282276241408360363573365136170788479211419
5621243555136911126874902092192949946122740478289839323850877528299
```

```
2534514715165653763355566476399262297744756561928502537832594734218
2517307460588495425327335034876035623574197164892871939912417875 10
1760146059302582569114268293510399289620117494651034552669370946 08
2600721953629679247304173351149017826405234920691693488656077272 71
7321458117956590402927706023192373539548257865928864628843485285 26
8253488737506097269589349345870441010628863169036189454430684276 98
7265151672812203818123360018740463991314217107541355432293841072 96
9963218087877517519730938666327606185430818902559606867930963592044
0693285453596654725391711933154388093864091888381364001602241517 81
6058591590858203539387803382962008214095868156703171131398628751 48
5021936911870618271275410173323557657830569071167524181420858474 8
2239818517986054041890461218875262875969424507868435159585420511 41
5402155032972364053696851382474958272200579413252055188843479051 70
8672450277259085077602688049592613470742753859783003757742153332 46
6163693146812507093761796527005877700992952606876662817164091207 16
0929333088153876394307528897793548006209454580722510791567660344 3
5158452475604370208396637254837179362732810682860828917094234144 54
0922811094955746198789087549897373580773575160574441968939726029 13
5186642217894480793273891829162452850073791032180272457137842772 97
9990906813060532284146009900776982261479285147844996815767905092 90
4902518037546080633012624107982680367900215366565252237358249396 82
6200850776531863214813557060146490877558550143228348970285583766 99
7779647939290602757065690019263992629523418240465414655118864079 12
0644642133237970866031724495084442786825978453544731453075165458 271
9785328342581584193429805310298248838279992858904391002987811706 65
3883327867798136547502704135599993330246261908110493918624488431 99
6867657835720100589712635101009135102376209988836426503163906995 76
2757322424853885490024202145883253006534680443779333601832870749 21
6132993888400861186195546433428316080696943967124913309819269238 74
9377903347230979800053618584852830251476386669491520745557574389 531
3634403036476079589862790854821583247276221166519228389752622455 18
5916297232380887456938367949566661418085653781050731923016749558 30
1487700411321137041324715954646454801602668610836035818675407632 78
9504750840652510467638346858596138181315314294827384254498721060 44
8651202908872525751811499313482564340523594454759127703088011207 47
9834059984371936315508711779844914253339393624254435464982285213 146
5679395144481043266751969278320276902316890565183002021921611562 73
0305318044269647852290119995273401760922282576431383895507787737 02
8707044986788962637552377371746708214121882256457360234115131319 68
5669424597551113103803639688207027020059215326006751952348578877 08
9066020299322599796081977456149576246920599061253264129937964305 44
1138602547587046153970927914363246231732613669437865493649580742 51
6219022588712996396721320351605517119414141717110642131269462774 27
1891165640547127919491059388341376154304013813170346121744354226 39
1457774762707226595075210567446711136152393181713277403347750784 96
0407474058604330741113454145980264093082096887711551056952267727 98
0894471940678585899327367325314070577765753530178232781734708415 0
9637524925074294025600649477725342334115624558413553279777257702 17
3194684120152478666161422865546961368346147006316544427156673697 48
2791910420988227524140343410014162834149124023888406883543634013 28
0852014026556633174153819036687401039005916322394590345138868020 89
2075292660094838716377934465573466872439766134738285752209251563 71
7200820716669555789525960503518737818293035394077344917516701624 21
5270942117672324738550617457502503114980560223810999489523564471 98
1291893579075710465768873114531252211914965894900529545905156757 01
4520794923429133942006513499514380451072347477300601196993067779 06
7768241072718112790898683845267507794223432720711533613422394998 77
9916835750955696876987508328964227218526545920275734334622918866 86
9393855679454583866531869143507797458830851816055666361789954391 93
9765308829860726400664329996603080380172286389032861993435157085 52
```

```
75999542576456755668125627838519924720958626701725752906423078766
52345726444284518192679325993697746970169591518863347765313474694
17719811527046002284285782083680394619067628709239124905289355815
53495698474235437727083899500067629333408636180919678168715058832
47272625671457731464980045304861714171718921058119116670938964076
23010038770315335340912289423737980597273106166450079769830699142
22510777643516737199084084262065973017309864654345375931609956856
36605082817146355981756738453072524089516972274778695131575441734
45472231952384246941765722568988931103741198664489157762273697918
24205586520760798995585844301144929901420553871053873765325608033
96818454616866102811535831816796143045658318071729440669411586376
38484611000438217867311835906684684503638545310225423627094604293
40655375560142986937850938251419954250093062755848365826452673187
55458188516506300240402277182482747117469916392793541064861603346
99171707268638654851571394219748711916357072843224598328820843029
85576234791759652491077509568894147027692006548333496695848885977
09088030768117692000539140847071329510172814621287091951518065110
05037815641412908539216981086765068508121346861797951925878916032
26673702693686799072977281717652804885975603078503181675441180048
87346117137728708121233361415629031555871144740387589966958056653
96605283683522524341760527733293448970924283352808620617571581842
01480630736873755675657074570658267308996196748386448730910525171
39088389512013313334954002840893279091764207499306144562633737996
89086148746897744623338024727100021387600130016419879141503114834
46085762748651935996918967596768225737910162215578861323125523373
53914384316443234312505699579981068857299605833398103507905397891
87614115302314027488009071360634872197827072857239706246380349762
20684388816133847768661702578175332556528927515481510496833441853
48265907541833005192252384516611550946805077160851658494458549888
65286782597173547819572439989821750440009606566410692416877215011
79898283811099201082576187883976384133236217473160385006466108189
09765360147538965823524330570083936135400462140752068307599894186
50346692035508762486276154311150368430667642908280074440988394532
11186140277272130722016617040548123348140458733435040503740064004
82650787112791443605886719566576176608430634671853980058310774936
75412683737907381699476437651809177981485180157006642169228631130
32856533729134304264161160306035702731954053893889233984443568070
06979519191328076164714789031733404130270278145014074786299403774
30064750337744016668851871570545373452670969830429217052959743467
35024484000458606484321251067182726114991255077410283770665338637
71712021128192363358233363403876630996813821037495121262939365053
14827665611036592860170977465471116753773512203563007505429472082
54699729397979575001983180678455749127031693750613907824125453314
46981557285147350720925140402302844893125259284342773874202006602
60416247000340305131846957597237094582875759383268955110782425590
23712422614027813712005738334729945237136283122876668672282294943
15773208438522866523199284912324852821420990768237411031053043174
93782721327678786576724507458080130356336506873670735907194248758
96733157905234047482399807770741145551313186080257755669293479046
72955421298940072512437665355353699182763645960452110921703011377
19237325705649508019783852013130745704131561280114092739151213741
92002153673145722018643941327099882185621756595659345447298506292
27984223394743205838065142255145878422778457516531452377600928320
87457901593950341777854034524798621560871703904246687644929911898
28041920726033073686580088906979567405733286321187486025369136222
71117039037671496617562142278710476367501977294877869715517055320
02156564923212299157588192624145247172814249204182576238257553922
37312588088773652228264200729028555887666187289565336372879637900
49016627899945584557250723205108426998644370704885795623223128219
64527054856931536462205074408600822016492969472723786966318972885
```

7535295496664529441238399636586701509654687188280246459962912518 94
3645395988734574902174365408201339206097043057588634843946599693 61
3751136699220860085078191900218784357761559301553955928822907142 74
3036869620845177168086924750642539902360039766027107635906562748 97
6085711230233926633986523984577192808211907510609451890427630793 92
4618288307997551831774785713343128232826112154502467372712011637 21
4179168910789031133699667063876845253007597562467527648836451906 66
2162762307388500978488336685856067574086110656005194140599859159 46
8110571271796360445501741440254596245878789797635421788371521087 0
6629766185284281404933313056206855422893076345920136087226007415 27
2346350294389161165166798128401472800711817348129165560394511057 37
2660435039248469519651186632701177283998712598037348685203878144 57
3609039891377985189672185579414701475789360767812442055555337574 74
7867044191714613342044252057531101079289522207989600625371130231 15
6827243427312892007384945416133386492731598170168987669608646428 44
8384961079043229172133327891404061443294353980689870450113388304 46
6253786599595563673921797718499910831107027710671906247241897718 93
1395703540974673149466121125395481783523771570739888297169104166 34
4151022080470229824351953195854031140283855677631141415962711583 89
2594746115769794070812354400524610529125384477524954718215981346 78
2825104315649470337194506231527911174645287746511674488932106999 79
4356821101491292586213970102087453236546414071425265451685312384 73
3794838677855371838484969386231958542673020678857691057456749439 78
9465782169150250744276503986200613658152009057903591828161480174 56
9984707839843964849371304922505525602538777338018562290369977133 70
2943549181034686707931106444345981136721567912431490071732713542 27
3746317932173009717553736350752874778473693619528085144244425311 66
4915782639172248632774607069418353175565060159440467166507873097 12
8511145168314255396588717990313657811934693040177688341451752385 33
6001257844076353671234950801711026805104627818593642357014391312 09
4881435804090498138460034997136665339915975252030399851619589805 83
0700192560727840369693729873556529024364319340585148901918365225 71
2922601242295107624031283464403286263713634470007263192351521020 74
7520098458750934980401237494797294662122948993842044193016904841 20
4390646281364098838187277975410993874855579862843014592070594313 2
9445612545199073257324237580094766758101266122854048507226973202 57
31849141493880004856742892

www.ingramcontent.com/pod-product-compliance
Lightning Source LLC
Chambersburg PA
CBHW071347210326
41597CB00015B/1563